COMPLEX ANALYSIS
AND ITS APPLICATIONS
Vol. II

INTERNATIONAL CENTRE FOR THEORETICAL PHYSICS, TRIESTE

COMPLEX ANALYSIS AND ITS APPLICATIONS

LECTURES PRESENTED AT
AN INTERNATIONAL SEMINAR COURSE
AT TRIESTE FROM 21 MAY TO 8 AUGUST 1975
ORGANIZED BY THE
INTERNATIONAL CENTRE FOR THEORETICAL PHYSICS, TRIESTE

In three volumes

VOL. II

INTERNATIONAL ATOMIC ENERGY AGENCY
VIENNA, 1976

THE INTERNATIONAL CENTRE FOR THEORETICAL PHYSICS (ICTP) in Trieste was established by the International Atomic Energy Agency (IAEA) in 1964 under an agreement with the Italian Government, and with the assistance of the City and University of Trieste.

The IAEA and the United Nations Educational, Scientific and Cultural Organization (UNESCO) subsequently agreed to operate the Centre jointly from 1 January 1970.

Member States of both organizations participate in the work of the Centre, the main purpose of which is to foster, through training and research, the advancement of theoretical physics, with special regard to the needs of developing countries.

COMPLEX ANALYSIS AND ITS APPLICATIONS
IAEA, VIENNA, 1976
STI/PUB/428
ISBN 92–0–130476–5

Printed by the IAEA in Austria
October 1976

FOREWORD

The International Centre for Theoretical Physics has maintained an interdisciplinary character in its research and training programmes in different branches of theoretical physics and related applied mathematics. In pursuance of this objective, the Centre has since 1964 organized extended research courses in various disciplines; most of the Proceedings of these courses have been published by the International Atomic Energy Agency.

In 1972 the ICTP held the first of a series of extended summer courses in mathematics and its applications. To date, the following courses have taken place: Global Analysis and its Applications (1972), Mathematical and Numerical Methods in Fluid Dynamics (1973), Control Theory and Topics in Functional Analysis (1974), and Complex Analysis and its Applications (1975). The present volumes consist of a collection of the long, basic courses (Volume I) and the individual lectures (Volumes II and III) given in Trieste at the 1975 Summer Course. The contributions are partly expository and partly research-oriented – in the spirit of the very wide range of interest of the participants. The programme of lectures was organized by Professors A. Andreotti (Pisa, Italy, and Oregon, United States of America), J. Eells (Warwick, United Kingdom) and F. Gherardelli (Florence, Italy).

Abdus Salam

CONTENTS OF VOLUME II

SEVERAL COMPLEX VARIABLES AND BANACH ALGEBRAS

G.R. ALLAN
School of Mathematics,
University of Leeds,
Leeds, United Kingdom

Abstract

SEVERAL COMPLEX VARIABLES AND BANACH ALGEBRAS.

This paper aims to present certain applications of the theory of holomorphic functions of several complex variables to the study of commutative Banach algebras. The material falls into the following sections: (A) Introduction to Banach algebras (this will not presuppose any knowledge of the subject); (B) Groups of differential forms (mainly concerned with setting up a useful language); (C) Polynomially convex domains in $\mathbb{C}^n$; (D) Holomorphic functional calculus for Banach algebras; (E) Some applications of the functional calculus.

A. INTRODUCTION TO BANACH ALGEBRAS

Let $\mathbb{C}$ be the field of complex numbers; an **algebra** A over $\mathbb{C}$ will always mean a **commutative** and **associative** linear algebra with an identity element $1 \neq 0$ (i.e. A is (i) a commutative ring with 1, and (ii) a complex vector space, such that $\lambda(xy) = (\lambda x)y = x(\lambda y)$ for all $\lambda \in \mathbb{C}$ and $x,y \in A$).

Example 1. $\mathbb{C}[X_1, ..., X_n]$, the algebra of all complex **polynomials** in n indeterminates $X_1, ..., X_n$.

Example 2. $\mathbb{C}[[X_1, ..., X_n]]$, the algebra of all formal **power series** in $X_1, ..., X_n$.

Example 3. Let U be an open subset of $\mathbb{C}^n$. Then each of the sets of functions C(U), $C^\infty(U)$, $\mathcal{O}(U)$, consisting respectively of all continuous, all smooth, all holomorphic $\mathbb{C}$-valued functions on U, is an example of an algebra when equipped with the obvious "pointwise" algebraic operations.

Example 4. Let K be a compact subset of $\mathbb{C}^n$. Then C(K), the algebra of all continuous $\mathbb{C}$-valued functions on K is another example (pointwise operations again).

Example 5. Let K be as in Example 4 and let P(K) be the subalgebra of C(K) consisting of all those functions uniformly approximable on K by complex polynomials (in $z_1, ..., z_n$).

Now let A be a complex algebra. An **algebra-norm** on A is a function $x \mapsto \|x\|$ of A into $\mathbb{R}$ (real numbers) such that, for all $x,y \in A$ and $\lambda \in \mathbb{C}$:

(i) $\|x\| \geqslant 0$ and $\|x\| = 0$ if and only if $x = 0$;

(ii) $\|\lambda x\| = |\lambda| \; \|x\|$;

(iii) $\|x+y\| \leqslant \|x\| + \|y\|$;

(iv) $\|xy\| \leqslant \|x\| \; \|y\|$;

(v) $\|1\| = 1$

If A is equipped with a given algebra-norm $\|.\|$, the pair $(A;\|.\|)$ is called a **normed algebra**; if, also, A is **complete** in the given norm then $(A; \|.\|)$ is a **Banach algebra.** (Of course, when no confusion seems possible, we shall write just A rather than $(A; \|.\|)$.)

Of the examples of algebras given above:

Example 1. $\mathbb{C}[X_1, ..., X_n]$ carries many different algebra-norms – but (provided $n \geqslant 1$) it is not a Banach algebra for any of them.

Example 2. $\mathbb{C}[[X_1, ..., X_n]]$: the same statement applies as in Example 1, but it is much less obvious (see Refs [1], [2]).

Example 3. $C(U)$, $C^\infty(U)$ *cannot* be given algebra-norms at all (we shall show this after Theorem for $C(U)$); the algebra $\mathcal{O}(U)$ can be given many different algebra-norms for certain U – but never a **Banach**-algebra norm (e.g., if U is connected and K is a compact subset of U with int $K \neq \emptyset$ then $\|f\|_K = \sup_{z\in K} |f(z)|$ $(f \in \mathcal{O}(U))$ defines an algebra-norm on $\mathcal{O}(U)$. On the other hand if, for example U is an open subset of the plane consisting of a union of infinitely many open discs with pairwise disjoint closures, then $\mathcal{O}(U)$ carries no algebra-norm).

Examples 4, 5. The algebras C(K), P(K) may be given Banach-algebra norms, by putting

$$\|f\|_K = \sup_{z\in K} |f(z)| \quad (f\in C(K))$$

We now give two more examples of Banach algebras:

Example 6. Let W be the algebra of all functions $f : \mathbb{R} \to \mathbb{C}$ representable by a Fourier series

$$f(x) = \sum_{n=-\infty}^{\infty} f_n e^{inx} \quad (x \in \mathbb{R})$$

where each $f_n \in \mathbb{C}$ and $\sum |f_n| < \infty$. The norm is $\|f\| = \sum |f_n|$ and the algebraic operations are pointwise on $\mathbb{R}$ (called the **Wiener algebra**).

Example 7. Let K be a compact subset of $\mathbb{C}$. Let R(K) be the uniform closure in C(K) of $\mathbb{C}$-valued rational functions without poles in K.

Throughout this section, A is some commutative Banach algebra.

The elementary theory of Banach algebras makes frequent use of the following trivial but vital technical lemma (we write A^{-1} for the group of invertible elements of A):

Lemma 1. *Let* A *be a Banach algebra, let* $x \in A$ *with* $\|1 - x\| < 1$; *then* $x \in A^{-1}$ *and*

$$x^{-1} = \sum_{n=0}^{\infty} (1-x)^n$$

Proof. Elementary calculation, after noting that

$$\sum_{0}^{\infty} \|(1-x)^n\| \leqslant \sum_{0}^{\infty} \|1-x\|^n < \infty$$

since $\|1 - x\| < 1$ and thus $\sum_{0}^{\infty} (1 - x)^n$ is norm-convergent in A since $(A; \|.\|)$ is **complete**.

Corollary 1. A^{-1} *is an open subset of* A.

Proof. Let $x \in A^{-1}$; then, for $y \in A$ with $\|y - x\| < \|x^{-1}\|^{-1}$, we have

$$y = x + (y - x) = x[1 + x^{-1}(y - x)]$$

Since $\|x^{-1}(y - x)\| \leqslant \|x^{-1}\|.\|y - x\| < 1$ and $x \in A^{-1}$, we see from Lemma 1 that $y \in A^{-1}$.

Corollary 2. *The mapping* $x \mapsto x^{-1}$ *is a homeomorphism of* A^{-1} *onto itself.*

Proof. Let $x \in A^{-1}$; by the calculation of Corollary 1, if $\|y - x\| < \|x^{-1}\|^{-1}$ then $y \in A^{-1}$ and

$$y^{-1} = x^{-1}[1 + x^{-1}(y - x)]^{-1} = \sum_{n=0}^{\infty} x^{-(n+1)}(x - y)^n$$

Thus

$$\|y^{-1} - x^{-1}\| = \|\sum_{n=1}^{\infty} x^{-(n+1)}(x - y)^n\|$$

$$\leqslant \sum_{n=1}^{\infty} \|x^{-1}\|^{n+1}\|y - x\|^n = \frac{\|x^{-1}\|^2\|y - x\|}{1 - \|x^{-1}\|.\|y - x\|}$$

Thus $y^{-1} \to x^{-1}$ as $y \to x$; which shows that $x \mapsto x^{-1}$ is a continuous mapping; but this mapping is its own inverse and is thus a homeomorphism.

Definition. Let A be a Banach algebra and let $x \in A$; the **spectrum** of x in A is

$$Sp_A(x) = \{\lambda \in \mathbb{C} : \lambda 1 - x \notin A^{-1}\}$$

Theorem 1. $Sp_A(x)$ *is a non-empty compact subset of* $\mathbb{C}$.

Proof. $Sp_A(x)$ is **closed** in $\mathbb{C}$, by Lemma 1, Corollary 1, and the continuity of $\lambda \mapsto \lambda 1 - x$ as a mapping of $\mathbb{C}$ into A.

Also, if $\lambda \in \mathbb{C}$ and $|\lambda| > \|x\|$, then $\lambda 1 - x = \lambda(1 - \lambda^{-1}x)$ and, since $\|\lambda^{-1}x\| = |\lambda|^{-1}\|x\| < 1$, we see from Lemma 1 that $\lambda 1 - x \in A^{-1}$, so that $\lambda \notin Sp_A(x)$. Thus $Sp_A(x) \subseteq \{\lambda \in \mathbb{C} : |\lambda| \leqslant \|x\|\}$ and, in particular, $Sp_A(x)$ is **bounded**. Thus $Sp_A(x)$ is compact; it remains to prove that $Sp_A(x) \neq \emptyset$.

Observe first that defining $f(\lambda) = (\lambda 1 - x)^{-1}$ defines a **continuous** function $f : \mathbb{C} \backslash Sp_A(x) \to A^{-1}$ (for f is the composition of $\lambda \mapsto \lambda 1 - x$ and $x \mapsto x^{-1}$ and we may use Lemma 1, Corollary 2). But it is trivial that, for any $\lambda, \mu \in \mathbb{C} \backslash Sp_A(x)$:

$$f(\lambda) - f(\mu) = -(\lambda - \mu)\ f(\lambda)\ f(\mu)$$

so that

$$(\lambda - \mu)^{-1}\ (f(\lambda) - f(\mu)) = - f(\lambda)\ f(\mu) \to - f(\mu)^2$$

as $\lambda \to \mu$, by the continuity of f at μ. Thus f is an A-valued **holomorphic** function on $\mathbb{C} \backslash \mathrm{Sp}_A(x)$. But also, for $\lambda \neq 0$, $f(\lambda) = (\lambda 1 - x)^{-1} = \lambda^{-1}(1 - \lambda^{-1} x)^{-1}$, so that, by Lemma 1, Corollary 2, $\|f(\lambda)\| \to 0$ as $|\lambda| \to \infty$. Thus, if $\mathrm{Sp}_A(x) = \emptyset$, then f would be an **entire** A-valued function with $\|f(\lambda)\| \to 0$ as $|\lambda| \to \infty$; by Liouville's theorem (for vector-valued functions) we could then deduce $f(\lambda) \equiv 0$ on $\mathbb{C}$ – obviously a contradiction since $f(\lambda) \in A^{-1}$. Hence $\mathrm{Sp}_A(x) \neq \emptyset$.

Definition. The **spectral radius** $r_A(x)$ of x in A is

$$r_A(x) = \sup\{|\lambda| \ : \ \lambda \in \mathrm{Sp}_A(x)\}$$

The above theorem contained a proof of:

Lemma 2. $r_A(x) \leqslant \|x\|$ $(x \in A)$.

(There is a more precise result:

$$r_A(x) = \lim_{n \to \infty} \|x^n\|^{1/n}$$

but we shall not need it in the sequel).

Theorem 2 (**Gelfand–Mazur**). *If* A *is a Banach algebra and is a field, then* $A \cong \mathbb{C}$.

Proof. Let $x \in A$; by Theorem 1, there is a complex number $\lambda \in \mathrm{Sp}_A(x)$, i.e. such that $\lambda 1 - x \notin A$ But if A is a field then $\lambda 1 - x = 0$. Thus, the mapping $\lambda \mapsto \lambda.1$ is an isometric isomorphism of $\mathbb{C}$ onto A.

Lemma 3. *The closure* $\bar{I}$ *of a proper ideal* I *of* A *is again a proper ideal.*

Proof. Since $I \neq A$ we have $I \cap A^{-1} = \emptyset$; but A^{-1} is open and so $\bar{I} \cap A^{-1} = \emptyset$; since $1 \in A^{-1}$ we hav $\bar{I} \neq A$. The fact that $\bar{I}$ is an ideal follows easily from the continuity of the algebraic operations in

Corollary. *Each maximal ideal of* A *is closed in* A.

Proof. Let M be a maximal ideal; by Lemma 3, $\bar{M}$ is a proper ideal, $\bar{M} \supseteq M$; by maximality of M, $M = \bar{M}$.

Definition. A **character** ϕ on A is an algebra homomorphism of A onto $\mathbb{C}$; we write Φ_A for the set of all characters on A.

Lemma 4. *Any character* ϕ *on* A *is continuous and* $\|\phi\| = 1$.

Proof. Let $x \in A$ with $\|x\| \leqslant 1$ and suppose that $|\phi(x)| > 1$ for some character ϕ. Set $y = \phi(x)^{-1} x$; then $\|y\| = |\phi(x)|^{-1}\ \|x\| < 1$, while $\phi(y) = 1$. By Lemma 1, $1 - y \in A^{-1}$ so that, for some $z \in A$, $z(1 - y) = 1$. But then $1 = \phi(1) = \phi(z)\ (1 - \phi(y)) = 0$, a contradiction.

Thus for any $x \in A$ with $\|x\| \leqslant 1$ and any $\phi \in \Phi_A$ we have $|\phi(x)| \leqslant 1$. This completes the pro

If ϕ is any character on A then

$$\ker \phi = \{x \in A : \phi(x) = 0\}$$

is easily seen to be a maximal ideal of A. It is an important fact that the converse is also true:

Theorem 3. *Let* M *be a maximal ideal of* A; *then* $M = \ker \phi$ *for some* $\phi \in \Phi_A$.

Proof. If I is any closed ideal of A it is not hard to prove that the quotient algebra A/I becomes a Banach algebra under the **quotient norm**

$$\|[x]\| = \inf_{i \in I} \|x + i\|$$

where we have written [x] for the coset x + I. (The fact that A/I is complete is standard elementary Banach space theory — true for any closed **subspace** I).

If M is maximal then M is closed (by Corollary to Lemma 3); thus A/M is a Banach algebra in the quotient norm — but is also a field since M is maximal (an elementary bit of pure algebra). Thus, by Theorem 2, $A/M \cong \mathbb{C}$ and the quotient mapping $x \mapsto x + M$ is thus a character ϕ with $\ker \phi = M$.

Corollary. *Let* $x \in A$; *then*

(i) $x \in A^{-1}$ *if and only if* $\phi(x) \neq 0$ *for every* $\phi \in \Phi_A$;

(ii) $\mathrm{Sp}_A(x) = \{\phi(x) : \phi \in \Phi_A\}$;

(iii) *if* p *is any complex polynomial then* $\mathrm{Sp}_A(p(x)) = \{p(\lambda) : \lambda \in \mathrm{Sp}_A(x)\}$.

Proof. (i) If $x \in A^{-1}$ then $1 = \phi(1) = \phi(x)\,\phi(x^{-1})$, so that $\phi(x) \neq 0$ for each $\phi \in \Phi_A$. Suppose that $x \notin A^{-1}$; then $Ax = \{ax : a \in A\}$ is a proper ideal of A (for $1 \notin Ax$) and so, by a standard Zorn's Lemma argument, there is a maximal ideal $M \supseteq Ax$. By Theorem 3, $M = \ker \phi$ for some $\phi \in \Phi_A$ so that $\phi(x) = 0$ for such a ϕ. Thus, if $\phi(x) \neq 0$ for every $\phi \in \Phi_A$ we must have $x \in A^{-1}$.

(ii) $\mathrm{Sp}_A(x) = \{\lambda \in \mathbb{C} : \lambda 1 - x \notin A^{-1}\}$
$= \{\lambda \in \mathbb{C} : \phi(\lambda 1 - x) = 0 \text{ for some } \phi\}$

by part (i); thus $\mathrm{Sp}_A(x) = \{\phi(x) : \phi \in \Phi_A\}$.

(iii) If p is a complex polynomial then, since a character ϕ is a homomorphism, $\phi(p(x)) = p(\phi(x))$. Hence

$$\begin{aligned} \mathrm{Sp}_A(p(x)) &= \{\phi(p(x)) : \phi \in \Phi_A\} \\ &= \{p(\phi(x)) : \phi \in \Phi_A\} \\ &= \{p(\lambda) : \lambda \in \mathrm{Sp}_A(x)\} \end{aligned}$$

We shall now determine the character sets in some particular cases.

(a) *The Wiener algebra W* (see Example 6)

Let $u \in W$ be defined by $u(x) = e^{ix}$; then $u \in W^{-1}$ with $u^{-1}(x) = e^{-ix}$ and clearly $\|u\| = \|u^{-1}\| = 1$. Thus, if ϕ is any character on W, we have $|\phi(u)| \leqslant \|u\| = 1$ (by Lemma 4) and also $|\phi(u)|^{-1} = |\phi(u^{-1})| \leqslant \|u^{-1}\| = 1$.

Thus $|\phi(u)| = 1$ — say $\phi(u) = e^{i\alpha}$ for some $\alpha \in \mathbb{R}$. But then for a general $f \in W$, say

$$f(x) = \sum_{n=-\infty}^{\infty} f_n e^{inx} \text{ where } \sum |f_n| < \infty$$

we have

$$f = \sum_{n=-\infty}^{\infty} f_n u^n$$

with norm-convergence in W. Thus

$$\phi(f) = \sum_{n=-\infty}^{\infty} f_n\, \phi(u)^n = \sum_{n=-\infty}^{\infty} f_n\, e^{in\alpha} = f(\alpha)$$

Thus ϕ is simply "evaluation at α"; clearly two real numbers α, α' give rise to the same character if and only if $\alpha - \alpha' \in 2\pi\mathbb{Z}$, so that we may identify (for the moment just as a set) Φ_W with $\mathbb{R}/2\pi\mathbb{Z}$, i.e. with the unit circle.

As a consequence of this determination of Φ_W, together with the Corollary to Theorem 3, w deduce the following classical theorem of Wiener:

If $f \in W$ *and* $f(x) \neq 0$ *(all* $x \in \mathbb{R}$*) then* $1/f \in W$.

(b) *The algebra* C(K) (see Example 4)

Let K be a compact subset of $\mathbb{C}^n$ – or, more generally, *any* compact Hausdorff topological space. For any $k \in K$ the mapping $\epsilon_k : C(K) \to \mathbb{C}$ defined by $\epsilon_k(f) = f(k)$ is clearly a character on C(K) (we call it "evaluation at k"). We shall see that these are the only characters on C(K).

First, let I be any proper ideal of C(K). Suppose that, for each $k \in K$ there is a function $f_k \in$ with $f_k(k) \neq 0$; then, by continuity of f_k at k, there is an open neighbourhood U_k of k such that $f_k(k') \neq 0$ (all $k' \in U_k$). By compactness of K, there is a finite subset $\{k_1, ..., k_m\}$ of K such that K is covered by $\{U_{k_1}, ..., U_{k_m}\}$. But then the function

$$f = \sum_{r=1}^{m} |f_{k_r}|^2 = \sum_{r=1}^{m} \overline{f_{k_r}} \,.\, f_{k_r} \in I$$

$f(x) \neq 0$ for all $x \in K$, and so $f \in C(K)^{-1}$ which contradicts I being a *proper* ideal of C(K). Thus fo some point $k \in K$ we have $I \subseteq \ker \epsilon_k$.

In particular, if M is a maximal ideal then $M = \ker \epsilon_k$ for some $k \in K$. Thus, finally, if ϕ is any character, then, by Theorem 3, $\ker \phi = \ker \epsilon_k$ for some $k \in K$. But then, for any $f \in C(K)$, $f - f(k).1 \in \ker \epsilon_k = \ker \phi$, so that $\phi(f) = f(k)\,\phi(1) = f(k) = \epsilon_k(f)$; thus $\phi = \epsilon_k$.

(c) *The algebra* P(K) (see Example 5)

Let K be a compact subset of $\mathbb{C}^n$. We define the **polynomial hull** $\hat{K}$ of K to be

$$\hat{K} = \{z \in \mathbb{C}^n : |p(z)| \leqslant \|p\|_K, \text{ for every complex polynomial p in n variables}\}$$

(**Note**: here we have written $z = (z_1, ..., z_n)$ and

$$\|p\|_K = \sup_{z \in K} |p(z)|$$

We shall use similar notation freely without further comment).

We say that K is **polynomially convex** if and only if $K = \hat{K}$. For any compact K, $\hat{K}$ is a com polynomially convex subset of $\mathbb{C}^n$ which contains K. (Verify!)

Let p be a polynomial; then $\|p\|_{\hat{K}} = \|p\|_K$, so that, for the algebra of polynomial functions c $\hat{K}$, the operation of restriction to K is an isometric isomorphism; thus restriction from $\hat{K}$ to K als gives an isometric isomorphism of $P(\hat{K})$ onto P(K). It follows that $\Phi_{P(K)}$ and $\Phi_{P(\hat{K})}$ may be identified and we thus consider just $\Phi_{P(\hat{K})}$.

Clearly ϵ_k, evaluation at k, is a character for any $k \in \hat{K}$. Now let ϕ be any character on $P(\hat{K})$. By a common abuse of notation we shall write simply z_r for the co-ordinate projection $z \mapsto z_r$ $(z \in \hat{K})$. Then $z_r \in P(\hat{K})$ and we define $a_r = \phi(z_r)$, r = 1, ..., n. Then $a = (a_1, ..., a_n) \in \mathbb{C}^n$; but also, by Lemma 4, for any polynomial p,

$$\|p\|_{\hat{K}} \geqslant |\phi(p)| = |p(a)|$$

so that $a \in \hat{K}$, since $\hat{K}$ is polynomially convex. Thus, since $\phi(p) = p(a) = \epsilon_a(p)$ for each polynomial p, we also must have $\phi(f) = \epsilon_a(f)$ for each $f \in P(\hat{K})$ by continuity of characters.

Thus $\Phi_{P(K)} \cong \Phi_{P(\hat{K})}$ has been identified with the set $\hat{K}$.

Note: for n = 1 and K a compact subset of $\mathbb{C}$, $\hat{K}$ is the union of K with the bounded components of $\mathbb{C} \backslash K$ (essentially the classical theorem of Runge). No similar simple description is available for subsets of $\mathbb{C}^n$ $(n > 1)$.

We shall briefly describe a standard way of topologizing Φ_A for any commutative Banach algebra A. If we consider $\Phi_A \subset A'$, the Banach dual space of A, then we may topologize Φ_A as a subspace of A′ in its "weak topology", $\sigma(A',A)$ (in Bourbaki notation); this relative topology is called the **Gelfand topology** on Φ_A. The Gelfand topology is the weakest topology on Φ_A for which each of the mappings $\phi \mapsto \phi(x)$ $(x \in A)$, of Φ_A into $\mathbb{C}$, is continuous.

If $\phi_0 \in \Phi_A$ then a neighbourhood base of ϕ_0 for the Gelfand topology is provided by sets of the form:

$$N(\phi_0;\ x_1, ..., x_n) \equiv \{\phi \in \Phi_A : |\phi(x_k) - \phi_0(x_k)| < 1 \ (k = 1, ..., n)\}$$

where n is a positive integer and $x_1, ..., x_n$ are elements of A.

Theorem 4. *For the Gelfand topology, Φ_A is a non-empty compact Hausdorff space.*

Proof (outline). For each $x \in A$, let $D_x = \{\lambda \in \mathbb{C} : |\lambda| \leqslant \|x\|\}$. Then D_x is a compact disc in $\mathbb{C}$ and so, by Tychonoffs theorem, the topological product

$$P = \prod_{x \in A} D_x$$

is a compact Hausdorff space.

We define $\theta : \Phi_A \to P$ by $\theta(\phi) = \{\phi(x)\}_{x \in A}$.

Then it is an easy exercise to see that θ is a homeomorphism of Φ_A onto a closed subset of P. The result follows.

We note that, with Φ_W, $\Phi_{C(K)}$, $\Phi_{P(K)}$ given their respective Gelfand topologies, the identifications of these spaces respectively with the unit circle in $\mathbb{C}$, K, $\hat{K}$ given above are topological and not merely set-theoretic (the proofs are easy exercises).

It is sometimes useful to change the viewpoint slightly; if A is as before and $x \in A$ then the **Gelfand transform** $\hat{x}$ of x is the function $\hat{x} \in C(\Phi_A)$ defined by $\hat{x}(\phi) = \phi(x)$ $(\phi \in \Phi_A)$. (The fact that $\hat{x}$ is continuous on Φ_A follows immediately from the definition of the Gelfand topology.)

Theorem 5. (i) *The mapping* $x \mapsto \hat{x}$ *is a continuous homomorphism of* A *into* $C(\Phi_A)$;
(ii) *for any* $x \in A$:

$$Sp_A(x) = \{\hat{x}(\phi) : \phi \in \Phi_A\}$$

Proof. (i) This is more or less trivial; the continuity follows because, for any $x \in A$:

$$\|\hat{x}\| = \sup\{|\hat{x}(\phi)| : \phi \in \Phi_A\}$$

$$= \sup\{|\phi(x)| : \phi \in \Phi_A\} \leqslant \|x\|$$

by Lemma 4.

(ii) Part (ii) of the Corollary to Theorem 3.

Example. *A non-normable function algebra.* Let U be open, $U \subseteq \mathbb{C}^n$.

We now fulfil the promise in Example 3 to show that C(U) cannot be given an algebra-norm. First we find the characters.

Let $U^* = U \cup \{\infty\}$ be the 1-point compactification of U; then $C(U^*)$ may be identified with a subalgebra of C(U). Let ϕ be any character on C(U); then $\phi|C(U^*)$ is a character on $C(U^*)$ (not the zero functional since $1 \in C(U^*)$) and so, by (b) above, since U^* is compact, there is a point $w \in U^*$ such that $\phi(f) = f(w)$ for each $f \in C(U^*)$. We shall now show that $w \in U$, i.e. that w is not the "point at infinity". We may, in many ways, find a function $g \in C(U)$ such that $g(z) \neq 0$ (all $z \in U$) and $|g(z)| \to \infty$ whenever either $|z| \to \infty$ in U or z approaches ∂U from within U; then $g^{-1} \in C(U^*)$ with $g^{-1}(\infty) = 0$, whereas $\phi(g^{-1}) = g^{-1}(w) \neq 0$ since g^{-1} is invertible in C(U); thus $w \neq \infty$, so that $w \in U$. We now show that $\phi(f) = f(w)$ for every $f \in C(U)$.

Let θ be an element of C(U) such that:(i) supp θ is a compact subset of U, and (ii) $\theta(w) = 1$. Then, for any $f \in C(U)$, θf has compact support in U, so that, in particular, θ and θf are both in $C(U^*)$. Thus $1 = \theta(w) = \phi(\theta)$, while $f(w) = (\theta f)(w) = \phi(\theta f) = \phi(\theta)\phi(f) = \phi(f)$. Thus the characters on C(U) are precisely the evaluation maps ϵ_z $(z \in U)$. (By a **character** we mean a non-zero homomorphism from $C(U) \to \mathbb{C}$; we are not, of course, asserting that C(U) is a Banach algebra).

Now suppose that C(U) can be given an algebra-norm $\|.\|$; then C(U) may be completed in this norm to give a Banach algebra B, say. Then Φ_B may be identified with the set of those characters on C(U) which are $\|.\|$-continuous on C(U); since Φ_B is compact for the Gelfand topology $\sigma(\Phi_B; B)$, it is also compact for the weaker topology $\sigma(\Phi_B; C(U))$ (which is in fact the *same* topology); but this is just the relativization to Φ_B of the topology $\sigma(U, C(U))$ on U, and this latter is the usual topology of U as a subset of $\mathbb{C}^n$.

Thus Φ_B has been identified with a compact subset $K \subset U$, namely $K = \{z \in U : \epsilon_z$ is $\|.\|$-continuous on $C(U)\}$.

But then, by Urysohn's lemma, we may easily find functions $u, v \in C(U)$ such that:

(i) $u(z) = v(z) = 0$ (all $z \in K$)

(ii) $uv = v$

(iii) $v \neq 0$

But, by (i), $\phi(u) = 0$ for each $\phi \in \Phi_B$ and so $\phi(1-u) = 1$ for each $\phi \in \Phi_B$. But then, by Corollary to Theorem 3, $1 - u \in B^{-1}$. But, by (ii), $v(1-u) = 0$ and so $v = v(1-u)\cdot(1-u)^{-1} = 0$, contrary to (iii). This contradiction proves the result.

We now meet, in a simple case, the notion of **holomorphic functional calculus.**

Theorem 6. *Let* $x \in A$ *and let* U *be an open subset of* $\mathbb{C}$ *with* $Sp_A(x) \subset U$. *Then there is a unique continuous unital homomorphism* $\Theta_x : \mathcal{O}(U) \to A$ *such that* $\Theta_x(z) = x$. *Moreover, for any* $f \in \mathcal{O}(U)$ *and any* $\phi \in \Phi_A$ *we have* $\phi(\Theta_x(f)) = f(\phi(x))$.

Proof. Let R(U) denote the subalgebra of $\mathcal{O}(U)$ consisting of all rational functions with no pole in U; then, for any $f \in R(U)$, we must define $\Theta_x(f) = f(x)$ if a definition of Θ_x is at all possible. Moreover it is clear that $f \mapsto f(x)$ $(f \in R(U))$ defines a unital homomorphism from R(U) into A such that $z \mapsto x$ and also that $\phi(f(x)) = f(\phi(x))$, for any $f \in R(U)$ and any $\phi \in \Phi_A$. By the classical approximation theorem of Runge, R(U) is **dense** in $\mathcal{O}(U)$, for the topology τ of uniform convergence on compact subsets of U, and the proof is therefore completed by showing that $f \mapsto f(x)$ is a continuous mapping from R(U) (with topology τ) into A. To show this we note that it is elementary to find Γ, a finite union of piecewise-smooth closed curves, such that (i) $\Gamma \subset U$, (ii) every point of $\mathbb{C}\backslash U$ has index 0 with respect to Γ, (iii) every point of $Sp_A(x)$ has index 1 with respect to Γ. Then it is again elementary that, for any $f \in R(U)$:

$$f(x) = \frac{1}{2\pi i} \int_\Gamma f(\xi)(\xi 1 - x)^{-1} d\xi \qquad (*)$$

and that,therefore,

$$\|f(x)\| \leqslant C \cdot \|f\|_\Gamma \quad (f \in R(U))$$

where C is independent of f. The result is thus proved; we note that the formula (*) must continue to hold for any $f \in \mathcal{O}(U)$, where f(x) is now taken to mean $\Theta_x(f)$.

Corollary. *If* $x \in A$ *and* $Sp_A(x)$ *is disconnected then* A *contains a non-trivial idempotent* e.

Proof. Let K be a non-empty, proper open-and-closed subset of $Sp_A(x)$ and let $L = Sp_A(x)\backslash K$. Then K, L are disjoint non-empty compact subsets of $\mathbb{C}$.

Let V, W be disjoint open neighbourhoods of K, L respectively and set $U = V \cup W$. Then the function E, defined to be 0 on W and 1 on V belongs to $\mathcal{O}(U)$ and satisfies $E^2 = E$. Thus, if we put $e = \Theta_x(E)$ we have $e^2 = e$ and $Sp_A(e) = \{0, 1\}$, so that $e \neq 0, 1$.

The rest of this paper is mainly devoted to a several-variable extension of Theorem 6, together with some applications. The original motive for this extension was precisely to give a several-variable version of the above Corollary. We shall see that the extension of Theorem 6 provides a profound tool in the theory of Banach algebras. First we need some definitions:

Let A be as before and let $x_1, ..., x_n \in A$, where $n \geqslant 1$; write $\underset{\sim}{x} = (x_1, ..., x_n) \in A^n$. Then the **joint spectrum**, $Sp_A(\underset{\sim}{x}) \equiv Sp_A(x_1, ..., x_n)$ is defined to be

$$Sp_A(\underset{\sim}{x}) = \{(\phi(x_1), ..., \phi(x_n)) : \phi \in \Phi_A\}$$

We note that, by an argument analogous to the proof of Theorem 3, Corollary, we have that, also,

$$Sp_A(\underset{\sim}{x}) = \{\underset{\sim}{\lambda} = (\lambda_1, ..., \lambda_n) \in \mathbb{C}^n : id_A(\lambda_k 1 - x_k : k = 1, ..., n) \neq A\}$$

where we have written $id_A(\lambda_k 1 - x_k : k = 1, ..., n)$ to mean the ideal of A generated by $\{\lambda_k 1 - x_k : k = 1, ..., n\}$.

Lemma 4. (i) $Sp_A(\underset{\sim}{x})$ *is a non-empty compact subset of* $\mathbb{C}^n$;

(ii) *If* $1 \leqslant m \leqslant n$ *and if* $\pi : \mathbb{C}^n \to \mathbb{C}^m$ *is projection onto the first* m *co-ordinates, then*

$$\pi(Sp_A(\underset{\sim}{x})) = Sp_A(x_1, ..., x_m)$$

(iii) $Sp_A(\underset{\sim}{x}) \subseteq \prod_{k=1}^{n} Sp_A(x_k) \subseteq \{\underset{\sim}{\lambda} \in \mathbb{C}^n : |\lambda_k| \leqslant r_A(x_k),\ k = 1, ..., n\}$

(iv) $Sp_A(p(x_1, ..., x_n)) = \{p(\lambda_1, ..., \lambda_n) : \underset{\sim}{\lambda} \in Sp_A(\underset{\sim}{x})\}$ *for any complex polynomial* p *in* n-*variables.*

Proof. (i) Follows from Theorem 4, since $\phi \mapsto (\phi(x_1), ..., \phi(x_n))$ is a continuous mapping of Φ_A onto $Sp_A(\underset{\sim}{x})$.

(ii), (iii), (iv) all follow more or less trivially from the definition of joint spectrum.

We shall ultimately see that an analogue of Theorem 6 holds in the case of several variables, with the joint spectrum $Sp_A(\underset{\sim}{x})$ playing the role of the spectrum. For the moment we content ourselves with a much simpler result.

Theorem 7. *Let* $x_1, ..., x_n \in A$ *and let* $D = \{\underset{\sim}{\lambda} \in \mathbb{C}^n : |\lambda_k| < r_k \ (k = 1, ..., n)\}$ *be an open polydisc in* $\mathbb{C}^n$ *with* $Sp_A(\underset{\sim}{x}) \subset D$. *Then there is a unique continuous unital homomorphism* $\Theta_{\underset{\sim}{x}} : \mathcal{O}(D) \to A$ *such that* $\Theta_{\underset{\sim}{x}}(z_k) = x_k \ (k = 1, ..., n)$. *Moreover, for any* $\phi \in \Phi_A$ *and any* $f \in \mathcal{O}(D)$ *we have*

$$\phi(\Theta_{\underset{\sim}{x}}(f)) = f(\phi(x_1), ..., \phi(x_n))$$

(As before, z_k is the kth co-ordinate projection $z_k(\underset{\sim}{\lambda}) = \lambda_k \ (k = 1, ..., n)$).

Proof. Choose $0 < r_k' < r_k \ (k = 1, ..., n)$ such that $Sp_A(\underset{\sim}{x}) \subset \{\underset{\sim}{\lambda} \in \mathbb{C}^n : |\lambda_k| < r_k' \ (k = 1, ..., n)\}$ Then, for any $f \in \mathcal{O}(D)$ define

$$\Theta_{\underset{\sim}{x}}(f) = \frac{1}{(2\pi i)^n} \underset{\substack{|\xi_k| = r_k' \\ (k = 1, ..., n)}}{\int \cdots \int} f(\underset{\sim}{\xi}) \cdot (\xi_1 1 - x_1)^{-1} \cdots (\xi_n 1 - x_n)^{-1} \, d\xi_1 \cdots d\xi_n$$

The proof that $\Theta_{\underset{\sim}{x}}$, so defined, has the required properties is very similar to that of Theorem 6 but technically a little simpler, since we have the set of **polynomials** (rather than just rational functions) dense in $\mathcal{O}(D)$, by Taylor's formula. We shall omit the details.

Our basic problem is to substitute for D a more general open neighbourhood which may approximate $Sp_A(\underset{\sim}{x})$ more closely.

The importance of **polynomial convexity** (introduced in the discussion of P(K) above) lies in the following result, with which we conclude the present section.

Lemma 5. *If* A *is generated, as a Banach algebra with identity, by* $x_1, ..., x_n$, *then* $Sp_A(\underset{\sim}{x})$ *is polynomially convex.*

(Note: the condition of this lemma means that polynomials in $x_1, ..., x_n$ are dense in A.)

Proof. Let $\underset{\sim}{\lambda} = (\lambda_1, ..., \lambda_n) \in \mathbb{C}^n \backslash Sp_A(\underset{\sim}{x})$. Then there are elements $y_1, ..., y_n$ in A such that

$$\sum_{k=1}^{n} y_k(\lambda_k 1 - x_k) = 1$$

Since polynomials in $x_1, ..., x_n$ are dense in A, we may approximate $y_1, ..., y_n$ as closely as we wish by suitable choice of such polynomials. In particular we may choose polynomials $p_1, ..., p_n$ such that

$$\left\| \sum_{k=1}^{n} p_k(x_1, ..., x_n) \cdot (\lambda_k 1 - x_k) - 1 \right\| < 1$$

Now define the complex polynomial q by

$$q(\mu_1, \ldots, \mu_n) = \sum_{k=1}^{n} p_k(\mu_1, \ldots, \mu_n)(\lambda_k - \mu_k) - 1$$

Then $|q(\underset{\sim}{\lambda})| = |-1| = 1$, but for any $\underset{\sim}{\mu} \in Sp_A(\underset{\sim}{x})$ we have $q(\underset{\sim}{\mu}) \in Sp_A(q(x_1, \ldots, x_n))$, by Lemma 4 (iv), and so $|q(\underset{\sim}{\mu})| \leqslant \|q(x_1, \ldots, x_n)\| < 1$. Thus $\underset{\sim}{\lambda} \notin \widehat{Sp_A(\underset{\sim}{x})}$ and therefore $Sp_A(\underset{\sim}{x})$ is polynomially convex.

B. GROUPS OF DIFFERENTIAL FORMS

It is useful to introduce a little abstract algebra; only the most elementary facts are presented.

By a **differential group** we mean a pair $(X; d)$, where X is an Abelian group (written additively) and $d : X \to X$ is a group endomorphism such that $d^2 = 0$.

The endomorphism d is the **differential** of X. We define subgroups:

$Z(X) = \ker d = \{x \in X : d(x) = 0\}$ (elements of Z(X) are called **closed** elements)

$B(X) = \operatorname{im} d = \{d(x) : x \in X\}$ (elements of B(X) are called **exact** elements).

Since $d^2 = 0$ we have $B(X) \subseteq Z(X)$ and may define

$$H(X) = Z(X)/B(X)$$

the **homology group** of X. If $x \in Z(X)$ we write $[x]$ for its canonical image in H(X) – called the **homology class** of x.

A **morphism** of differential groups X, X′ is a group homomorphism $f : X \to X'$ such that

$$\begin{array}{ccc} X & \xrightarrow{f} & X' \\ d\downarrow & & \downarrow d' \\ X & \xrightarrow{f} & X' \end{array}$$

commutes. (Of course d, d′ are the differentials of X, X′ respectively.) If $f : X \to X'$ is a morphism of differential groups then clearly $fZ(X) \subseteq Z(X')$, $fB(X) \subseteq B(X')$ and there is thus a naturally induced group homomorphism $H(f) : H(X) \to H(X')$. More precisely, H is a covariant functor from the category of differential groups and their morphisms to the category of Abelian groups and their homomorphisms.

The Abelian group X will be called **graded** (by the integers) if for each integer n there is a subgroup X_n with

$$X = \sum_{n=-\infty}^{\infty} \oplus X_n$$

In this context, the elements of X_n are called **homogeneous of degree n.** A differential map $D : X \to X$ will be said to **respect the grading** provided $dX_n \subseteq X_{n+1}$ for each n. The pair $(X; d)$ is then a **graded differential group.** If X, X′ are graded differential groups then a **morphism** $f : X \to X'$ is defined to be a morphism of the underlying differential groups such that $f(X_n) \subseteq X'_n$ for each n.

If (X, d) is graded then the homology group H(X) inherits a natural grading:

$$H(X) = \sum_{n=-\infty}^{\infty} H_n(X)$$

where

$$H_n(X) \cong (Z(X) \cap X_n)/B(X) \cap X_n$$

for each n. The group $H_n(X)$ is the nth **homology group** of X

If $f : X \to X'$ is a morphism of graded differential groups, then the map $H(f) : H(X) \to H(X')$ is also graded, i.e. H(f) maps $H_n(X)$ into $H_n(X')$ for each n.

Example: We now turn to our prime example of a differential group. Let U be an open subset of $\mathbb{C}^n$; write $C^\infty(U)$ for the algebra of all smooth (i.e. infinitely differentiable) $\mathbb{C}$-valued functions on For p = 0, 1, 2, ... we shall write $\overline{\mathscr{E}}^p(U)$ for the group of all complex-conjugate smooth differential forms on U of type (0, p). Thus a typical element of $\overline{\mathscr{E}}^p(U)$ looks like

$$\omega = \sum \omega_{i_1 \dots i_p} \, d\bar{z}_{i_1} \wedge \dots \wedge d\bar{z}_{i_p}$$

where each $\omega_{i_1 \dots i_p} \in C^\infty(U)$ and the summation is over all p-tuples $(i_1, \dots, i_p)$ of positive integers such that $1 \leqslant i_1 < i_2 < \dots < i_p \leqslant n$.

Thus

$$\begin{cases} \overline{\mathscr{E}}^0(U) = \mathscr{C}^\infty(U) \\ \overline{\mathscr{E}}^p(U) = 0 \ (p \geqslant n+1) \end{cases}$$

The $\bar{\partial}$ operator maps $\overline{\mathscr{E}}^p(U) \to \overline{\mathscr{E}}^{p+1}(U)$ for each p and extends uniquely, by additivity to a group endomorphism of

$$\overline{\mathscr{E}}(U) \equiv \sum_{p=0}^{n} \oplus \overline{\mathscr{E}}^p(U) \equiv \sum_{p=-\infty}^{\infty} \oplus \overline{\mathscr{E}}^p(U)$$

(where we put $\overline{\mathscr{E}}^p(U) = 0$ for $p < 0$).

By well-known properties of $\bar{\partial}$ (see e.g. Part I of the paper by M.J. Field in Volume I of these Proceedings), we have $\bar{\partial}^2 = 0$; thus $(\overline{\mathscr{E}}(U), \bar{\partial})$ is a graded differential group.

The homology groups $H^p(\overline{\mathscr{E}}(U))$ are called the **Dolbeault cohomology groups** of U.

By the Cauchy-Riemann criteria, we have $H^0(\overline{\mathscr{E}}(U)) = \mathscr{O}(U)$ for any open $U \subset \mathbb{C}^n$.

We now give an important example of a morphism of graded differential groups.

Let U be open in $\mathbb{C}^n$, V open in $\mathbb{C}^m$ and let $F_1, \dots, F_m \in \mathscr{O}(U)$. For each $z \in U$ define $F(z) \in \mathbb{C}^m$ by $F(z) = (F_1(z), \dots, F_m(z))$ and suppose that $F(U) \subseteq V$; such an F is called a **holomorphic mapping** of U into V. If, now, $\omega \in \overline{\mathscr{E}}^p(V)$, say

$$\omega = \sum \omega_{i_1 \dots i_p} \, d\bar{\xi}_{i_1} \wedge \dots \wedge d\bar{\xi}_{i_p}$$

where each $\omega_{i_1 \dots i_p} \in C^\infty(V)$ and we write $\xi = (\xi_1, \dots, \xi_m)$ for the current variable in $\mathbb{C}^m$, then we define

$$F^*(\omega) = \sum (\omega_{i_1 \dots i_p} \circ F) \, \bar{\partial} \bar{F}_{i_1} \wedge \dots \wedge \bar{\partial} \bar{F}_{i_p}$$

where, of course, summation in each case is over all p-tuples $(i_1, \dots, i_p)$ of positive integers such that $1 \leqslant i_1 < \dots < i_p \leqslant n$. Then $F^*(\omega) \in \mathscr{E}^p(U)$ and F^* extends uniquely by additivity to a group homomorphism from $\mathscr{E}(V)$ into $\mathscr{E}(U)$. A simple calculation (essentially the "chain rule") shows that $\bar{\partial}(F^*\omega) = F^*(\bar{\partial}\omega)$, so that F^* is thus a morphism of the graded differential groups $\mathscr{E}(V)$, $\mathscr{E}(U)$.

We conclude this section with a special case of the so-called first Cousin problem.

Lemma 6. *Let* U, V *be open subsets of* $\mathbb{C}^n$ *and let* $H^{p+1}(\mathscr{E}(U \cup V)) = 0$ *for some* $p \geqslant 0$. *Then, given* $\omega \in \mathscr{E}^p(U \cap V)$ *with* $\bar{\partial}\omega = 0$, *there exist* $\alpha \in \mathscr{E}^p(U)$, $\beta \in \mathscr{E}^p(V)$, *with* $\bar{\partial}\alpha = 0$, $\bar{\partial}\beta = 0$, *such that* $\omega = \alpha - \beta$ *on* $U \cap V$.

It seems worth stating the case p = 0 separately:

Corollary. *Let* U, V *be open subsets of* $\mathbb{C}^n$ *and let* $H^1(\mathscr{E}(U \cup V)) = 0$. *Then, given* $f \in \mathscr{O}(U \cap V)$, *there exist* $g \in \mathscr{O}(U)$, $h \in \mathscr{O}(V)$ *such that* $f = g - h$ *on* $U \cap V$.

Proof of Lemma 6. We first observe that we can find $\theta \in C^\infty(U \cup V)$ such that $\theta \equiv 1$ on a neighbourhood of U\V and $\theta \equiv 0$ on a neighbourhood of V\U. (This is a special case of a well-known fact about existence of partitions of unity).

Now let $\omega \in \mathscr{E}^p(U \cap V)$ with $\bar{\partial}\omega = 0$. We define $\alpha_1 \in \mathscr{E}^p(U)$, $\beta_1 \in \mathscr{E}^p(V)$ by:

$$\alpha_1(z) = \begin{cases} (1 - \theta(z)) \, \omega(z) & (z \in U \cap V) \\ 0 & (z \in U \setminus V) \end{cases}$$

$$\beta_1(z) = \begin{cases} -\theta(z) \, \omega(z) & (z \in U \cap V) \\ 0 & (z \in V \setminus U) \end{cases}$$

The fact that α_1, β_1 are **smooth** forms follows from the stated properties of θ; clearly $\omega = \alpha_1 - \beta_1$ on $U \cap V$. However, α_1, β_1 need not be closed forms.

Now observe that, on $U \cap V$,

$$0 = \bar{\partial}\omega = \bar{\partial}\alpha_1 - \bar{\partial}\beta_1$$

and we may thus uniquely define $\eta \in \mathscr{E}^{p+1}(U \cup V)$ by putting:

$$\eta = \begin{cases} \bar{\partial}\alpha_1 \text{ on } U \\ \bar{\partial}\beta_1 \text{ on } V \end{cases}$$

Moreover, $\bar{\partial}\eta = 0$, being equal to $\bar{\partial}^2\alpha_1$ on U and to $\bar{\partial}^2\beta_1$ on V. By our assumption that $H^{p+1}(\mathscr{E}(U \cup V)) = 0$, there is thus a form $\gamma \in \mathscr{E}^p(U \cup V)$ such that $\eta = \bar{\partial}\gamma$. We now define $\alpha = \alpha_1 - \gamma$ on U and $\beta = \beta_1 - \gamma$ on V; then $\alpha - \beta = \alpha_1 - \beta_1 = \omega$ on $U \cap V$, while $\bar{\partial}\alpha = \bar{\partial}\alpha_1 - \bar{\partial}\gamma = \bar{\partial}\alpha_1 - \eta = 0$ on U and, similarly, $\bar{\partial}\beta = 0$ on V. The proof is thus complete.

C. POLYNOMIALLY CONVEX DOMAINS IN $\mathbb{C}^n$

Recall that, for a compact set $K \subset \mathbb{C}^n$, its polynomial hull is

$$\hat{K} = \{z \in \mathbb{C}^n : |p(z)| \leqslant \|p\|_K \text{ (all polynomials p)}\}$$

and that K is polynomially convex if and only if $K = \hat{K}$.

We now define a **polynomially convex domain** to be an **open** subset U of $\mathbb{C}^n$ such that for any compact $K \subset U$ we have also $\hat{K} \subset U$.

An **open polydisc** is a set of the form

$$D = \{z \in \mathbb{C}^n : |z_k - a_k| < r_k \ (k = 1, ..., n)\}$$

where the **centre** $\underset{\sim}{a} = (a_1, ..., a_n) \in \mathbb{C}^n$ and the **polyradius** $\underset{\sim}{r} = (r_1, ..., r_n)$ is an n-tuple of positive real numbers.

A **polynomial polyhedron** is a set of the form

$$P = \{z \in D : |p_k(z)| < 1 \ (k = 1, ..., \nu)\}$$

where D is an open polydisc in $\mathbb{C}^n$ and $p_1, ..., p_\nu$ are finitely many polynomials. An open polydisc is thus a special kind of polynomial polyhedron. It is a very simple exercise to show that any polynomial polyhedron is a polynomially convex domain.

An easy compactness argument proves the following:

Lemma 7. *Let* K *be a compact polynomially convex subset of* $\mathbb{C}^n$ *and let* U *be open,* $U \supset K$. *Then there is a polynomial polyhedron* P *with* $K \subset P \subset U$.

We shall need to know that $H^p(\overline{\mathscr{E}}(P)) = 0$ for all $p \geqslant 1$, where P is any polynomial polyhedron. Although we only use the result in the case p = 1, the method of proof involves proving the theorem simultaneously for p = 1, 2, 3, ...

Our starting point is the case of an open polydisc; this result will be stated without proof; a proof may be found, e.g. in Ref. [3] (Chapter I, D5).

Theorem 8. *If* D *is an open polydisc in* $\mathbb{C}^n$, *then* $H^p(\overline{\mathscr{E}}(D)) = 0 \ (p \geqslant 1)$.

We shall now show how to extend this result to a general polynomial polyhedron; the method is due to Oka and the key idea lies in the following lemma.

Suppose that U is open in $\mathbb{C}^n$, that $\phi \in \mathscr{O}(U)$ and let $\Delta = \{z \in \mathbb{C} : |z| < 1\}$. We define $U_\phi = \{z \in U : |\phi(z)| < 1\}$.

Lemma 8. *If* $H^p(\overline{\mathscr{E}}(U \times \Delta)) = 0$ *for all* $p \geqslant 1$ *then*

(i) $H^p(\overline{\mathscr{E}}(U_\phi)) = 0$ *(all* $p \geqslant 1$*)*;

(ii) *for any* $f \in \mathscr{O}(U_\phi)$, *there is a function* $F \in \mathscr{O}(U \times \Delta)$ *such that* $f(z) = F(z, \phi(z)) \ (z \in U_\phi)$.

Proof. Define $\mu : U \to \mathbb{C}^{n+1}$ by $\mu(z) = (z, \phi(z))$; observe that $U_\phi = \mu^{-1}(U \times \Delta)$. Then μ is certainly a holomorphic mapping and we may thus define the morphism of graded differential groups $\mu^* : \overline{\mathscr{E}}(U \times \Delta) \to \overline{\mathscr{E}}(U_\phi)$, as in the section just before Lemma 6. By hypothesis, $H^p(\overline{\mathscr{E}}(U \times \Delta)) = 0 \ (p \geqslant 1)$ and the lemma would thus be proved if we could show that, for each $p \geqslant 0$, $H(\mu^*)$ maps $H^p(\overline{\mathscr{E}}(U \times \Delta))$ *onto* $H^p(\overline{\mathscr{E}}(U_\phi))$. (Recall that $H^0(\overline{\mathscr{E}}(V)) = \mathscr{O}(V)$ for any open set V, so that part (ii) of the lemma is just the case p = 0). But it is then sufficient to show that, for each $p \geqslant 0$ and each p-form ω on U_ϕ with $\bar{\partial}\omega = 0$, there exists $\Omega \in \overline{\mathscr{E}}^p(U \times \Delta)$ with $\bar{\partial}\Omega = 0$ and $\mu^*(\Omega) = \omega$.

Thus, let $\omega \in \overline{\mathscr{E}}^p(U_\phi)$, with $\bar\partial\omega = 0$. Define $A = U_\phi \times \Delta$, $B = \{(z,w) \in U \times \Delta : w \neq \phi(z)\}$; then A, B are open subsets of $U \times \Delta$ and $A \cup B = U \times \Delta$. On $A \cap B$ define the p-form η by

$$\eta(z,w) = (w - \phi(z))^{-1}\ \omega(z)$$

Then, since $(w - \phi(z))^{-1} \in \mathscr{O}(A \cap B)$ and $\bar\partial\omega = 0$, we deduce that $\bar\partial\eta = 0$ on $A \cap B$. But then, by Lemma 6, there exist $\alpha \in \overline{\mathscr{E}}^p(A)$, $\beta \in \overline{\mathscr{E}}^p(B)$ with $\bar\partial\alpha = 0$, $\bar\partial\beta = 0$ and $\eta = \alpha - \beta$ on $A \cap B$. Thus, on $A \cap B$,

$$\omega(z) - (w - \phi(z))\,\alpha(z,w) = -\,(w - \phi(z))\,\beta(z,w)$$

There is thus a well-defined element $\Omega \in \overline{\mathscr{E}}^p(U \times \Delta)$ defined by

$$\Omega(z,w) = \begin{cases} \omega(z) - (w - \phi(z))\,\alpha(z,w) \text{ on A} \\ -\,(w - \phi(z))\,\beta(z,w) \text{ on B} \end{cases}$$

Clearly $\bar\partial\Omega = 0$ and $\mu^*(\Omega) = \omega$. The proof is thus complete.

Corollary 1. *Let* D *be an open polydisc in* $\mathbb{C}^n$, *let* $p_1, \ldots, p_\nu$ *be given polynomials and let*

$$P = \{z \in D : |p_k(z)| < 1 \quad (k = 1, \ldots, \nu)\}$$

Then (i) $H^p(\overline{\mathscr{E}}(P)) = 0$ $(p \geqslant 1)$; (ii) *for any* $f \in \mathscr{O}(P)$ *there is a function* $F \in \mathscr{O}(D \times \Delta^\nu)$ *with*

$$f(z) = F(z, p_1(z), \ldots, p_\nu(z)) \quad (z \in P)$$

Proof. This is a simple induction on ν, using Lemma 8. The case $\nu = 1$ follows from Lemma 8 since $H^p(D \times \Delta) = 0$ $(p \geqslant 1)$ by Theorem 8. We omit further detail.

Corollary 2. *Let* Q *be a polynomial polyhedron, let* $p_1, \ldots, p_\nu$ *be given polynomials and let*

$$P = \{z \in Q : |p_k(z)| < 1 \ (k = 1, \ldots, \nu)\}$$

Then for any $f \in \mathscr{O}(P)$ *there is a function* $F \in \mathscr{O}(Q \times \Delta^\nu)$ *with*

$$f(z) = F(z, p_1(z), \ldots, p_r(z)) \quad (z \in P)$$

Proof. This is similar to Corollary 1, except that the start of the induction uses Corollary 1 itself to give $H^1(Q \times \Delta) = 0$ (for $Q \times \Delta$ is a polynomial polyhedron in $\mathbb{C}^{n+1}$).

Corollary 3 **(Oka - Weil theorem).**

Let P *be a polynomial polyhedron in* $\mathbb{C}^n$; *then the set of polynomials is dense in* $\mathscr{O}$(P) *(for the topology of uniform convergence on compact subsets of* P).

Proof. Let P be as in Corollary 1, let $f \in \mathscr{O}(P)$ and let $F \in \mathscr{O}(D \times \Delta^\nu)$ be such that

$$f(z) = F(z, p_1(z), \ldots, p_\nu(z))$$

Then substitution of $p_1(z), \ldots, p_\nu(z)$ into the Taylor series of F on $D \times \Delta^\nu$ gives the desired sequence of approximations to f – by polynomials in $\{z_1, \ldots, z_n, p_1(z), \ldots, p_\nu(z)\}$, hence by polynomials in $\{z_1, \ldots, z_n\}$.

Corollary 4. *With the notation of Corollary 2, let* $F \in \mathcal{O}(Q \times \Delta^\nu)$ *be such that*

$$F(z, p_1(z), ..., p_\nu(z)) = 0 \text{ (all } z \in P)$$

Then there exist $H_1, ..., H_\nu \in \mathcal{O}(Q \times \Delta^\nu)$ *such that*

$$F(z, w) = \sum_{k=1}^{\nu} H_k(z, w)(w_k - p_k(z))$$

for all $(z, w) \equiv (z_1, ..., z_n, w_1, ..., w_\nu) \in Q \times \Delta^\nu$.

Proof. This is again by induction on ν, but we shall give some details.

Case $\nu = 1$ (this case makes no use of any cohomology vanishing): We have $F \in \mathcal{O}(Q \times \Delta)$ such that $F(z, p(z)) = 0$ for any $z \in Q_p \equiv \{z \in Q : |p(z)| < 1\}$, and we wish to find $H \in \mathcal{O}(Q \times \Delta)$ such that $F(z, w) = H(z, w)(w - p(z))$ on $Q \times \Delta$. Any such H is clearly **unique** on any open set (if such an H exists at all), so that it is sufficient to define an H on a neighbourhood of each point of $Q \times \Delta$.

Thus, let $(z_0, w_0) \in Q \times \Delta$ and suppose that $w_0 = p(z_0)$ (for otherwise the definition of H is trivial). Now the mapping $\theta(z, w) = (z, w - p(z)$ is $(1-1)$ holomorphic of $\mathbb{C}^n$ onto $\mathbb{C}^n$, the inverse mapping being $\theta^{-1}(z, w) = (z, w + p(z))$, which is clearly also holomorphic. Now $\theta(z_0, w_0) = (z_0, 0)$, so that $F \circ \theta^{-1}$ is holomorphic on the open neighbourhood $\theta(Q \times \Delta)$ of $(z_0, 0)$ and $(F \circ \theta^{-1})(z, w) = 0$ whenever $w = 0$; thus the Taylor expansion of $F \circ \theta^{-1}$ at $(z_0, 0)$ contains w as a factor, i.e. $F(\theta^{-1}(z, w)) = wH'(z, w)$ on some open $W \ni (z_0, 0)$, where $H' \in \mathcal{O}(W)$. Thus, for $(z, w) \in \theta^{-1}(w)$ we have $F(z, w) = (w - p(z)) H(z, w)$, where $H = H' \circ \theta \in \mathcal{O}(\theta^{-1}(w))$. Thus the result is proved for $\nu = 1$.

Case $\nu > 1$: We shall use notation such as $V_{(p_1, ..., p_\nu)}$ to mean $\{z \in V : |p_k(z)| < 1 \ (k = 1, ..., \nu)\}$.

We are given $F \in \mathcal{O}(Q \times \Delta^\nu)$ with

$$F(z, p_1(z), ..., p_\nu(z)) = 0 \quad (z \in Q_{(p_1, ..., p_\nu)})$$

But $Q_{(p_1, ..., p_\nu)} = (Q_{p_1})_{(p_2, ..., p_\nu)}$; we define $G \in \mathcal{O}(Q_{p_1} \times \Delta^{\nu \ 1})$ by

$$G(z, w_2, ..., w_\nu) = F(z, p_1(z), w_2, ..., w_\nu)$$

(for $(z, w_2, ..., w_\nu) \in Q_{p_1} \times \Delta^{\nu-1}$); then we have

$$G(z, p_2(z), ..., p_\nu(z)) = 0 \quad (z \in (Q_{p_1})_{p_2, ..., p_\nu})$$

Thus, by the induction hypothesis, we can find $H'_2, ..., H'_\nu$ in $\mathcal{O}(Q_{p_1} \times \Delta^{\nu-1}$ such that

$$G(z, w_2, ..., w_\nu) = \sum_{k=2}^{\nu} H'_k(z, w_2, ..., w_\nu)(w_k - p_k(z)), \text{ on } Q_{p_1} \times \Delta^{\nu-1}$$

By Corollary 2, there are $H_k \in \mathcal{O}(Q \times \Delta^\nu)$ $(k = 2, ..., \nu)$ such that

$$H'_k(z, w_2, ..., w_\nu) = H_k(z, p_1(z), w_2, ..., w_\nu)$$

for $(z, w_2, ..., w_\nu) \in Q_{p_1} \times \Delta^{\nu-1}$. Thus

$$F(z, w_1, ..., w_\nu) - \sum_{k=2}^{\nu} H_k(z, w_1, ..., w_\nu)(w_k - p_k(z))$$

is holomorphic on $Q \times \Delta^\nu$ and vanishes whenever $w_1 = p_1(z)$. Thus, by case $\nu = 1$ above, this function has the form $H_1(z, w_1, ..., w_\nu)(w_1 - p_1(z))$ for some $H_1 \in \mathcal{O}(Q \times \Delta^\nu)$. The proof is thus complete.

D. HOLOMORPHIC FUNCTIONAL CALCULUS

We start by recalling Theorem 7; if A is a commutative Banach algebra and $x_1, ..., x_n \in A$, then for any open polydisc $D \supset Sp_A(\underset{\sim}{x})$ there is a unique continuous unital homomorphism $\Theta_{\underset{\sim}{x}} : \mathcal{O}(D) \to A$ such that $\Theta_{\underset{\sim}{x}}(z_k) = x_k$. As explained in Section A, we aim to replace D by an arbitrary open neighbourhood of $Sp_A(\underset{\sim}{x})$. The next step is as follows:

Lemma 9. *Let* P *be a polynomial polyhedron* $P \supset Sp_A(\underset{\sim}{x})$. *Then there is a unique continuous unital homomorphism* $\Theta_{\underset{\sim}{x}} : \mathcal{O}(P) \to A$ *with* $\Theta_{\underset{\sim}{x}}(z_k) = x_k$ $(k = 1, ..., n)$. *Moreover, for any* $f \in \mathcal{O}(P)$ *and any* $\phi \in \Phi_A$ *we have* $\phi(\Theta_{\underset{\sim}{x}}(f)) = f(\phi(x_1), ..., \phi(x_n))$.

Proof. Let $P = \{z \in D : |p_k(z)| < 1 \ (k = 1, ..., \nu)\}$, where D is an open polydisc in $\mathbb{C}^n$ and $p_1, ..., p_\nu$ are polynomials. Define

$$\rho : \mathcal{O}(D \times \Delta^\nu) \to \mathcal{O}(P)$$

by putting

$$\rho(F)(z) = F(z, p_1(z), ..., p_\nu(z)) \quad (z \in P)$$

Then ρ is clearly a continuous homomorphism and, by Lemma 8, Corollary 1, ρ is **surjective.** Thus, since $\mathcal{O}(D \times \Delta^\nu)$, $\mathcal{O}(P)$ are Fréchet spaces, ρ is an open mapping (by the open mapping theorem).

Also we have defined, by Theorem 7, a continuous unital homomorphism:

$$\Theta_{\underset{\sim}{x}, p_1(\underset{\sim}{x}), ..., p_\nu(\underset{\sim}{x})} : \mathcal{O}(D \times \Delta^\nu) \to A$$

such that

$$\begin{cases} \Theta_{\underset{\sim}{x}, p_1(\underset{\sim}{x}), ..., p_\nu(\underset{\sim}{x})}(z_k) = x_k \quad (k = 1, ..., n) \\ \Theta_{\underset{\sim}{x}, p_1(\underset{\sim}{x}), ..., p_\nu(\underset{\sim}{x})}(w_k) = p_k(\underset{\sim}{x}) \quad (k = 1, ..., \nu) \end{cases}$$

where $\{z_1, ..., z_n, w_1, ..., w_\nu\}$ are the co-ordinate projections on $\mathbb{C}^{n+\nu}$. Note that

$$Sp_A(\underset{\sim}{x}, p_1(\underset{\sim}{x}), ..., p_\nu(\underset{\sim}{x})) \subseteq Sp_A(\underset{\sim}{x}) \times \prod_{k=1}^{\nu} p_k(Sp_A(\underset{\sim}{x})) \subseteq P \times \Delta^\nu \subseteq D \times \Delta^\nu$$

But, by Lemma 8, Corollary 4, $\ker \rho$ is generated as an ideal by $\{w_k - p_k(z) : k = 1, ..., \nu\}$, and thus

$$\ker \rho \subseteq \ker \Theta_{\underset{\sim}{x}, p_1(\underset{\sim}{x}), ..., p_\nu(\underset{\sim}{x})}$$

It follows that there is a unique continuous unital homomorphism $\Theta_{\underset{\sim}{x}} : \mathcal{O}(P) \to A$ such that

$$\Theta_{\underset{\sim}{x}} \circ \rho = \Theta_{\underset{\sim}{x}, p_1(\underset{\sim}{x}), \ldots, p_\nu(\underset{\sim}{x})}$$

Thus we have $\Theta_{\underset{\sim}{x}}(z_k) = x_k$ $(k = 1, \ldots, n)$. The uniqueness of $\Theta_{\underset{\sim}{x}}$ follows since polynomials are dense in $\mathcal{O}(P)$, by Lemma 8, Corollary 3. By the same Corollary we also deduce that for any $f \in \mathcal{O}(P)$ and $\phi \in \Phi_A$ we have

$$\phi(\Theta_{\underset{\sim}{x}}(f)) = f(\phi(x_1), \ldots, \phi(x_n))$$

since the result is clear when f is a polynomial.

Now we come to the functional calculus theorem itself. (It would be possible to strengthen the statement to include a form of uniqueness; see Ref. [9].)

Theorem 9. *Let* U *be an open subset of* $\mathbb{C}^n$ *with* $Sp_A(\underset{\sim}{x}) \subset U$. *Then there is a continuous unital homomorphism* $\Theta_{\underset{\sim}{x}} : \mathcal{O}(U) \to A$ *such that:*

(i) $\Theta_{\underset{\sim}{x}}(z_k) = x_k$ $(k = 1, \ldots, n)$;

(ii) *for any* $f \in \mathcal{O}(U)$ *and* $\phi \in \Phi_A$, $\phi(\Theta_{\underset{\sim}{x}}(f)) = f(\phi(x_1), \ldots, \phi(x_n))$.

Proof. The method of deducing Theorem 9 from Lemma 9 is usually known as "the Arens–Calderón method", since it is essentially contained in Ref. [5].

Let $D = \{z \in \mathbb{C}^n : |z_k| \leqslant \|x_k\|\ (k = 1, \ldots, n)\}$; thus D is a compact polydisc with $Sp_A(\underset{\sim}{x}) \subseteq D$ (by Lemma 4). If $\underset{\sim}{\lambda} = (\lambda_1, \ldots, \lambda_n) \in D \backslash U$, then $\underset{\sim}{\lambda} \notin Sp_A(\underset{\sim}{x})$ so that there are elements $y_1, \ldots, y_n \in A$ for which

$$\sum_{k=1}^{n} y_k(\lambda_k 1 - x_k) = 1$$

It follows from Lemma 1 that, for all $\underset{\sim}{\mu}$ in some open neighbourhood $V_{\underset{\sim}{\lambda}}$ of $\underset{\sim}{\lambda}$, the element

$$\sum_{k=1}^{n} y_k(\mu_k 1 - x_k)$$

is invertible in A and that the inverse lies in the closed subalgebra $B_{\underset{\sim}{\lambda}}$ of A generated by $\{1, x_1, \ldots, x_n, y_1, \ldots, y_n\}$; thus $V_{\underset{\sim}{\lambda}} \cap Sp_{B_{\underset{\sim}{\lambda}}}(x_1, \ldots, x_n) = \emptyset$. Since $D \backslash U$ is compact it is covered by finitely many of the neighbourhoods $V_{\underset{\sim}{\lambda}}$ $(\underset{\sim}{\lambda} \in D \backslash U)$ and we thus obtain a finitely generated subalgebra B of A, generated, say, by $\{1, x_1, \ldots, x_n, b_1, \ldots, b_\nu\}$ such that

$$Sp_B(x_1, \ldots, x_n) \subset U$$

By Lemma 5, $Sp_B(x_1, \ldots, x_n, b_1, \ldots, b_\nu)$ is polynomially convex and so, by Lemma 7, there is an open polynomial polyhedron P with

$$Sp_B(x_1, \ldots, x_n, b_1, \ldots, b_\nu) \subset P \subset U \times \mathbb{C}^\nu$$

By Lemma 9, there is a (unique) continuous unital homomorphism $\Theta_{\underset{\sim}{x},\underset{\sim}{b}} : \mathcal{O}(P) \to A$ such that

$$\Theta_{\underset{\sim}{x},\underset{\sim}{b}}(z_k) = \begin{cases} x_k \ (k = 1, \ldots, n) \\ b_{k-n} \ (k = n+1, \ldots, n+\nu) \end{cases}$$

We define $i : \mathcal{O}(U) \to \mathcal{O}(P)$ by "ignoring co-ordinates" $i(f)(z_1, \ldots, z_{n+\nu}) = f(z_1, \ldots, z_n)$ $(f \in \mathcal{O}(U))$. Then clearly i is a continuous homomorphism. We now define

$$\Theta_{\underset{\sim}{x}} : \mathcal{O}(U) \to A$$

by

$$\Theta_{\underset{\sim}{x}} = \Theta_{\underset{\sim}{x},\underset{\sim}{b}} \circ i$$

That $\Theta_{\underset{\sim}{x}}$ has the required properties now follows from Lemma 9.

As an immediate application we deduce the **Shilov idempotent theorem** [4], which provides a strong generalization of the Corollary to Theorem 6. Observe that the statement makes no mention of complex variable theory; no more direct proof is known, however.

Corollary (**Shilov idempotent theorem**). *Let* E *be an open-and-closed subset of* Φ_A. *There is a unique idempotent element* $e \in A$ *such that* $\hat{e} = \chi_E$ *(the characteristic function of* E*).*

Proof. Let $F = \Phi_A \backslash E$; thus $\Phi_A = E \cup F$, where E, F are disjoint compact subsets of Φ_A. It is a simple consequence of the definition of the Gelfand topology that there are elements $x_1, \ldots, x_n \in A$ such that $\hat{\underset{\sim}{x}}(E) \cap \hat{\underset{\sim}{x}}(F) = \emptyset$ (where, for example, $\hat{\underset{\sim}{x}}(E) = \{(\hat{x}_1(\phi), \ldots, \hat{x}_n(\phi)) : \phi \in E\}$); clearly $\hat{\underset{\sim}{x}}(E)$, $\hat{x}(F)$ are compact subsets of $\mathbb{C}^n$.

Now let V, W be open neighbourhoods of $\hat{\underset{\sim}{x}}(E)$, $\hat{\underset{\sim}{x}}(F)$ respectively with $V \cap W = \emptyset$ and set $U = V \cup W$. Define $f \in \mathcal{O}(U)$ by $f(\underset{\sim}{\lambda}) \equiv 1$ $(\lambda \in V)$, $f(\underset{\sim}{\lambda}) \equiv 0$ $(\lambda \in W)$; then $f^2 = f$. Now let $e = \Theta_{\underset{\sim}{x}}(f)$; then $e^2 = e$ and, for any $\phi \in \Phi_A$,

$$\hat{e}(\phi) = \phi(\Theta_{\underset{\sim}{x}}(f)) = f(\phi(x_1), \ldots, \phi(x_n))$$
$$= \begin{cases} 1 \ (\phi \in E) \\ 0 \ (\phi \in F) \end{cases}$$

so that $\hat{e} = \chi_E$.

For uniqueness, suppose $d = d^2 \in A$ and that $\hat{d} = \hat{e}$. Then $(d-e)^3 = d^3 - 3d^2e + 3de^2 - e^3 = d^3 - e^3 = d - e$; thus $(d-e)\,x = 0$, where $x = 1 - (d-e)^2$. But, for any $\phi \in \Phi_A$, $\phi(x) = 1 - (\phi(d) - \phi(e))^2 = 1$, so that $x \in A^{-1}$ (Corollary to Theorem 3) and thus $d - e = 0$. The proof is complete.

E. SOME APPLICATIONS OF FUNCTIONAL CALCULUS TO THE STUDY OF COMMUTATIVE BANACH ALGEBRAS

The importance of the functional calculus theorem lies in the fact that it provides homomorphisms (i.e. the maps $\Theta_{\underset{\sim}{x}}$) from algebras of holomorphic functions which have an intensively studied structure, into a commutative Banach algebra where, from many viewpoints, direct knowledge of the structure is hard to obtain.

This was strikingly illustrated in the Shilov idempotent theorem (Corollary to Theorem 9) where a trivial fact about $\mathcal{O}(U)$ (that it contains any locally constant function) led to the highly non-trivial fact that a Banach algebra with a disconnected character space Φ_A contains a non-trivial idempotent. Naturally, we hope that more subtle information about $\mathcal{O}(U)$ may lead to further facts about A. We shall aim to illustrate that this is so.

Our first illustration is a remarkable result of H. Rossi – the "local peak set theorem". First some terminology. Let A be a commutative Banach algebra with 1 and let Φ_A be the space of characters in its Gelfand topology. A subset $K \subset \Phi_A$ is a **peak set** for A if and only if there is an element $x \in A$ such that

(i) $\hat{x}(\phi) = 1 \quad (\phi \in K)$
(ii) $|\hat{x}(\phi)| < 1 \quad (\phi \in \Phi_A \backslash K)$

Note that a peak set is necessarily compact. A closed subset K of Φ_A will be called a **local peak set** if and only if there is an element $x \in A$ and an open neighbourhood U of K in Φ_A such that

(i) $\hat{x}(\phi) = 1 \quad (\phi \in K)$
(ii) $|\hat{x}(\phi)| < 1 \quad (\phi \in U \backslash K)$

Clearly any peak set is also a local peak set. The theorem of Rossi gives the converse:

Theorem 10 (Rossi [6]) – *Any local peak set is a peak set.*

Proof. We reduce at once to the case in which K consists of a single point – for let A_K be the closed subalgebra of A, $A_K = \{x \in A : \hat{x}/K \text{ is constant}\}$; then it is almost trivial that Φ_{A_K} is the quotient space of Φ_A obtained by identifying K to a point [K], say, and that {[K]} is a local peak set for A_K. It is also clear that proof of the theorem for A_K would immediately imply its truth for A itself.

Thus we suppose that $K = \{\phi_0\}$ and that for some open neighbourhood U_0 of ϕ_0 and element $x_0 \in A$ we have $\hat{x}_0(\phi_0) = 1$, $|\hat{x}_0(\phi)| < 1 \ (\phi \in U_0 \backslash \{\phi_0\})$.

By definition of the Gelfand topology we may suppose that

$$U_0 = \{\phi \in \Phi_A : |\hat{x}_k(\phi)| < 2 \ \ (k = 1, \ldots, n)\}$$

where $x_k \in A$, $\hat{x}_k(\phi_0) = 0 \ (k = 1, \ldots, n)$.

Now let $\underset{\sim}{x} = (x_0, \ldots, x_n)$, $\hat{\underset{\sim}{x}} = (\hat{x}_0, \ldots, \hat{x}_n)$, $\sigma \equiv \sigma_A(\underset{\sim}{x}) \equiv \{\hat{\underset{\sim}{x}}(\phi) : \phi \in \Phi_A\}$; observe that $P_0 \equiv \hat{\underset{\sim}{x}}(\phi_0) = (1, 0, 0, \ldots, 0)$. Let

$$V_0 = \{\underset{\sim}{z} = (z_0, \ldots, z_n) \in \mathbb{C}^{n+1} : |z_k| < 1 \ (k = 1, \ldots, n)\}$$

Note that $z_0 = 1$ at P_0, but that $|z_0| < 1$ whenever $\underset{\sim}{z} \in \overline{V}_0 \cap \sigma$, $\underset{\sim}{z} \neq P_0$ (for, if $z \in \overline{V}_0 \cap \sigma$ then $\underset{\sim}{z} = \hat{\underset{\sim}{x}}(\phi)$ for some ϕ with $|\hat{\underset{\sim}{x}}_k(\phi)| \leq 1 < 2 \ (k = 1, \ldots, n)$ and so with $\phi \in U_0$).

Thus (bdy V_0) $\cap\ \sigma$ is a compact subset of $\{\underset{\sim}{z} \in \mathbb{C}^{n+1} : |z_0| < 1\}$; thus there exists an open set $V_1 \supset \sigma \backslash V_0$ such that $V_0 \cap V_1 \subset \{\underset{\sim}{z} \in \mathbb{C}^{n+1} : |z_0| < 1\}$. Then $V = V_0 \cup V_1$ is an open neighbourhood of σ in $\mathbb{C}^{n+1}$.

Using the Arens-Calderón method (see proof of Theorem 9) we now choose $\{x_{n+1}, \ldots, x_\nu\}$ in A such that, if $B = \overline{\text{alg}}_A\{x_0, x_1, \ldots, x_\nu\}$ then $\text{Sp}_B(\underset{\sim}{x}) \subset V$. Now $\text{Sp}_B(x_0, \ldots, x_\nu)$ is polynomially convex (Lemma 5) and so we may choose a polynomial polyhedron P in $\mathbb{C}^{\nu+1}$ such that

$$\text{Sp}_B(x_0, \ldots, x_\nu) \subset P \subset V \times \mathbb{C}^{\nu-n}$$

Set $P_0 = (V_0 \times \mathbb{C}^{\nu-n}) \cap P$, $P_1 = (V_1 \times \mathbb{C}^{\nu-n}) \cap P$; then $P = P_0 \cup P_1$. On $P_0 \cap P_1$ we have $|z_0| < 1$ so that we may define a holomorphic branch of log $(z_0 - 1)$. By Lemma 6, Corollary, and Lemma 8, Corollary 1, there are functions $h_0 \in \mathcal{O}(P_0)$, $h_1 \in \mathcal{O}(P_1)$ with

$$(z_0 - 1)^{-1} \log(z_0 - 1) = h_1(z) - h_0(z) \text{ on } P_0 \cap P_1$$

i.e. $\quad (z_0 - 1) \exp((z_0 - 1)\, h_0(z)) = \exp((z_0 - 1)\, h_1(z)) \text{ on } P_0 \cap P_1$

Thus there is $h \in \mathcal{O}(P)$ defined by

$$h(z) = \begin{cases} \exp((z_0 - 1)h_1(z)) & (z \in P_1) \\ (z_0 - 1)\exp((z_0 - 1)h_0(z)) & (z \in P_0) \end{cases}$$

It is now an elementary geometrical argument (in the complex plane) to show that, for a sufficiently small $\epsilon > 0$, the function H, defined by

$$H(z) = -\epsilon(h(z) - \epsilon)^{-1} \quad (z \in P)$$

is holomorphic on a neighbourhood of $Sp_B(x_0, ..., x_\nu)$ and peaks precisely $(\hat{x}_0(\phi_0), ..., \hat{x}_\nu(\phi_0))$ relative to $Sp_B(x_0, ..., x_\nu)$. Thus, putting $y = \Theta_{(x_0, ..., x_\nu)}(H)$, we have that $\hat{y}(\phi_0) = 1$, $|\hat{y}(\phi)| < 1$ $(\phi \in \Phi_A \backslash \{\phi_0\})$, i.e. $\{\phi_0\}$ is a peak set, as required.

Our final application is the Arens-Royden theorem (see Refs [7], [8]). Suppose that K is any compact Hausdorff space; let C = C(K), the algebra of all continuous complex-valued functions on K, let C^{-1} be the multiplicative group of units and let exp C be the subgroup of C^{-1} consisting of those elements having continuous logarithms defined on K. Then a theorem from topology gives that $C^{-1}/\exp C \cong H^1(X, \mathbb{Z})$. We shall show that the same result holds not merely for C, but for *any* commutative Banach algebra A with $\Phi_A = K$. To avoid presupposing knowledge of topology we shall simply prove that $A^{-1}/\exp A \cong C^{-1}/\exp C$. First we need some definitions.

Let A be, as usual, a commutative Banach algebra with 1 and let A^{-1} be the group of units (invertible elements). For any $x \in A$ we define

$$\exp x = \sum_{n=0}^{\infty} \frac{x^n}{n!}$$

(the series is norm-convergent in A) and let $\exp A = \{\exp x : x \in A\}$. Then exp A is a subgroup of A^{-1} (in fact exp A is the topological component containing 1). We write $K = \Phi_A$ and C = C(K) as above.

Theorem 11 (**Arens – Royden**)

$$\frac{A^{-1}}{\exp A} \cong \frac{C^{-1}}{\exp C}$$

Proof. Let $G: A \to C$ be the Gelfand representation of A (i.e. $G(x) = \hat{x}$ $(x \in A)$). Then clearly $G(A^{-1}) \subseteq C^{-1}$, $G(\exp A) \subseteq \exp C$, so that there is a naturally induced homomorphism $\overline{G}: A^{-1}/\exp A \to C^{-1}/\exp C$, defined by $\overline{G}([x]) = [G(x)]$ (where, for $x \in A^{-1}$, [x] denotes the coset $x \cdot \exp A$, etc.).

We shall show that $\overline{G}$ effects the required isomorphism between $A^{-1}/\exp A$ and $C^{-1}/\exp C$. The proof falls into two parts:

(i) $\overline{G}$ *is* (1 – 1): for this we must show that if $x \in A^{-1}$ and $\hat{x} \in \exp C$, then $x \in \exp A$.

(ii) $\overline{G}$ *maps onto* $C^{-1}/\exp C$: we must show that if $f \in C^{-1}$ then there exists $x \in A^{-1}$ such that $f/\hat{x} \in \exp C$.

Proof of (i): Let $x_1 \in A^{-1}$ be such that $x_1 \in \exp C$, i.e. there exists $f \in C$ with $\hat{x}_1(\phi) = \exp f(\phi)$ (all $\phi \in K$).

We shall show now that there exist elements $x_2, \ldots, x_n \in A$ such that whenever $\phi_1, \phi_2 \in K$ and $\hat{x}_k(\phi_1) = \hat{x}_k(\phi_2)$ $(k = 1, \ldots, n)$ then also $f(\phi_1) = f(\phi_2)$. Thus, suppose that $\phi_1^0 \neq \phi_2^0$; then we can find $a \in A$ with $\hat{a}(\phi_1^0) \neq \hat{a}(\phi_2^0)$ and hence also $\hat{a}(\phi_1) \neq \hat{a}(\phi_2)$ for all (ϕ_1, ϕ_2) in some neighbour W of (ϕ_1^0, ϕ_2^0) in $K \times K$. For a diagonal point $(\phi^0, \phi^0) \in K \times K$ we choose a neighbourhood V of ϕ^0 in K such that $|f(\phi) - f(\phi^0)| < \pi$ (all $\phi \in V$); then if $(\phi_1, \phi_2) \in V \times V$ and if $\hat{x}_1(\phi_1) = \hat{x}_1(\phi_2)$ then $\exp f(\phi_1) = \exp f(\phi_2)$ and $|f(\phi_1) - f(\phi_2)| < 2\pi$, so that $f(\phi_1) = f(\phi_2)$. Thus for (ϕ^0, ϕ^0) we take $W = V \times V$, $a = x_1$. Then for each $(\phi_1^0, \phi_2^0) \in K \times K$ we have found a neighbourhood W and an element $a \in A$ such that if $(\phi_1, \phi_2) \in W$ and $\hat{a}(\phi_1) = \hat{a}(\phi_2)$ then $f(\phi_1) = f(\phi_2)$. The compactness $K \times K$ now yields the required finite set of elements $x_2, \ldots, x_n \in A$.

Define $\alpha : K \to \sigma \equiv Sp_A(x_1, \ldots, x_n)$ by $\alpha(\phi) = (\phi(x_1), \ldots, \phi(x_n))$; then there is a unique functi $F : \sigma \to \mathbb{C}$ such that $F \circ \alpha = f$ and it is elementary that F is continuous. Also $e^{F(z)} = z_1$ $(z \in K)$. We shall now show that F may be extended to a holomorphic function $\tilde{F}$, on a neighbourhood U of σ, such that $e^{\tilde{F}(z)} = z_1$ on U.

For each $z^0 \in \sigma$ we choose $\delta(z^0) > 0$ such that (if $B(z^0; \delta(z^0))$ is the open ball, centre z^0 and radius $\delta(z^0)$) then

(a) $|F(z) - F(z^0)| < \pi$ $(z \in \sigma \cap B(z^0; \delta(z^0)))$

(b) there is a unique holomorphic function ϕ_{z^0} on $B(z^0; \delta(z^0))$ such that both $\phi_{z^0}(z^0) = F($ and $\exp(\phi_{z^0}(z)) = z_1$ on $B(z^0; \delta(z^0))$

(c) $|\phi_{z^0}(z) - \phi_{z^0}(z^0)| < \pi$ on $B(z^0; \delta(z^0))$

Then (a) and (c) together imply that $F(z) = \phi_{z^0}(z)$ for all $z \in B(z^0; \delta(z^0)) \cap \sigma$.
Define

$$U = \bigcup_{z^0 \in \sigma} B(z^0; \tfrac{1}{3}\delta(z^0))$$

Then U is an open neighbourhood of σ. We claim that there is a well-defined $\tilde{F} \in \mathcal{O}(U)$ defined by putting $\tilde{F}(z) = \phi_{z_0}(z)$ on $B(z^0; \frac{1}{3}\delta(z^0))$ for each $z^0 \in \sigma$. Assuming, for the moment, that $\tilde{F}$ is well-defined, it is evident that $\tilde{F}/\sigma = F$ and that $\exp(\tilde{F}(z)) = z_1$ on U. If we now put $y = \Theta_{(x_1, \ldots, x_n)}(\tilde{F})$ we have that $\exp y = \Theta_{(x_1, \ldots, x_n)}(\exp \tilde{F}) = x_1$ and we have proved that $x_1 \in \exp A$, as required.

It remains to show that $\tilde{F}$ was well-defined on U. Thus, suppose that, for some $z^0, z^1 \in \sigma$ we have $B(z^0; \frac{1}{3}\delta(z_0)) \cap B(z^1; \frac{1}{3}\delta(z_1)) \neq \emptyset$. Suppose, without loss of generality, that $\delta(z_1) \leq \delta(z_0)$; then $B(z_1; \frac{1}{3}\delta(z_1)) \subseteq B(z^0; \delta(z^0))$ and so, by the uniqueness assertion for $B(z^1; \frac{1}{3}\delta(z^1))$ (since $\phi_{z^0}(z^1) = F(z^1) = \phi_{z^1}(z^1)$) we must have $\phi_{z^0}(z) = \phi_{z^1}(z)$ on $B(z^1; \frac{1}{3}\delta(z^1))$ and so, in particular, on $B(z^0; \frac{1}{3}\delta(z^0)) \cap B(z^1; \frac{1}{3}\delta(z^1))$. Thus $\tilde{F}$ was well-defined and the proof of (i) is complete.

Proof of (ii): Let $f \in C^{-1}$ and let $\epsilon = \inf\{|f(\phi)| : \phi \in K\}$; then if $g \in C$ with $\|g - f\|_K < \epsilon$ we have $g \in C^{-1}$ and $f/g \in \exp C$ (for $gf^{-1} = 1 + (g - f)f^{-1}$, and $\|(g - f)f^{-1}\|_K < 1$).

In particular, by the Stone-Weierstrass theorem, we may find $x_1, \ldots, x_{2n} \in A$ such that, for

$$g = \sum_{k=1}^{n} \hat{x}_k \overline{\hat{x}_{k+n}}$$

we have $\|g - f\|_K < \epsilon$ and thus $fg^{-1} \in \exp C$.

Define $\alpha : K \to \sigma \equiv Sp_A(x_1, \ldots, x_{2n})$ by $\alpha(\phi) = (\phi(x_1), \ldots, \phi(x_{2n}))$ and define G on $\mathbb{C}^{2n}$ by

$$G(z_1, \ldots, z_{2n}) = \sum_{k=1}^{n} z_k \overline{z_{k+n}}$$

Then G is a smooth function and $G(\underset{\sim}{z}) \neq 0$ on σ and hence also $G(\underset{\sim}{z}) \neq 0$ for $\underset{\sim}{z}$ in some open set $U \supset \sigma$.

By the Arens-Calderón method we choose elements $x_{2n+1}, \dots, x_\nu$ in A such that, for $B = \overline{\mathrm{alg}_A\{x_1, \dots, x_\nu\}}$ we have $\mathrm{Sp}_B(x_1, \dots, x_{2n}) \subset U$. Then $\mathrm{Sp}_B(x_1, \dots, x_\nu)$ is polynomially convex and so we may choose a polynomial polyhedron P with

$$\mathrm{Sp}_B(x_1, \dots, x_\nu) \subset P \subset U \times \mathbb{C}^{\nu-2n}$$

By an abuse of notation we regard G as defined on P, by the formula

$$G(z_1, \dots, z_\nu) = \sum_{k=1}^{n} z_k \bar{z}_{k+n}$$

Then G is smooth on P and $G(\underset{\sim}{z}) \neq 0$ $(\underset{\sim}{z} \in P)$. Also, if $\beta : K \to \mathrm{Sp}_A(x_1, \dots, x_\nu)$ is defined by $\beta(\phi) = (\phi(x_1), \dots, \phi(x_\nu))$ $(\phi \in K)$ then $G \circ \beta = g$ (note that $\mathrm{Sp}_A(x_1, \dots, x_\nu) \subseteq \mathrm{Sp}_B(x_1, \dots, x_\nu)$).

Define $\omega \in \overline{\mathscr{E}}^1(P)$ by

$$\omega = \frac{1}{G} \sum_{k=1}^{\nu} \frac{\partial G}{\partial \bar{z}_k} \, d\bar{z}_k$$

Then a trivial calculation shows that $\bar{\partial}\omega = 0$ and so since, by Lemma 8, Corollary 1, $H^1(\overline{\mathscr{E}}(P)) = 0$, there exists $k \in C^\infty(P)$ such that $\omega = \bar{\partial}k$. Set $h = e^{-k} G \in C^\infty(P)$; then $\bar{\partial}h = e^{-k}(\bar{\partial} G - G \bar{\partial}k) = 0$, so that $h \in \mathscr{O}(P)$ and $G h^{-1} = e^k$ on P. But then if we put $x = \Theta_{(x_1, \dots, x_\nu)}(h)$ we have $\hat{x} = h \circ \beta$ and so

$$g/\hat{x} = e^{k \circ \beta} \in \exp C$$

Thus $f/\hat{x} = (f/g)(g/\hat{x}) \in \exp C$, and the proof is complete.

REFERENCES

[1] ALLAN, G.R., Embedding the algebra of formal power series in a Banach algebra, Proc. London Math. Soc. (3) **25** (1972) 329–340.
[2] HAGHANY, G., Norming formal power series in n variables, to appear in Proc. London Math. Soc.
[3] GUNNING, R.C., ROSSI, H., Analytic Functions of Several Complex Variables, Prentice-Hall (1965).
[4] SHILOV, G.E., On the decomposition of a normed ring as a direct sum of ideals, Mat. Sb. **32** (1954) 353–364 and A.M.S. Translations (2) **1** (1955) 37–48.
[5] ARENS, R., CALDERON, A.P., Analytic functions of several Banach algebra elements, Ann. Math. **62** (1955) 204–216.
[6] ROSSI, H., The local maximum modulus principle, Ann. Math. **72** (1960) 1–11.
[7] ARENS, R., The group of invertible elements of a commutative Banach algebra, Studia Math. (Seria Spec.) **Z. 1** (1963) 21–23.
[8] ROYDEN, H., Function algebras, Bull. Am. Math. Soc. **69** (1963) 281–298.
[9] BOURBAKI, N., Théories spectrales, Hermann, Paris (1967).

ANALYTIC CONVEXITY
Some comments on an example of de Giorgi and Piccinini

A. ANDREOTTI, M. NACINOVICH
Istituto Matematica,
Università di Pisa,
Pisa, Italy

Abstract

ANALYTIC CONVEXITY: SOME COMMENTS ON AN EXAMPLE OF DE GIORGI AND PICCININI.
1. Introduction. 2. C^∞ and analytical convexity for a complex of differential equations with constant coefficients. 3. Reduction of the analytic convexity to the C^∞ convexity. 4. The example of de Giorgi-Piccinini.

1. INTRODUCTION

De Giorgi was the first to make the following observation: Consider the Laplace operator in two variables x, y

$$\Delta = \frac{\partial^2}{\partial x^2} + \frac{\partial^2}{\partial y^2}$$

as operating on functions of three variables x, y, t, and consider the equation

(1) $$\Delta u = f$$

for u and f functions of x, y, t.

Then we have the following facts:

(a) $\forall f \in C^\infty(\mathbb{R}^3)$ there exist $u \in C^\infty(\mathbb{R}^3)$ such that (1) holds;

(b) $\forall R$ and every $\varepsilon > 0$ and for every f complex valued real analytic in $B(R) = \{(x,y,t) \in \mathbb{R}^3 \mid x^2 + y^2 + t^2 < R\}$ we can find u complex valued, real analytic in $B(R-\varepsilon)$ such that (1) holds;

(c) There exist some f real analytic in $\mathbb{R}^3$ such that the equation (1) does not admit any global solution real analytic in $\mathbb{R}^3$.

Facts (a) and (b) can be proved directly by means, for instance, of the Poisson integral.

The surprising phenomenon is fact (c) and Piccinini has given an explic example of this phenomenon.

Let us make first the following remark. If we set $z = x + iy$ then

$$\Delta = 4 \frac{\partial^2}{\partial \bar{z}\, \partial z}$$

Therefore to validate fact (c) it is enough to show that there exists a complex valued real analytic function f on $\mathbb{R}^3$ such that the equation

$$(2) \qquad \frac{\partial u}{\partial \bar{z}} = f$$

does not admit any global complex valued real analytic solutions.

(Indeed, if $\Delta u = f$ could be solved with a global real analytic function u then $v = 4 \frac{\partial u}{\partial z}$ would be global real analytic and would satisfy equation (2)).

This phenomenon is similar to the phenomenon of Hans Lewy and is closely related to the theory of convexity for differential operators with constant coefficients.

2. C^∞ AND ANALYTIC CONVEXITY FOR A COMPLEX OF DIFFERENTIAL EQUATIONS WITH CONSTANT COEFFICIENTS

(a) Let Ω be an open set in $\mathbb{R}^n$, and let

$$(1) \qquad \mathcal{E}^{p_o}(\Omega) \xrightarrow{A_o(D)} \mathcal{E}^{p_1}(\Omega) \xrightarrow{A_1(D)} \cdots \xrightarrow{A_{d-1}(D)} \mathcal{E}^{p_d}(\Omega) \to 0$$

be a complex of differential operators with constant coefficients. Here

(α) $\mathcal{E}(\Omega)$ denote the space of C^∞ functions on Ω and

$\mathcal{E}^{p_i}(\Omega) = \mathcal{E}(\Omega) \times \ldots \times \mathcal{E}(\Omega)$ p_i-times;

(β) $A_i(D) = (a^{(i)}_{\alpha\beta}(D))$ is a matrix of type (p_{i+1}, p_i) with entries differential operators with constant coefficients;

(γ) The fact that the above sequence is a complex means that

$$A_{i+1}(D)\, A_i(D) \equiv 0 \ , \ \forall i$$

<u>Example</u>. Let $A_o(\xi)$ be a (p_1, p_o) matrix with polynomial entries. Consider the ring $\mathscr{P} = \mathbb{C}[\xi_1, \ldots, \xi_n]$ of polynomials in n variables and consider ${}^tA_o$ as a $\mathscr{P}$-homomorphism

$$\mathscr{P}^{p_1} \xrightarrow{{}^tA_o(\xi)} \mathscr{P}^{p_o}$$

Then by Hilbert's theorem we can continue this map by a finite sequence of maps

$$0 \longrightarrow \mathscr{P}^{p_d} \xrightarrow{{}^tA_{d-1}(\xi)} \mathscr{P}^{p_{d-1}} \longrightarrow \cdots \xrightarrow{{}^tA_2(\xi)} \mathscr{P}^{p_2} \xrightarrow{{}^tA_1(\xi)} \mathscr{P}^{p_i} \xrightarrow{{}^tA_o(\xi)} \mathscr{P}^{n_o} \longrightarrow N - 10$$

(and $d \leqslant n$ or $d = 2$ if $n = 1$) to give an exact sequence.

Replacing the matrices ${}^tA_i(\xi)$ by their transposed $A_i(\xi)$ and the variables ξ_j by $\frac{\partial}{\partial x_j}$ we get a complex (1) of differential operators with constant coefficients which has moreover the property to be exact on every open convex set Ω (Poincaré lemma).

The condition that the complex (1) be exact on convex Ω's characterize the complexes obtained from a "Hilbert resolution". Thus, in the sequel <u>we will assume that (1) is a "Hilbert complex"</u>.

We will say that Ω (not necessarily convex) is <u>C^∞ convex</u> for (1) iff (1) is acyclic.

If $\mathscr{E}_o$ denotes the sheaf of germs of C^∞ solutions of $A_o(D)u = 0$, C^∞-convexity can be stated by the conditions

$$H^j(\Omega, \mathscr{E}_o) = 0 \ , \ \forall\, j \geqslant 1$$

(b) Let us replace the space $\mathcal{E}(\Omega)$ of C^∞ functions on Ω by the space $\mathcal{A}(\Omega)$ of complex valued real analytic functions. We get a complex

$$(1)_{\mathcal{A}} \qquad \mathcal{A}^{p}(\Omega) \xrightarrow{A_o(D)} \mathcal{A}^{p_1}(\Omega) \xrightarrow{A_1(D)} \cdots \xrightarrow{A_{d-1}(D)} \mathcal{A}^{p_d}(\Omega) \longrightarrow 0 .$$

We will say that Ω is analytically convex if $(1)_{\mathcal{A}}$ is acyclic.

The first question that arises is the following. Let $\mathcal{A}_o$ be the sheaf of germs of complex valued real analytic solutions of $A_o(D)u = 0$. Then is Ω-analytically convex for $(1)_{\mathcal{A}}$ equivalent to

$$H^j(\Omega, \mathcal{A}_o) = 0 \ , \ \forall\, j \geq 1 \ ?$$

The answer to this question is affirmative and it is a consequence of the following two statements:

(α) (analytic Poincaré lemma). Let $\mathcal{A}$ denote the sheaf of germs of complex valued real analytic functions. Then the sequence

$$\mathcal{A}^{p_o}_{x_o} \xrightarrow{A_o} \mathcal{A}^{p_1}_{x_o} \xrightarrow{A_1} \cdots \xrightarrow{A_{d-1}} \mathcal{A}^{p_d}_{x_o} \longrightarrow 0$$

is exact for every $x_o \in \mathbb{R}^n$.

(β) For every open set $\Omega \subset \mathbb{R}^n$ we have

$$H^j(\Omega, \mathcal{A}) = 0 \ , \ \forall\, j \geq 1$$

The second of these statements is a consequence of a theorem of Grauert: "every open set in $\mathbb{R}^n$ has a fundamental system of neighborhoods in $\mathbb{C}^n$ which are open sets of holomorphy".

The first of these statements will be proved later; it is a sort of Cauchy-Kovalewska theorem for overdetermined systems with constant coefficients.

Assuming these facts for the moment, let us go back to the example of de Giorgi and Piccinini. We take $\Omega = \mathbb{R}^3$ and there, with respect to the coordinates $z = x + iy$ and t, we consider the complexes

$$0 \longrightarrow \mathcal{E}_o(\mathbb{R}^3) \rightarrowtail \mathcal{E}(\mathbb{R}^3) \xrightarrow{\frac{\partial}{\partial \bar{z}}} \mathcal{E}(\mathbb{R}^3) \longrightarrow 0$$

and

$$0 \longrightarrow \mathcal{A}_o(\mathbb{R}^3) \rightarrowtail \mathcal{A}(\mathbb{R}^3) \xrightarrow{\frac{\partial}{\partial \bar{z}}} \mathcal{A}(\mathbb{R}^3) \longrightarrow 0$$

We have

$$H^1(\mathbb{R}^3, \mathcal{E}_o) = 0 \ , \ H^1(\mathbb{R}^3, \mathcal{A}_o) \neq 0$$

as we do have the exact cohomology sequences

$$0 \longrightarrow \mathcal{E}_o(\mathbb{R}^3) \longrightarrow \mathcal{E}(\mathbb{R}^3) \xrightarrow{\frac{\partial}{\partial \bar{z}}} \mathcal{E}(\mathbb{R}^3) \longrightarrow H^1(\mathbb{R}^3, \mathcal{E}_o) \longrightarrow 0$$

$$0 \longrightarrow \mathcal{A}_o(\mathbb{R}^3) \longrightarrow \mathcal{A}(\mathbb{R}^3) \xrightarrow{\frac{\partial}{\partial \bar{z}}} \mathcal{A}(\mathbb{R}^3) \longrightarrow H^1(\mathbb{R}^3, \mathcal{A}_o) \longrightarrow 0$$

and as the equation on $\mathbb{R}^3$

$$\frac{\partial u}{\partial \bar{z}} = f$$

is always solvable for $f \in \mathcal{E}(\mathbb{R}^3)$ with $u \in \mathcal{E}(\mathbb{R}^3)$ (as the complex is Hilbert and $\mathbb{R}^3$ is convex) but is not always solvable for $f \in \mathcal{A}(\mathbb{R}^3)$ with $u \in \mathcal{A}(\mathbb{R}^3)$. Therefore we do have complexes (Hilbert) which are C^∞ convex for some Ω open in $\mathbb{R}^n$ but are not analytically convex.

3. REDUCTION OF THE ANALYTIC CONVEXITY TO THE C^∞ CONVEXITY

a) Let

$$\mathcal{E}^{p_o}(\Omega) \xrightarrow{A_o(D)} \mathcal{E}^{p_1}(\Omega) \xrightarrow{A_1(D)} \cdots \xrightarrow{A_{d-1}(D)} \mathcal{E}^{p_d}(\Omega) \longrightarrow 0$$

be a Hilbert complex defined for all open sets $\Omega \subset \mathbb{R}^n$. We consider $\mathbb{R}^n \subset \mathbb{C}^n$ as the real part of $\mathbb{C}^n$ and we consider for $\tilde{\Omega}$ open in $\mathbb{C}^n$ the same complex of differential operators as acting on functions of 2n variables the real and imaginary part of the compelx coordinates in $\mathbb{C}^n$.

$$(\tilde{H}) : \mathcal{E}^{p_o}(\tilde{\Omega}) \xrightarrow{A_o(D)} \mathcal{E}^{p_2}(\tilde{\Omega}) \xrightarrow{A_1(D)} \dots \xrightarrow{A_{d-1}(D)} \mathcal{E}^{p_d}(\tilde{\Omega}) \longrightarrow 0$$

where $D_i = \frac{\partial}{\partial x_i}$, $z_j = x_j + iy_j$ being the complex coordinates in $\mathbb{C}^n$.

Also on $\mathbb{C}^n$ we consider the Dolbeault complex

$$(\mathcal{D}) : C^{oo}(\tilde{\Omega}) \xrightarrow{\bar{\partial}} C^{o1}(\tilde{\Omega}) \xrightarrow{\bar{\partial}} \dots \xrightarrow{\bar{\partial}} C^{on}(\tilde{\Omega}) \longrightarrow 0$$

where $C^{os}(\tilde{\Omega})$ denote the space of C^∞ form of type (o,s) in $\tilde{\Omega}$ and where $\bar{\partial}$ is the exterior differentiation with respect to antiholomorphic local coordinates.

From these two complexes we form a third complex; the tensor product $(\tilde{H}) \otimes (\mathcal{D})$.

This is (by definition) the simple complex associated to the double complex $K = \{K^{r,s} , \bar{\partial}, A_*\}$ where

$$K^{r,s} = C^{or}(\tilde{\Omega}) \otimes_{\mathcal{E}(x)} \mathcal{E}^{p_s}(\tilde{\Omega})$$
$$= (C^{or}(\tilde{\Omega}))^{p_s}$$

and where the differential operators are induced by $\bar{\partial}$ and A_* .

This complex is a complex of differential operators with constant coefficients; rather large. It looks like this:

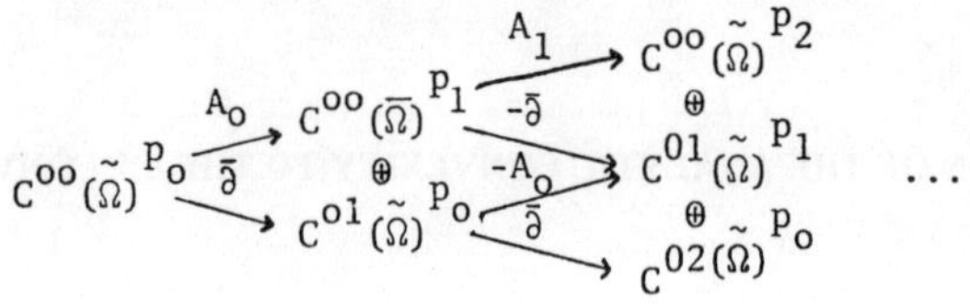

<u>Theorem 1.</u> (a) <u>The $(\mathcal{D})$-suspended complex $(\tilde{H}) \otimes \mathcal{D}$ is a Hilbert complex. In particular it is acyclic on an open convex sets $\tilde{\Omega}$. (b) If $\tilde{\Omega}$ is an open set of holomorphy in $\mathbb{C}^n$ the cohomology of that complex is natural isomorphic to the cohomology of the complex</u>

$$(2)\qquad \Gamma(\tilde{\Omega},\mathcal{O})^{p_o} \xrightarrow{A_o(\frac{\partial}{\partial z})} \Gamma(\tilde{\Omega},\mathcal{O})^{p_1} \xrightarrow{A_1(\frac{\partial}{\partial z})} \cdots \xrightarrow{A_{d-1}(\frac{\partial}{\partial z})} \Gamma(\tilde{\Omega},\mathcal{O})^{p_d} \longrightarrow 0$$

where $\Gamma(\tilde{\Omega},\mathcal{O})$ denotes the space of holomorphic functions on $\tilde{\Omega}$.

Outline of proof. According to the previous remarks about Hilbert complexes, statement (a) reduces to the following algebraic verification. Let

$$\mathcal{E}(\tilde{\Omega})^{q_o} \xrightarrow{B_o(D)} \mathcal{E}(\tilde{\Omega})^{q_1} \xrightarrow{B_1(D)} \cdots \xrightarrow{B_{\ell-1}(D)} \mathcal{E}(\tilde{\Omega})^{q_\ell} \longrightarrow 0$$

be the considered suspended complex where $B_j(D)$ are matrices with entries polynomial in $\frac{\partial}{\partial x_j}, \frac{\partial}{\partial x_{n+j}}$ $(z_j = x_{j+i}\, x_{n+j})$. If $\mathcal{P} = \mathbb{C}[\xi_1,\ldots,\xi_{2n}]$ is the polynomial ring in $2n$ variables, statement (a) is equivalent to the exactness of the sequence of $\mathcal{P}$-homomorphisms

$$0 \longrightarrow \mathcal{P}^{q_\ell} \xrightarrow{{}^tB_{\ell-1}(\xi)} \mathcal{P}^{q_{e-1}} \xrightarrow{{}^tB_{\ell-2}(\xi)} \cdots \xrightarrow{{}^tB_1(\xi)} \mathcal{P}^{q_2} \xrightarrow{{}^tB_o(\xi)} \mathcal{P}^{x_o}$$

This is an algebraic verification and it is left to the reader. Statement (b) is an immediate consequence of the spectral sequence of the double complex $\{K^{r,s}, \bar{\partial}, A_*\}$.

In particular if $\tilde{\Omega}$ is open convex then $\tilde{\Omega}$ is a domain of holomorphy and thus the sequence (2) must be exact.

Corollary. The analytic complex $(1)_{\mathcal{A}}$

$$\mathcal{A}^{p_o}(\Omega) \xrightarrow{A_o} \mathcal{A}^{p_1}(\Omega) \xrightarrow{A_1} \cdots \xrightarrow{A_{d-1}} \mathcal{A}^{p_d}(\Omega) \longrightarrow 0$$

admits the Poincaré (analytic) lemma.

Indeed let $x_0 \in \Omega$ and let $f \in \mathcal{A}^{p_s}(w)$ $(s \geq 1)$ where w is a connected neighborhood of x_0. Then f extends to a holomorphic function F defined in some connected neighborhood $\tilde{w}$ of w in C^n. Assume that $A_s f = 0$ then $A_s F = 0$. We can find a convex neighborhood V of x_o in C^n with $V \subset \tilde{w}$. There, there exist G holomorphic in V, $G \in \Gamma(V,\mathcal{O})^{p_{s-1}}$ such that $A_{s-1} G = F$. Restricting to $V \cap R^n$ we get for $g = G|_{V \cap R^n}$, $A_{s-1}\, g = f$.

(b) Let $\Omega \subset \mathbb{R}^n$ and let $\tilde{\Omega}$ be an open set of holomorphy in $\mathbb{C}$ with $\tilde{\Omega} \cap \mathbb{R}^n = \Omega$.

Let $\tilde{\mathcal{E}}_o$ denote the sheaf of germs of solutions of the first operator of the suspended complex. We do have

$$H^j(\tilde{\Omega}, \tilde{\mathcal{E}}_o) \simeq H^j(\tilde{\Omega}, \Gamma(\tilde{\Omega}, \mathcal{O})^*, A_*)$$

and thus, by restriction a natural map

$$H^j(\tilde{\Omega}, \tilde{\mathcal{E}}_o) \to H^j(\Omega, \mathcal{A}_o)$$

One can then prove the following.

Theorem 2. We have a natural isomorphism

$$H^j(\Omega, \tilde{\mathcal{E}}_o) = \varinjlim_{\tilde{\Omega} \cap \mathbb{R}^n = \Omega} H^j(\tilde{\Omega}, \tilde{\mathcal{E}}_o) \simeq H^j(\Omega, \mathcal{A}_o)$$

which shows that the analytic convexity of Ω is the limit of the C^∞ convexity of the suspended complex.

(c) What we are interested in are the cohomology groups $H^j(\Omega, \mathcal{A}_o)$. Instead of suspending the given complex from $\mathbb{R}^n$ to $\mathbb{C}^n$ via the Dolbeault complex, one can suspend the complex from $\mathbb{R}^n$ to $\mathbb{R}^{n+1}$ via a complex given for instance, by a single differential operator

$$(\mathcal{L}) : \mathcal{E}(\tilde{\Omega}) \xrightarrow{L(D)} \mathcal{E}(\tilde{\Omega}) \longrightarrow 0,$$

$\tilde{\Omega}$ open in $\mathbb{R}^{n+1}$. What is required from the previous considerations are the following conditions:

(1) $(\tilde{H}) \otimes (\mathcal{L})$ should be a Hilbert complex. This amounts to a "transversality" condition of $\mathcal{L}$ with respect to $\mathbb{R}^n$ into $\mathbb{R}^{n+1}$ ($\mathbb{R}^n$ should be non-characteristic for L in $\mathbb{R}^{n+1}$).

(2) The first operator B_o of the suspended complex $(\tilde{H}) \otimes (\mathcal{L})$ should be elliptic, i.e. $B_o u = 0 \Rightarrow u$ real analytic. Then one obtains a theorem of the following type

$$\varinjlim_{\tilde{\Omega} \cap \mathbb{R}^n = \Omega} H^j(\tilde{\Omega}, \tilde{\mathcal{E}}_o) = \bigoplus_1^k H^j(\Omega, \mathcal{A}_o)$$

where k is the "transversal order of the operator L on $\mathbb{R}^n$" .

1. THE EXAMPLE OF DE GIORGI-PICCININI

We consider $\mathbb{R}^3$, where (x_1, y_1, x_2) are cartesian coordinates, imbedded in $\mathbb{R}^4 = \mathbb{C}^2$ where $z_1 = x_1 + iy_1$, $z_2 = x_2 + iy_2$ are complex coordinates.

On $\mathbb{R}^3$ we consider the complex

$$\mathcal{A}(\mathbb{R}^3) \xrightarrow{\frac{\partial}{\partial \bar{z}_1}} \mathcal{A}(\mathbb{R}^3) \longrightarrow 0$$

then $\mathcal{A}_o$, the sheaf of germs of analytic solutions of the homogeneous equation $\frac{\partial u}{\partial \bar{z}_1} = 0$ is the sheaf of germs of functions which are real analytic, holomorphic in z_1 and thus admitting power series expansions in z_1 and x_2 .

If $\mathcal{O}$ is the sheaf of germs of holomorphic functions in $\mathbb{C}^2$, the restriction map on the points of $\mathbb{R}^3$ gives a homomorphism

$$\mathcal{O}_{\mathbb{R}^3} \longrightarrow \mathcal{A}_o, \quad (\mathcal{O}_{\mathbb{R}^3} = \text{topological restriction of } \mathcal{O} \text{ to } \mathbb{R}^3)$$

which is surjective and has zero kernel. Thus an isomorphism; therefore

$$H^1(\mathbb{R}^3, \mathcal{A}_o) = H^1(\mathbb{R}^3, \mathcal{O}_{\mathbb{R}^3})$$

This is nothing else but the statement of Theorem 2 when we suspend the complex from $\mathbb{R}^3$ to $\mathbb{C}^2$ via the complex

$$(\mathcal{L}) : \qquad \mathcal{E}(\tilde{\Omega}) \xrightarrow{\frac{\partial}{\partial \bar{z}_2}} \mathcal{E}(\tilde{\Omega}) \longrightarrow 0$$

At this point the statement of de Giorgi-Piccinini amounts to show that

$$H^1(\mathbb{R}^3, \mathcal{O}_{\mathbb{R}^3}) \neq 0$$

Now one can prove directly the following more comprehensive statement.

Theorem 3. We have $\dim_{\mathbb{C}} H^1(\mathbb{R}^3, \mathcal{O}_{\mathbb{R}^3}) = \infty$

This already gives more precision to the statement of de Giorgi-Piccinini in the sense that not only is it true that, for infinitely many analytic functions $f \in \mathcal{A}(\mathbb{R}^3)$, the equation $\frac{\partial u}{\partial \bar{z}_1} = f$ is not analytically solvable, but it is true also that the same happens for infinitely many cohomology classes (each one containing infitely many functions).

Outline of the proof of Theorem 3. (α) The proof is based on the followin

Lemma. Consider in $\mathbb{C}^2$ a compact region K with non-empty interior defin by the following inequalities:

$$y_1 \geqslant 0 \ , \ y_2 \geqslant 0 \ , \ H(z_1, z_2) \leqslant 0$$

where H is a pluri-harmonic function.

Let S be the part of ∂K on which $y_1 y_2 = 0$.

Let $p_1, \dots, p_k$ be distinct points in $\mathring{K}$ and let us consider the (2,1 $\bar{\partial}$-closed form

$$\chi = \sum_1^k c_i K_{p_i} \ , \ c_i \in \mathbb{C}$$

where

$$K_{p_i} = \frac{(\bar{z}_1 - \bar{z}_1(p_i))d\bar{z}_2 - (\bar{z}_2 - \bar{z}_2(p_i))d\bar{z}_1}{(|z_1 - z_1(p_i)|^2 + |z_2 - z_2(p_i)|^2)^2} dz_1 \, dz_2$$

If in some neighborhood V of S (not containing the points p_i) we have $\chi = \bar{\partial}\mu$ where μ is a (2,0) C^∞ form on V, then $c_i = 0$ for $1 \leqslant i \leqslant k$.

(β) From the exact sequence

$$0 \longrightarrow \mathcal{O}_{\mathbb{C}^2-\mathbb{R}^3} \longrightarrow \mathcal{O} \longrightarrow \mathcal{O}_{\mathbb{R}^3} \longrightarrow 0$$

we deduce that (as $\mathbb{C}^2$ is a domain of holomorphy)

$$H^1(\mathbb{R}^3, \mathcal{O}_{\mathbb{R}^3}) \simeq H^2_\phi(\mathbb{C}^2-\mathbb{R}^3, \mathcal{O})$$

where ϕ denotes the family of closed subsets of $\mathbb{C}^2$ contained in $\mathbb{C}^2 - \mathbb{R}^3$.

(γ) We can construct[1] on the first quadrant of $\mathbb{R}^2(y_1, y_2)$ a function

$$y_2 = k(y_1)$$

having the following properties:

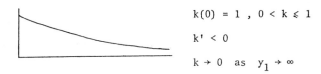

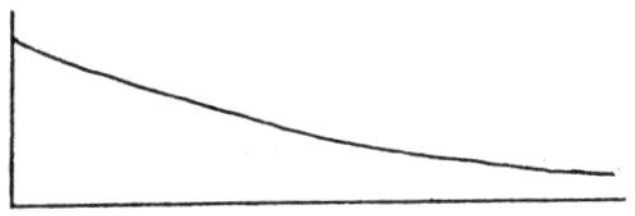

$k(0) = 1 \;,\; 0 < k \leqslant 1$

$k' < 0$

$k \to 0$ as $y_1 \to \infty$

such that, for every $\varepsilon > 0$, $\delta > 0$ one can construct a region $K(\varepsilon,\delta)$ as in the lemma contained in

$$|z_2| < \varepsilon \;,\; |x_2| < \varepsilon \quad \text{and} \quad K(\varepsilon,\delta) \cap \mathbb{R}^2(y_1 y_2) \supset \{\varepsilon y_2 \leqslant k(\varepsilon y_1) - \delta\}$$

(δ) We select a sequence of points $\{p_N\} \in \mathbb{R}^2(y_1, y_2)$ with the property that for any $\varepsilon > 0$, $\varepsilon\, y_2(p_N) < k(\varepsilon y_1(p_N))$ if $N \geqslant N(\varepsilon)$ is sufficiently large $(y_1(p_N) \to +\infty)$.

(ε) We remark that $\mathcal{O} \simeq \Omega^2$ the sheaf of germs of holomorphic 2-forms. We choose $c_N \in \mathbb{C}$, $c_N \neq 0$, $\Sigma |c_N| < \infty$ and consider the distribution

$$\Sigma\, c_N\, \delta_{p_N}$$

as an element of $H^2_\phi(\mathbb{C}^2 - \mathbb{R}^3, \Omega^2)$.

---

[1] Using elementary properties of conformal mappings in the plane of the z_1 variable.

We have

$$\Sigma c_N \delta_{p_N} = \bar{\partial} \Sigma c_N K_{p_N}$$

(up to a non-zero constant).

If

$$\Sigma c_N \delta_{p_N} = \bar{\partial}\mu \quad \text{with} \quad \text{supp } \mu \in \phi$$

then in some neighborhood V of $\mathbb{R}^3$ in $\mathbb{C}^2$ we must have

$$\Sigma c_N K_{p_N} = \bar{\partial} \Theta$$

for some C^∞ form Θ of type $(2,0)$ in V.

Combining this fact with (δ) and the lemma, we get a contradiction and indeed the proof that the space generated by the classes $\{\Sigma c_N \delta_{p_N}\}$ is already infinite dimensional.

Remark 1. Another more conceptual proof can be obtained along the following lines. From the statement at the end of Section 3, which in this case amounts to

$$H^1(\mathbb{R}^3, \mathcal{A}_o) = H^1(\mathbb{R}^3, \mathcal{O}_{\mathbb{R}^3}) = \varinjlim_{\tilde{\Omega} \supset \mathbb{R}^3} H^1(\tilde{\Omega}, \mathcal{O})$$

we see that a sufficient condition for $H^1(\mathbb{R}^3, \mathcal{A}_o)$ to vanish is that $\mathbb{R}^3$ admits a fundamental system of neighborhoods in $\mathbb{C}^2$ which are open sets of holomorphy. Now one can show that this condition is also necessary. Therefore, the fact that $H^1(\mathbb{R}^2, \mathcal{A}_o) \neq 0$ can be concluded from the fact that $\mathbb{R}^3$ does not admit a fundamental system of neighborhood in $\mathbb{C}^2$ which are open sets of holomorphy. This last statement is indeed a not too diffi exercise.

Remark 2. The argument given above cannot be extended to the case of more than 2 variables. This is not surprising as one can easily show by direct argument (using flabby resolutions of the sheaf $\tilde{\mathcal{E}}_0$) that for Ω convex in $\mathbb{R}^n$ one has always

$$H^j(\Omega, \tilde{\mathcal{E}}_0) \simeq H^j(\Omega, \mathcal{A}_0) = 0 \quad \text{if} \quad j \geqslant 2$$

It is worth noticing that if $\Omega \subset \mathbb{R}^n$ is not convex we may well have $H^j(\Omega, \mathcal{A}_0) \neq 0$ for $j \geqslant 2$ as can be seen on examples.

For the case of a simple complex consisting of a single differential operator, Hörmander has obtained necessary and sufficient conditions for solvability. We hope to have clarified conditions in the light of the foregoing.

BIBLIOGRAPHY

DE GIORGI, E., "Solutions analytiques des équations aux dérivées partielles à coefficients constants", Séminaire Goulaouic-Schwartz, Ecole Polytechnique (1971-1972), exposé 29.

HÖRMANDER, L., On the existence of real analytic solutions of partial differential equations with constant coefficients, Inv. Math. **21** (1973) 151–182.

PICCININI, L., Non-surjectivity of the Cauchy-Riemann operator on the space of the analytic functions in $\mathbb{R}^n$, generalization to parabolic operators, Bull. U.M.I. No.4, Vol. 7 (1973) 12–28.

COMPLEX TORI AND JACOBIANS

M. CORNALBA*
Department of Mathematics,
University of California,
Berkeley, California,
United States of America

Abstract

COMPLEX TORI AND JACOBIANS.

1. Generalities. 2. Cohomology of X. 3. Line bundles on complex tori. 4. The Jacobian of a curve. 5. The Riemann-Roch theorem for complex tori. 6. Projective embeddings. 7. Meromorphic functions. 8. Geometry of the theta-division.

This paper is meant to be an introduction to the geometry of complex tori and to the theory of theta-vanishing in Jacobians. The treatment of complex tori is close to the point of view of Ref. [3]. *Some of the arguments in Section 8 are adapted from Lewittes' paper* [2].

1. GENERALITIES

One of the main goals of this paper is the study of compact connected complex Lie groups, or, as they are usually called, <u>complex tori</u>. These are compact, connected complex manifolds X with holomorphic maps

$$(x,y) \to xy \qquad x,y \in X$$
$$x \to x^{-1}$$

satisfying the usual group axioms. They are very easy to describe explicitly, as a consequence of

<u>Proposition 1.1</u>. <u>A compact, connected, complex Lie group is abelian</u>.

To see this consider the function

$$\phi_x(y) = xyx^{-1}y^{-1}$$

When x is the identity, $\phi_x(y)$ is equal to the identity for every y Therefore, when x is close to the identity ϕ_x maps X into a co-

* Present address: Istituto Matematico, Università di Pavia, Pavia, Italy.

ordinate patch. Vector-valued holomorphic mappings on compact manifolds are constant, hence $\phi_x(y)$ is equal to the identity for every y. This proves that X is commutative. From now on the product in X will be written additively. We are now in a position to describe X. Let

$$\tilde{X} \overset{\phi}{\to} X$$

be the universal covering map. X is an abelian, simply connected Lie group, and is therefore isomorphic to a $\mathbb{C}^N$. ϕ is a group homomorphism and its kernel is a discrete subgroup of $\mathbb{C}^N$ of maximal rank (a lattice), a free abelian group on $2N$ generators.

Any complex torus is thus isomorphic to the quotient of a $\mathbb{C}^N$ by a lattice.

Example 1.2. When X has dimensions 1, it is the quotient of $\mathbf{C}$ by the subgroup Λ generated by two complex numbers ω_1, ω_2 that are linearly independent over $\mathbb{R}$.

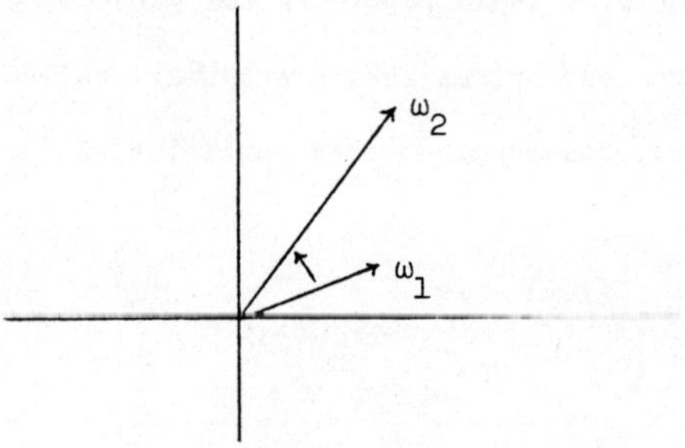

Notice that ω_1, ω_2 have been numbered so that the angle between them is less than π. Any other basis of Λ with the above property is of the form

$$\omega_1' = a\omega_1 + b\omega_2 \qquad \omega_2' = c\omega_1 + d\omega_2$$

where

$$\begin{pmatrix} a & b \\ c & d \end{pmatrix}$$

is an integral matrix with determinant +1.

Two lattices Λ, Λ' in $\mathbf{C}$ (or $\mathbf{C}^N$) are said to be equivalent if they correspond to each other under a $\mathbf{C}$-linear transformation. This amounts to saying that $\mathbf{C}^N/\Lambda$ and $\mathbf{C}^N/\Lambda'$ are isomorphic tori.

Returning to our example, we may divide by ω_1 to normalize our basis ω_1, ω_2. This gives us a new lattice which is equivalent to Λ and is generated by 1 and $\omega = \omega_2/\omega_1$. As a consequence of the assumptions, ω has positive imaginary part. It is now clear that $(1,\omega)$ and $(1,\omega')$ generate equivalent lattices if and only if

$$\omega' = \frac{a\omega + b}{c\omega + d}$$

where

$$\begin{pmatrix} a & b \\ c & d \end{pmatrix}$$

is an integral matrix with determinant 1. Thus the isomorphism classes of 1-dimensional tori correspond to points in the upper-half plane modulo linear fractional transformations with integral entries and determinant 1.

Remark 1.3. The addition law on X is determined by the complex structure, up to translation. In fact, let X' be another complex torus and f a holomorphic mapping of X into X' carrying the identity of X into the identity of X'. Then f is a group homomorphism. This can be shown by the same method used to prove Proposition 1.1. Consider the function

$$f(x+y) - f(x) - f(y) = \phi(x,y)$$

$\phi(0,y)$ is identically zero, so for x close to zero and every y, $\phi(x,y)$ lies in a coordinate patch and must be a constant function of y. But $\phi(x,0) = 0$, so $\phi(x,y)$ vanishes identically.

2. COHOMOLOGY OF X

Let X be the compact torus $\mathbf{C}^N/\Lambda$. Choose a basis $u_1, \ldots, u_{2N}$ for Λ. Then, from the real point of view, X is the product of the $2N$ 1-spheres $\mathbb{R}u_i/\mathbb{Z}u_i$, $i = 1,\dots,2N$. As a consequence, the fundamental group (or the first homology group) of X can be identified with Λ, and the first cohomology group of X with

$$\mathrm{Hom}(\Lambda,\mathbb{Z})$$

Moreover, it follows from repeated applications of the Künneth formula that the cohomology algebra $H^*(X,\mathbb{Z})$ is just the exterior algebra on $H^1(X,\mathbb{Z}) = \mathrm{Hom}(\Lambda,\mathbb{Z})$. If we introduce <u>real</u> coordinates $x_1,\dots,x_{2N}$ in $\mathbb{C}^N$ by the requirement

$$z = \Sigma x_i u_i$$

then the differentials $dx_1,\dots,dx_{2N}$ are translation-invariant, hence are pull-backs of differentials on X that we will denote by the same symbols. It is then apparent that, under the deRham isomorphism, the cohomology classes of $dx_1,\dots,dx_{2N}$ correspond to the dual basis of $u_1,\dots,u_{2N}$; in fact, this simply means that

$$\int_0^{u_i} dx_j = \delta_{ij}$$

Moreover, since cup-product and exterior multiplication of forms correspond to each other under the deRham isomorphism, $H^q(X,\mathbb{C})$ is generated freely by the <u>integral</u> classes

$$dx_{i_1} \wedge \dots \wedge dx_{i_q} \ , \ i_1 < i_2 < \dots < i_q$$

In other terms, $H^q(X,\mathbb{C})$ is isomorphic to the group of translation-invariant q-forms on $\mathbb{C}^N$.

So far we have not taken into account the complex structure of $\mathbb{C}^N$. If we do so, and introduce complex coordinates $z_1,\dots,z_N$, another basis for the invariant m-forms is given by forms of the kind:

$$dz_{i_1} \wedge \dots \wedge dz_{i_r} \wedge d\bar{z}_{j_1} \wedge \dots \wedge d\bar{z}_{j_q}$$

$$r+q = m \ , i_1 < \dots < i_r \ , j_1 < \dots < j_q$$

Therefore any invariant m-form ϕ can be written uniquely as

$$\phi = \sum_{r+q=m} \phi^{r,q}$$

where $\phi^{r,q}$ has pure type (r,q). This means that the mth complex cohomology group of X has a direct sum decomposition

$$H^m(X,\mathbb{C}) = \sum_{r+q=m} H^{r,q}(X)$$

where $H^{r,q}$ is the group of the cohomology classes of closed forms of pure type (r,q), and as a consequence:

$$H^{r,q} = \overline{H^{q,r}}$$

This is the so-called Hodge decomposition, which is valid for all compact Kähler manifolds, not only complex tori [4].

To complete the picture we only have to identify each $H^{r,q}$ with the cohomology group $H^q(X,\Omega^r)$, where Ω^r stands for the sheaf of holomorphic r-forms on X. The identification is made via the Dolbeault isomorphism:

$$H^q(X,\Omega^r) = \frac{\bar{\partial}\text{-closed } (r,q)\text{-forms on } X}{\bar{\partial}\text{-exact } (r,q)\text{-forms on } X}$$

It is clear that every class in $H^{r,q}$, viewed as an invariant (r,q)-form, gives a class in $H^q(X,\Omega^r)$. That this homomorphism is injective can be seen as follows. Choose a translation-invariant measure $d\mu$ on $\mathbb{C}^N$; this measure can be viewed as being induced by a measure on X that will be denoted by the same symbol. We may assume that $d\mu$ has been normalized so that X has volume one. Then to each form ϕ on X we may associate an invariant one, $I(\phi)$, by averaging:

$$I(\phi) = \int_{g\in X} T_g^*(\phi)d\mu(g)$$

where T_g is translation by g. The normalization assumption ensures that $I(\phi) = \phi$ when ϕ is invariant. It is also clear that I commutes with exterior differentiation, and therefore by averaging a $\bar{\partial}$-exact form we get zero. But then two invariant forms cannot differ by a $\bar{\partial}$-exact one unless they are the same.

The surjectivity statement is slightly more subtle. First of all remark that, since X is a group, its tangent bundle T and all the associated bundles, in particular $(T^*)^{\wedge r}$, are trivial. It is therefore sufficient to prove our surjectivity statement for $H^q(X,\mathcal{O})$, where $\mathcal{O}$ stands for the sheaf of germs of holomorphic functions on X. Let

$$\Delta = \sum_i \frac{\partial^2}{\partial z_i \partial \bar{z}_i}$$

be the Laplace operator on $\mathbb{C}^N$ (or X). Δ operates on forms

$$\Sigma\, a_{i_1\ldots i_r j_1\ldots j_q}\, dz_{i_1} \wedge \ldots \wedge dz_{i_r} \wedge d\bar{z}_{j_1} \ldots \wedge d\bar{z}_{j_q}$$

by operating on each coefficient separately.

We claim that, for any $\bar{\partial}$-closed $(0,q)$ form ϕ, the form

$$I(\phi) - \phi$$

is $\bar{\partial}$-exact or, in other terms, that a $\bar{\partial}$-closed form whose average is zero is $\bar{\partial}$-exact. This will follow from the auxiliary

Lemma 2.1. Let (c_{ij}) be a $2N \times 2N$ symmetric positive matrix. Write

$$E = \Sigma c_{ij} \frac{\partial^2}{\partial x_i \partial x_j}$$

and let α _be a smooth function on_ X _such that_ $I(\alpha) = 0$. _Then there is a unique smooth_ β _such that_ $E\beta = \alpha$, $I(\beta) = 0$.

We postpone the proof of (2.1) to a later time and try to conclude from it. We let

$$\phi = \sum_{i_1<\dots<i_q} \phi_{i_1\dots i_q} d\bar{z}_{i_1} \wedge \dots \wedge d\bar{z}_{i_q}$$

be a $\bar{\partial}$-closed $(0,q)$-form such that $I(\phi) = 0$. Lemma 2.1 applies to $E = \Delta$ and we may write $\phi = \Delta\psi$ for some $(0,q)$-form

$$\psi = \sum_{i_1<\dots<i_q} \psi_{i_1\dots i_q} d\bar{z}_{i_1} \wedge \dots \wedge d\bar{z}_{i_q}$$

ψ is $\bar{\partial}$-closed; in fact $\Delta\bar{\partial}\psi = \bar{\partial}\Delta\psi = 0$ and $I(\bar{\partial}\psi) = 0$, hence $\bar{\partial}\psi = 0$ by the uniqueness part of Lemma 2.1. That ψ is $\bar{\partial}$-closed means that

$$(2.2) \qquad 0 = \sum_j (-1)^{j-1} \frac{\partial}{\partial\bar{z}_{i_j}} \psi_{i_1\dots\hat{i}_j\dots i_{q+1}}, \quad i_1 < \dots < i_j < \dots < i_{q+1}$$

For any permutation σ of the numbers $1,2,\dots,r$, and any form

$$\alpha = \sum_{j_1<\dots<j_r} \alpha_{j_1\dots j_r} d\bar{z}_{j_1} \wedge \dots \wedge d\bar{z}_{j_r}$$

set
$$\alpha_{j_{\sigma(1)}\dots j_{\sigma(r)}} = \mathrm{Sgn}(\sigma)\, \alpha_{j_1\dots j_r}$$

where $\mathrm{Sgn}(\sigma)$ stands for the signature of σ. Now define

$$\gamma_{i_1\dots i_{q-1}} = \sum_i \frac{\partial}{\partial z_i} \psi_{ii_1\dots i_{q-1}}$$

$$\gamma = \sum_{i_1<\dots<i_{q-1}} \gamma_{i_1\dots i_{q-1}} d\bar{z}_{i_1} \wedge \dots \wedge d\bar{z}_{i_{q-1}}$$

Then
$$(\bar{\partial}\gamma)_{i_1\ldots i_q} = \sum_j (-1)^{j-1} \frac{\partial}{\partial \bar{z}_{i_j}} \gamma_{i_1\ldots \hat{i}_j \ldots i_q}$$

$$= \sum_j \sum_i (-1)^{j-1} \frac{\partial^2}{\partial z_i \partial \bar{z}_{i_j}} \psi_{i i_1 \ldots \hat{i}_j \ldots i_q}$$

$$= \sum_i \frac{\partial^2}{\partial z_i \partial \bar{z}_i} \psi_{i_1\ldots i_q} = \phi_{i_1\ldots i_q}$$

as follows from (2.2).

It remains to prove Lemma 2.1. α has a Fourier series that converges absolutely together with all of its term by term derivatives:

$$\alpha = \Sigma \alpha_{h_1\ldots h_{2N}} e^{2\pi\sqrt{-1}\,(\Sigma_i h_i x_i)}$$

where the h are integers. The assumption $I(\alpha) = 0$ simply means that $\alpha_{0\ldots 0} = 0$. Consider another Fourier series with undetermined coefficients and no constant term:

$$\beta = \Sigma \beta_{h_1\ldots h_{2N}} e^{2\pi\sqrt{-1}(\Sigma_i h_i x_i)}$$

and compute $E\beta$ by formally differentiating term by term. The result is

$$E\beta = -4\pi^2 \sum_{h_1\ldots h_{2N}} \beta_{h_1\ldots h_{2N}} \Big(\sum_{j,k} c_{jk} h_j h_k\Big) e^{2\pi\sqrt{-1}(\Sigma_i h_i x_i)}$$

$\sum_{j,k} c_{jk} h_j h_k$ is always positive by hypothesis, therefore

$$\beta = \Sigma \beta_{h_1\ldots h_{2N}} e^{2\pi\sqrt{-1}(\Sigma_i h_i x_i)}$$

$$\beta_{h_1\ldots h_{2N}} = - \frac{1}{4\pi^2 \Sigma c_{jk} h_j h_k} \alpha_{h_1\ldots h_{2N}}$$

is the unique formal Fourier series such that $I(\beta) = 0$, $E\beta = \alpha$ (formally). It follows easily from the convergence properties of α that β converges absolutely together with all of its term-by-term derivatives.

3. LINE BUNDLES ON COMPLEX TORI

Let L be a line bundle on an N-dimensional complex torus $X = \mathbb{C}^N/\Lambda$. Denote by π the quotient map $\mathbb{C}^N \to X$. $\pi^* L$ is a line bundle on $\mathbb{C}^N$. But:

Proposition 3.1. Any line bundle (any vector bundle, for that matter) on $\mathbb{C}^N$ is isomorphic to a trivial one.

This is an easy consequence of Cousin's theorem [1]:

Proposition 3.2. $H^1(\mathbb{C}^N, \mathcal{O}) = 0$.

To deduce (3.1) from (3.2) choose a covering $\{U_i\}$ of $\mathbb{C}^N$ by open sets on which L is trivial; we may and will suppose that the U_i's are balls (or, if we wanted to do the analogue of this on another complex manifold, we would want $U_i \cap U_j$ to be simply connected for any choice of i and j). Let f_{ij} be transition functions for L. Because of the simply-connectedness of $U_i \cap U_j$ we may write

$$f_{ij} = e^{2\pi\sqrt{-1}\,\gamma_{ij}}$$

and the cocycle condition $f_{ij}f_{jk} = f_{ik}$ implies that

$$\gamma_{ij} + \gamma_{jk} = \gamma_{ik} + n_{ijk}$$

where the n_{ijk}'s are integers and satisfy the cocycle condition

$$n_{jke} - n_{ike} + n_{ije} - n_{ijk} = 0$$

Therefore $\{n_{ijk}\}$ determines an integral cohomology class of degree 2, the Chern class of L. In the specific case of $\mathbb{C}^N$, $H^2(\mathbb{C}^N,\mathbb{Z}) = 0$, therefore $\{n_{ijk}\}$ is a coboundary and we may modify the γ_{ij}'s in such a way that γ_{ij} is a cocycle. But then γ_{ij} is a coboundary by (3.2), i.e. we may write

$$\gamma_{ij} = \zeta_j - \zeta_i$$

$$f_{ij} = e^{2\pi\sqrt{-1}\,\zeta_j}\, e^{-2\pi\sqrt{-1}\,\zeta_i}$$

and L is trivial.

In more fancy language, the above argument means that on any complex manifold M there is an exact sequence of sheaves:

$$0 \to \mathbb{Z} \to \mathcal{O} \xrightarrow{e^{2\pi\sqrt{-1}}} \mathcal{O}^* \to 0$$

where $\mathcal{O}^*$ stands for the sheaf of non-vanishing holomorphic functions, whence a long exact sequence

$$\ldots \to H^1(M,\mathbb{Z}) \to H^1(M,\mathcal{O}) \to H^1(M,\mathcal{O}^*) \overset{c}{\to} H^2(M,\mathbb{Z}) \to \ldots$$

The group $H^1(M,\mathcal{O}^*)$ is the group of isomorphism classes of holomorphic line bundles on M, and the coboundary map c associates to each line bundle L its Chern class c(L). When $M = \mathbb{C}^N$, $H^2(M,\mathbb{Z}) = H^1(M,\mathcal{O}) = 0$, so $H^1(M,\mathcal{O}^*) = 1$. In general, two line bundles that have the same Chern class differ by an element of $H^1(M,\mathcal{O})/H^1(M,\mathbb{Z})$, that is, a line bundle with transition functions f_{ij} (with respect to a suitable covering) of the form

$$f_{ij} = e^{2\pi\sqrt{-1}\,\gamma_{ij}}$$

$$\gamma_{ij} + \gamma_{jk} = \gamma_{ik}$$

For later applications, it is interesting to find an explicit formula for the deRham representative of the Chern class of a line bundle.

Let M be a complex manifold, and L a line bundle on M with transition functions f_{ij} with respect to the covering $\{U_i\}$. Suppose that $U_i \cap U_j$ is simply connected and write

$$f_{ij} = e^{2\pi\sqrt{-1}\,\gamma_{ij}}$$

The Chern cocycle $\{n_{ijk}\}$ is the Cech coboundary of the cochain $\{\gamma_{ij}\}$; applying exterior differentiation to the latter gives a 1-cocycle of holomorphic 1-forms $\left\{\frac{1}{2\pi\sqrt{-1}}\, d \log f_{ij}\right\}$. Using a partition of unity we may write

$$\frac{1}{2\pi\sqrt{-1}}\, d \log f_{ij} = \omega_j - \omega_i \quad \text{in} \quad U_i \cap U_j$$

for some C^∞ 1-forms $\{\omega_i\}$ on $\{U_i\}$. A deRham representative of $c(L)$ is the global 2-form which is equal to $d\omega_i$ in U_i for every i. This makes sense since

$$d\omega_i = d\omega_j \quad \text{in} \quad U_i \cap U_j$$

Suppose now that L has a hermitian metric. Let α_i be the squared length of the section of L on U_i that corresponds to the constant function 1 under the given local trivializations. Then

$$\alpha_j = |f_{ij}|^2\, \alpha_i$$

Now
$$\frac{1}{2\pi\sqrt{-1}}\, d \log f_{ij} = \frac{1}{2\pi\sqrt{-1}}\partial \log |f_{ij}|^2$$

$$= \frac{1}{2\pi\sqrt{-1}}\, [\partial \log \alpha_j - \partial \log \alpha_i]$$

hence a deRham representative of $c(L)$ is the global (1,1)-form with local expression

$$\frac{1}{2\pi\sqrt{-1}}\, \bar{\partial}\, \partial \log \alpha_i$$

To conclude, remark that the cocycle $\left\{ \frac{1}{2\pi\sqrt{-1}} \, d \log f_{ij} \right\}$ defines a class in $H^1(M,\Omega^1)$, the so-called Atiyah Chern class of L. On a Kähler manifold, under the Hodge decomposition this corresponds to the usual Chern class of L.

Now we return to our original problem. We noticed that π^*L is trivial. Choose one specific isomorphism $\pi^*L \xrightarrow{\phi} \mathbb{C}^N \times \mathbb{C}$. Any other trivialization of π^*L is obtained by composing ϕ and an automorphism of $\mathbb{C}^N \times \mathbb{C}$, i.e. multiplication by a nowhere vanishing holomorphic function on $\mathbb{C}^N$. L can be viewed as the quotient of $\mathbb{C}^N \times \mathbb{C}$ modulo the identifications

$$(z,\xi) \equiv (z+u, f_u(z)\cdot\xi) \qquad \begin{array}{l} z \in \mathbb{C}^N \\ \xi \in \mathbb{C} \\ u \in \Lambda \end{array}$$

where the f_u are nowhere vanishing holomorphic functions satisfying the cocycle conditions

$$f_u(z+v)f_v(z) = f_{u+v}(z) \tag{3.3}$$

Multiplication by a nowhere vanishing function $g(z)$ has the effect of replacing the cocycle $\{f_u\}$ with the new cocycle

$$f_u(z)g(z+u)g(z)^{-1}$$

We now want to describe explicitly all the line bundles on X, i.e. we want to put the cocycle $\{f_u\}$ in standard form.

We first try functions f_u that are exponentials of linear functions

$$f_u = e^{2\pi\sqrt{-1}[a_u(z)+b_u]}$$

where a_u is a linear form and b_u a number. The cocycle condition (3.3) translates into

$$a_u(z) + a_v(z) = a_{u+v}(z) \tag{3.4}$$

$$a_u(v) + b_u + b_v \equiv b_{u+v} \pmod{\mathbb{Z}} \tag{3.5}$$

(3.4) enables us to extend $a_u(z)$ by linearity to a bilinear form $A(z,w)$ such that $a_u(z) = A(z,u)$. $A(z,w)$ is $\mathbb{C}$-linear in the first variable and $\mathbb{R}$-linear only in the second. Interchanging the roles of u and v in (3.5) implies that, for any $u,v \in \Lambda$,

$$A(u,v) - A(v,u) \in \mathbb{Z} \tag{3.6}$$

Set

$$E(u,v) = A(u,v) - A(v,u)$$

$$c_u = b_u - \frac{1}{2} A(u,u)$$

Then formula (3.5) translates into:

$$c_u + c_v \equiv c_{u+v} + \frac{1}{2} E(u,v) \pmod{\mathbb{Z}}$$

This shows that $\mathrm{Im}(c_u)$ is additive in u and therefore extends to an $\mathbb{R}$-linear form on $\mathbb{C}^N$. Therefore, after multiplication by the exponential of a suitable $\mathbb{C}$-linear form, we may suppose that c_u is real for any u. If we set

$$\rho(u) = e^{2\pi\sqrt{-1}\, c_u}$$

the following hold:

$$|\rho(u)| = 1$$

(3.7)

$$\rho(u)\rho(v) = \rho(u+v)\, e^{\pi\sqrt{-1}\, E(u,v)} = \pm\rho(u+v)$$

Incidentally, it is clear that, for any given antisymmetric integral E, numbers $\rho(u)$ with the above properties exist.

At this point our cocycle $\{f_u\}$ has the form:

$$f_u = \rho(u)e^{2\pi\sqrt{-1}\,[A(z,u)+\frac{1}{2}A(u,u)]}.$$

We may still change the trivialization of $\mathbf{C}^N \times \mathbf{C}$ by multiplying by

$$g(z) = e^{2\pi\sqrt{-1}\,\mathcal{L}(z,z)}$$

where $\mathcal{L}(z,w)$ is a symmetric $\mathbf{C}$-bilinear form. The effect of this is to replace f_u with:

$$f_u(z)g(z+u)g(z)^{-1} = f_u(z)\, e^{2\pi\sqrt{-1}[2\mathcal{L}(z,u)\,+\mathcal{L}(u,u)]}$$

Notice that, by (3.6), $E(z,w)$ is real, and therefore the imaginary part of $A(z,w)$ is symmetric, hence:

$$E(\sqrt{-1}z,\sqrt{-1}w) = A(\sqrt{-1}z,\sqrt{-1}w) - A(\sqrt{-1}w,\sqrt{-1}z)$$

$$= \sqrt{-1}[A(z,\sqrt{-1}w) - A(w,\sqrt{-1}z)]$$

$$= \sqrt{-1}[A(\sqrt{-1}w,z) - A(\sqrt{-1}z,w)] = E(z,w)$$

It follows that $E(z,\sqrt{-1}w)$ is a real symmetric form and that

$$H(z,w) = -E(z,\sqrt{-1}w) + \sqrt{-1}\,E(z,w)$$

is hermitian. Moreover

$$\frac{1}{2\sqrt{-1}}[H(z,w) - H(w,z)] = E(z,w)$$

Therefore we may take $\mathcal{L}$ to be

$$\mathcal{L}(z,w) = \frac{1}{4\sqrt{-1}} H(z,w) - \frac{1}{2} A(z,w)$$

and reduce our cocycle $\{f_u\}$ to its final form

$$f_u = \rho(u)\, e^{\pi H(z,u) + \frac{\pi}{2} H(u,u)} \tag{3.8}$$

For any hermitian H whose imaginary part E is integral on $\Lambda \times \Lambda$ and any ρ such that (3.7) holds, we will define $L(H,\rho)$ to be the line bundle on X given by the cocycle (3.8). Note that

$$L(H_1,\rho_1) \otimes L(H_2,\rho_2) = L(H_1 + H_2, \rho_1\rho_2)$$

The main result of this section is the

Theorem 3.9. (Appell-Humbert)

(i) *Any line bundle on* X *is of the form* $L(H,\rho)$.

(ii) $L(H_1,\rho_1)$ *and* $L(H_2,\rho_2)$ *are isomorphic if and only if* $H_1 = H_2$, $\rho_1 = \rho_2$.

(iii) $E = \operatorname{Im}(H)$ *is the Chern class of* $L(H,\rho)$.

Notice that the statement of (3.9) can be divided in two parts: (iii) and the fact that the group $H^1(X,\mathcal{O})/H^1(X,\mathbb{Z})$ is isomorphic for the group of characters

$$\Lambda \xrightarrow{\rho} \{z \in \mathbb{C} \mid |z| = 1\}$$

We will prove them in this order.

A metric on $\mathbb{C}^N \times \mathbb{C}$ that is invariant under the action

$$(z,\xi) \to (z+u, \rho(u)\, e^{\pi H(z,u)+\frac{\pi}{2}H(u,u)}\xi)$$

is given by

$$|(z,\xi)|^2 = e^{-\pi H(z,z)}|\xi|^2$$

This metric induces one on $L(H,\rho)$ which can be used to compute $c(L(H,\rho))$.

If we write $H(z,w) = \Sigma h_{ij} z_i \bar{w}_j$, the Chern form of the above metric is

$$\frac{1}{2\pi\sqrt{-1}}\,\bar{\partial}\,\partial \log e^{-\pi H(z,z)} = \frac{1}{2\sqrt{-1}}\,\Sigma h_{ij}\, dz_i \wedge d\bar{z}_j$$

which corresponds exactly to E.

We now prove the remaining part of the theorem of Appell-Humbert. Let ω be a translation-invariant (0,1)-form on $\mathbb{C}^N$. ω corresponds to a class in $H^1(X,\mathcal{O})$, which in turn determines a line bundle on X with zero Chern class. We claim that this line bundle is $L(0,\rho)$, where

$$\rho(u) = e^{2\pi\sqrt{-1}\int_0^u (\omega + \bar{\omega})} \tag{3.10}$$

This is a straightforward application of the explicit formulae that give the Dolbeault isomorphism and is left to the reader. That $\rho(u)$ is a character follows from the fact that ω is translation-invariant. Granting (3.10), it remains to prove (ii) when $H = 0$. But it will be shown in Section 5 that if $\rho \neq 1$, then $L(0,\rho)$ has no nonzero sections and therefore is not trivial. This concludes the proof.

The sections of $L(H,\rho)$ correspond to functions on $\mathbb{C}^N$ which satisfy the functional equation

$$\theta(z+u) = \rho(u)\, e^{\pi H(z,u)+\frac{\pi}{2}H(u,u)}\,\theta(z)$$

These are called <u>theta-functions</u> relative to the hermitian form H and the <u>multiplicator</u> ρ. Among our goals in the next section will be that of determining the number of linearly independent theta-functions relative to a fixed quadratic form and multiplicator (Riemann-Roch problem).

Remark 3.11. We want to explicitly describe the line bundle

$$T_a^* L(H,\rho)$$

where T_a denotes translation by a on X. It is clear that if θ is a section of $L(H,\rho)$ (we are assuming here that $L(H,\rho)$ has sections, but the argument is just the same if this is not the case), viewed as a theta-function, then $\theta(z+a)$ represents a section of $T_a^* L(H,\rho)$ and satisfies the functional equation

$$\theta(z+a+u) = \rho(u) e^{\pi H(z,u) + \frac{\pi}{2} H(u,u) + \pi H(a,u)} \theta(z+a)$$

We are allowed to multiply $\theta(z+a)$ by a never vanishing function, for example $e^{\pi H(z,a)}$. Then $\hat{\theta}(z) = e^{\pi H(z,a)} \theta(z+a)$ represents a section of $T_a^* L(H,\rho)$ and satisfies the functional equation

$$\hat{\theta}(z+u) = \rho(u)\, e^{2\pi\sqrt{-1}\, E(a,u)}\, e^{\pi H(z,u) + \frac{\pi}{2} H(u,u)} \hat{\theta}(z)$$

Therefore it is a theta function for the hermitian form H and the multiplicator $\rho(u)\, e^{2\pi\sqrt{-1}\, E(a,u)}$.

Let us notice that if a,b are points of X then

$$T_a^* L(H,\rho) \otimes T_b^* L(H,\rho) = T_{a+b}^* L(H,\rho) \otimes L(H,\rho)$$

This follows by comparing hermitian forms and multiplicators. It is clear that the hermitian forms of the two sides are the same. As for multiplicators, on the left-hand side we have

$$\rho(u)\, e^{2\pi\sqrt{-1}\, E(a,u)} \rho(u)\, e^{2\pi\sqrt{-1}\, E(b,u)} = \rho(u)^2\, e^{2\pi\sqrt{-1}\, E(a+b,u)}$$

as on the right-hand side. The above formula means that the mapping

$$a \to T_a^* L(H,\rho) \otimes L(-H,\rho^{-1})$$

of X into the group of line bundles on X with zero Chern class is a group homomorphism. The kernel of this homomorphism is the set of the classes modulo Λ of all those a such that $E(a,u) \in \mathbb{Z}$ for any $u \in \Lambda$. If H is non-degenerate (positive, for example), i.e. if E is non-degenerate, the kernel is finite and the mapping

$$a \to T_a^* L(H,\rho) \otimes L(-H,\rho^{-1})$$

is onto, for dimension reasons; in fact the set of isomorphism classes of line bundles with zero Chern class is the complex torus $H^1(X,\mathcal{O})/H^1(X,\mathbb{Z})$, which has the same dimension as X.

4. THE JACOBIAN OF A CURVE

Let C be a smooth compact Riemann surface (a curve). Let g be the genus of C. $K = K_C$ will denote the canonical bundle of C, i.e. the line bundle whose associated sheaf is the sheaf Ω_C^1 of holomorphic 1-forms on C. In what follows we will freely use the language of divisors and line bundles associated to them; more details on this formalism will be given in Section 7. We will write $h^i(,)$ for the dimension of $H^i(,)$, and $\mathcal{O}(L)$ for the sheaf of holomorphic sections of the line bundle L. Let $D = \Sigma m_i p_i$ be a divisor on C ($p_i \in C$). We will denote by L_D the line bundle associated to D and set

$$\deg(L_D) = \deg(D) = \Sigma m_i$$

This degree depends only on the linear equivalence class of D. From the theory of curves we will assume (at least) the following facts:

(a) Every line bundle on C is of the form L_D for some divisor D.

(b) (Riemann-Roch for curves)

(i) $h^0(C,L) - h^1(C,L) = \chi(C,L) = \deg(L) + 1 - g$

(ii) (Duality) $H^1(C,L^{-1} \otimes K)$ is dual to $H^0(C,L)$, the pairing being the cup product:

$$H^1(C,L^{-1} \otimes K) \times H^0(C,L) \to H^1(C,K) \cong \mathbb{C}$$

(c) The topological structure of C as a sphere with g handles attached.

Notice that from the short exact sequence

$$0 \to \mathbb{C} \to \mathcal{O} \xrightarrow{d} \Omega^1_C \to 0$$

there follows a short exact sequence

$$0 \to H^0(C,K) \to H^1(C,\mathbb{C}) \to H^1(C,\mathcal{O}) \to 0$$

which naturally splits ($\overline{H^1(C,\mathcal{O})}$ can be identified with $H^0(C,K)$. This is the simplest case of the Hodge decomposition.

If $\deg(L) < 0$, L has no sections, because the number of zeroes of a section of a line bundle is equal to the degree of the line bundle itself. Therefore Riemann-Roch tells us the dimension of $H^1(C,L)$. Dually, $H^1(C,L) = 0$ when $\deg(L) > \deg(K) = 2g-2$. So we know $h^0(C,L)$ (or, which is the same, $h^1(C,L)$), when $\deg(L)$ is not between 0 and $2g-2$. When $\deg(L) = 0$, two cases are possible; either L is trivial or it has no sections (for a non-trivial section of L cannot have zeroes). Dually, when $\deg(L) = 2g-2$, either $L = K$ or $h^1(C,L) = 0$ (i.e. $h^0(C,L) = g-1$).

There remains the range of values $0 < \deg(L) < 2g-2$. First of all, notice that if r is a point of C, then

$$h^0(C,L_{D+r}) \leqslant h^0(C,L_D) + 1$$

This follows from the cohomology exact sequence of

$$0 \to \mathcal{O}(L_D) \to \mathcal{O}(L_{D+r}) \to \mathbf{C}_r \to 0$$

where $\mathbf{C}_r$ stands for the sheaf which is concentrated at r and has stalk C at r. Now assume that there is on C a line bundle of degree 1 with two linearly independent sections s_0, s_1; each one of these has only one zero. Moreover, s_0, s_1 never vanish simultaneously; in fact, if this were the case, s_1/s_0 would be a holomorphic function on C, hence a constant. It follows that s_0, s_1 define a mapping of C into $\mathbf{P}^1$ (the "meromorphic function" s_1/s_0). More details on this construction can be found in Section 6. The assumption $\deg(L) = 1$ implies that this mapping is 1-1, hence an isomorphism. This shows that $g = 0$.

Putting the above remarks together, and using duality, we see that for a curve of positive genus the possible dimensions for $H^0(C,L)$ are as shown in Fig.1.

We will see later that "generically" the dimension of $H^0(C,L)$ is minimum possible, given the degree of L.

From now on C will be a curve of positive genus. We can now introduce the <u>Jacobian variety</u> of C, written $J(C)$; we will do this in two equivalent ways. Set

$$J(C) = \text{group of (equivalence classes of) line bundles of degree } 0 \text{ on } C = H^1(C,\mathcal{O})/H^1(C,\mathbf{Z}).$$

deg L	$h^0(C,L)$
1	0,1
2	0,1,2
.	.
.	.
.	.
g-1	0,1,...,g-1
g	1,...,g-1
.	
.	
.	
2g-3	g-2,g-1

FIG. 1. Possible dimensions for $H^0(H,L)$ for a curve of positive genus.

Since $H^1(C,\mathbf{Z})$ is a lattice in $H^1(C,\mathcal{O})$, $J(C)$ is a complex torus. Choose a base point $q \in C$. Every point of $J(C)$ is of the form

$$\text{class of } (L_{r_1+\ldots+r_g-gq})$$

as follows from the table of Figure 1. There is a holomorphic mapping ϕ_q (or ϕ, for short) of C into $J(C)$, defined as follows:

$$\phi_q(r) = \text{class of } (L_{r-q})$$

Remark 4.1. Choose a basis $\omega_1 \ldots \omega_g$ for the holomorphic differentials on C. Then $J(C) = H^1(C,\mathcal{O})/H^1(C,\mathbf{Z})$ is isomorphic to $\mathbf{C}^g$ modulo the lattice consisting of all the vectors of the form

$$(\int_\gamma \omega_1, \ldots, \int_\gamma \omega_g) \qquad \gamma \in H_1(C,\mathbf{Z})$$

the isomorphism being induced by

$$H^1(C,\mathcal{O}) \ni \omega \mapsto (\int_C \omega \wedge \omega_1, \ldots, \int_C \omega \wedge \omega_g)$$

We will need the following version of Abel's theorem.

Theorem 4.2. With the above identification

$$\phi_q(r) = \text{class of } \left(\int_q^r \omega_1, \ldots, \int_q^r \omega_g\right)$$

(independent of the path used to join r and q), or, equivalently, the isomorphism between $J(C)$ and the torus constructed above is:

$$\text{class of } (L_{r_1+\ldots+r_g-gp}) \to \ldots$$

$$\ldots \to \text{class of } \left(\int_q^{r_1}\omega_1+\ldots+\int_q^{r_2}\omega_1, \ldots, \int_q^{r_1}\omega_g+\ldots+\int_q^{r_q}\omega_q\right)$$

Proof: Consider a line bundle of the form L_{r-q}. For simplicity, we shall assume that r and q are the points 1 and 0 in a suitable coordinate patch P with coordinate z. Consider a covering $\mathcal{U} = \{U,V\}$ of C consisting of the complement U of the segment joining r and q, and of a small neighborhood V of σ. Relative to $\mathcal{U}$, a transition function for L_{r-q} is $\left(\frac{z}{z-1}\right)$. This function has a single-valued logarithm in $U \cap V$; upon crossing σ from the lower half-plane to the upper half-plane, $\log\left(\frac{z}{z-1}\right)$ increases by $2\pi\sqrt{-1}$. A Dolbeault representative of the cohomology class of the 1-cocycle $\frac{1}{2\pi\sqrt{-1}} \log\left(\frac{z}{z-1}\right)$ can be obtained as follows. Let χ be a smooth function which is equal to 1 in a neighborhood of $\bar{V}$ in P and vanishes outside a neighborhood of $\bar{V}$ in P. Then the cohomology class of $\frac{1}{2\pi\sqrt{-1}} \log\left(\frac{z}{z-1}\right)$ is represented by

$$\frac{1}{2\pi\sqrt{-1}}\,\bar{\partial}\left(\chi \log\left(\frac{z}{z-1}\right)\right) = \omega$$

Now

$$\int_C \omega \wedge \omega_i = \frac{1}{2\pi\sqrt{-1}} \int_C d\left(\chi \log\left(\frac{z}{z-1}\right)\omega_i\right)$$

$$= \frac{1}{2\pi\sqrt{-1}}\left[\underset{\text{upper}}{\int_0^1 \log\left(\frac{z}{z-1}\right)\omega_i} - \underset{\text{lower}}{\int_0^1 \log\left(\frac{z}{z-1}\right)\omega_i}\right] = \int_0^1 \omega_i$$

where "upper" and "lower" denote upper and lower determination of $\log \frac{z}{z-1}$ along σ. This concludes the proof of (4.2).

Remark 4.3. ϕ is an embedding. To begin with, let us show that it is 1-1. If not, there are two distinct point r_1, r_2 such that $r_1 - q$ and $r_2 - q$ are linearly equivalent, i.e. r_1 and r_2 are linearly equivalent. But this means that $h^0(C, L_{r_1}) = 2$, which is absurd. It remains to show that ϕ has non-zero differential everywhere, i.e. that given a point r of C, there is a global holomorphic differential that does not vanish at r. If this is not true, the mapping

$$H^0(C, K \otimes L_{-r}) \to H^0(C, K)$$

is an isomorphism, or , dually,

$$H^1(C, \mathcal{O}) \to H^1(C, L_r)$$

is an isomorphism. By Riemann-Roch this implies that $h^0(C, L_r) = 2$, absurd.

Remark 4.4. $J(C)$ has one more piece of structure. On $H^1(C, \mathcal{O})$ there is a positive definite hermitian form:

$$H(\omega, \omega') = \frac{2}{\sqrt{-1}} \int_C \omega \wedge \bar{\omega}'$$

whose imaginary part restricts to the negative of the intersection pairing on $H^1(C, \mathbf{Z})$. For any multiplicator ρ for H, $L(H, \rho)$ is a line bundle on $J(C)$. We shall see later that the vector space of its sections is one-dimensional and we will explicitly write down a generator, "the" Riemann theta function of C (well defined up to translation).

5. THE RIEMANN-ROCH THEOREM FOR COMPLEX TORI

Let $X = \mathbb{C}^N/\Lambda$ be a complex torus. Suppose that there exists a positive definite hermitian form H on $\mathbb{C}^N$ whose imaginary part E is integral on $\Lambda \times \Lambda$ ("usually" there is no such form, as will be made clear later). Let ρ be any multiplicator relative to H. The datum of such a positive H and a multiplicator ρ is called a polarization of X. We want to compute the dimension of the space of sections of $L(H,\rho)$.

We will use the following

Lemma 5.1. It is possible to choose a basis

$$u_1 \ldots u_N \quad \tilde{u}_1 \ldots \tilde{u}_N$$

for Λ such that

$$E(u_i,u_j) = E(\tilde{u}_i,\tilde{u}_j) = 0$$

$$-E(u_i,\tilde{u}_j) = E(\tilde{u}_j,u_i) = d_i\delta_{ij}$$

where the d_i are positive integers and $d_1|d_2|\ldots|d_N$. In other terms, the matrix of E relative to the above basis can be written in block form as

$$\begin{pmatrix} 0 & -\Delta \\ \Delta & 0 \end{pmatrix}$$

where $\Delta = \begin{pmatrix} d_1 & & 0 \\ & \ddots & \\ 0 & & d_N \end{pmatrix}$ and $d_1|\ldots|d_N$.

Proof. Choose a basis for Λ and let $Q = (q_{ij})$ be the matrix of E relative to this basis. Q is skew-symmetric. A change of basis changes Q into MQ^tM, where M is a unimodular integral matrix. MQ^tM is said to be unimodularly congruent to Q. We must show that any non-degenerate, skew-symmetric Q is unimodularly congruent to a matrix of the form $\begin{pmatrix} 0 & -\Delta \\ \Delta & 0 \end{pmatrix}$. Elementary unimodular congruences are:

(i) Interchanging two rows and the two corresponding columns. This is accomplished by choosing an M of the form

$$\begin{pmatrix} 1 & & & & & & \\ & \ddots & & & & & \\ & & 0 & \cdots & 1 & & \\ & & \vdots & \ddots & \vdots & & \\ & & 1 & \cdots & 0 & & \\ & & & & & \ddots & \\ & & & & & & 1 \end{pmatrix}$$

(ii) Changing the sign of a row and the corresponding column. M has to be of the form

$$\begin{pmatrix} 1 & & & & & \\ & \ddots & & & 0 & \\ & & 1 & & & \\ & & & -1 & & \\ & 0 & & & \ddots & \\ & & & & & 1 \end{pmatrix}$$

(iii) Replacing a row by that same row plus k times another (k any integer) then doing the same for columns. The matrix M is

$$\begin{pmatrix} 1 & & & \\ & \ddots & & \\ 0 & k & \ddots & \\ & 0 & & 1 \end{pmatrix} \quad \text{or} \quad \begin{pmatrix} 1 & 0 & k & 0 \\ & \ddots & & \\ & & \ddots & \\ 0 & & & 1 \end{pmatrix}$$

We will show that Q is congruent to a matrix:

$$\begin{pmatrix} 0 & -d_1 & & & & & \\ d_1 & 0 & & & & & \\ & & 0 & -d_2 & & & \\ & & d_2 & 0 & & & \\ & & & & \ddots & & \\ & & & & & 0 & -d_N \\ & & & & & d_N & 0 \end{pmatrix}$$

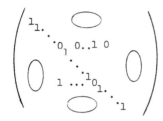

Repeated applications of (i) then show that this matrix is congruent to

$$\begin{pmatrix} 0 & -\Delta \\ \Delta & 0 \end{pmatrix}$$

The proof is by induction on N. The case $N = 2$ is clear. Repeated applications of (i) enable us to assume that q_{12} is not zero and is the smallest in absolute value among the non-zero entries of matrices congruent to Q. Repeated applications of (iii) then enable us to replace $q_{13},\dots,q_{1N}$ with their residues modulo q_{12}. These are zero by the very choice of q_{12}. Q is now of the form

$$\begin{pmatrix} 0 & q_{12} & 0 & \cdots & \cdots & 0 \\ -q_{12} & 0 & q_{23} & \cdots & \cdots & \\ \vdots & \vdots & & & & \\ \vdots & \vdots & & & & \end{pmatrix}$$

Applying (i) to the first two rows and columns gives a Q of the form

$$\begin{pmatrix} 0 & -q_{12} & q_{23} & \cdots & \cdots & \\ q_{12} & 0 & 0 & \cdots & \cdots & 0 \\ -q_{23} & 0 & & \cdots & \cdots & \\ \vdots & \vdots & \vdots & & & \end{pmatrix}$$

Repeated applications of (iii) enable us to assume that $q_{23},\dots,q_{2N}$ also vanish. Applying rule (ii), if necessary, we may assume that q_{12} is negative and set $d_1 = -q_{12}$. Now induction proves (5.1) except for one point. We must show that d_1 divides

d_2. Suppose that this is not the case. Using (iii), Q is congruent to

$$\begin{pmatrix} 0 & -d_1 & 0 & -d_2 & & & \\ d_1 & 0 & 0 & 0 & & & \\ 0 & 0 & 0 & -d_2 & & & \\ d_2 & 0 & d_2 & 0 & & & \\ & & & & 0 & -d_3 & \\ & & & & d_3 & 0 & \\ & & & & & & \ddots \end{pmatrix}$$

Another application of (iii) enables us to replace d_2 by its residue mod d_1, which is not zero. This contradicts the minimality of $d_1 = -q_{12}$.

Remark 5.2. It should be noticed that we have not used the positivity of H, but only the fact that H (or E) is non-degenerate. Moreover, the lemma is obvious for the Jacobian of a curve. A basis $\gamma_1 \ldots \gamma_{2g}$ of $H_1(C,\mathbb{Z})$ whose intersection matrix is

$$(I(\gamma_i,\gamma_j)) = \begin{pmatrix} 0 & I \\ -I & 0 \end{pmatrix}$$

is shown as:

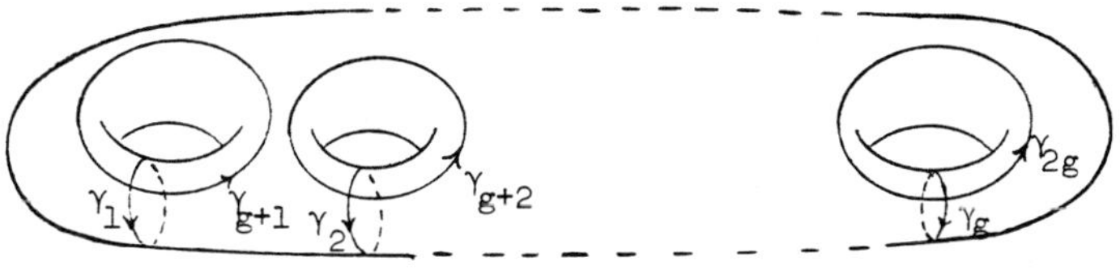

Notice that with the notations of Lemma 5.1, $u_1,\ldots,u_N$ constitute a basis of $\mathbb{C}^N$. In fact, denote by U the vector space over $\mathbb{R}$ that they generate. $U \cap \sqrt{-1}\, U$ is a complex vector space on which E, and hence H, vanish. But H is positive, so $U \cap \sqrt{-1}\, U = \{0\}$, $U + \sqrt{-1}\, U = \mathbb{C}^N$. We also set

$$\tilde{u}_i = \Sigma t_{ji} u_j \quad , \quad T = (t_{ij}) \quad , \quad h_{ij} = H(u_i, u_j) .$$

h_{ij} is a real number, so $B(\Sigma z_i u_i, \Sigma w_j u_j) = \Sigma h_{ij} z_i w_j$ is a $\mathbb{C}$-bilinear, symmetric form which is real on U. Moreover,

$$H(z,u) = B(z,u)$$

whenever u belongs to U, by linearity.

We are considering theta-functions θ that satisfy the functional equation

$$\theta(z+u) = \rho(u)\, e^{\pi H(z,u) + \frac{\pi}{2} H(u,u)}\, \theta(z) \qquad u \in \Lambda \tag{5.3}$$

It is easily checked that the function

$$\tilde{\theta}(z) = e^{-\frac{\pi}{2} B(z,z)}\, \theta(z)$$

satisfies the related functional equation

$$\tilde{\theta}(z+u) = \rho(u)\, e^{\pi[H(z,u) - B(z,u) + \frac{1}{2}(H(u,u) - B(u,u))]}\, \tilde{\theta}(z)$$

In particular

$$\begin{aligned} &\tilde{\theta}(z+u_i) = \rho(u_i)\, \tilde{\theta}(z) \\ &\tilde{\theta}(z+\tilde{u}) = \rho(\tilde{u})\, e^{-2\pi\sqrt{-1}\, \Sigma z_i \mu_i d_i - \pi\sqrt{-1}\, \Sigma \mu_i d_i t_{ij} \mu_j}\, \tilde{\theta}(z) \end{aligned} \tag{5.4}$$

where we are writing $z = \Sigma z_i u_i$, $\tilde{u} = \Sigma \mu_j \tilde{u}_j$, and μ_j is an integer for any index j. (Notice that $H(z,\tilde{u}_j) - B(z,\tilde{u}_j) = 2\sqrt{-1}\ \Sigma z_i\ E(u_i,\tilde{u}_j))$. On U, ρ is a homomorphism; therefore for any $u \in U$ we may write

$$\rho(u) = e^{2\pi\sqrt{-1}\lambda(u)}$$

where λ is a $\mathbf{C}$-linear form on $\mathbf{C}^N$ which is real on U. Then $e^{-2\pi\sqrt{-1}\lambda(z)}\tilde{\theta}(z)$ is periodic with respect to U, and $\tilde{\theta}(z)$ has a Fourier series expansion:

(5.5)
$$\tilde{\theta}(z) = \Sigma a_{n_1 \ldots n_N}\, e^{2\pi\sqrt{-1}(\lambda(z) + \Sigma n_i z_i)}$$

The functional equation (5.3) implies some restrictions on the coefficients $a_{n_1 \ldots n_N}$, namely

(5.6)
$$a_{m_1+k_1 d_1, \ldots, m_N + k_N d_N} = \rho(\tilde{u})\, e^{-2\pi\sqrt{-1}\lambda(\tilde{u})}\, e^{\pi\sqrt{-1}\ \Sigma k_i d_i t_{ij} k_j + 2\pi\sqrt{-1}\ \Sigma m_i t_{ij} k_j}\, a_{m_1 \ldots m_N}$$

where we have written $\tilde{u} = -\Sigma k_i \tilde{u}_i$. Therefore the Fourier coefficients $a_{n_1 \ldots n_N}$ are uniquely determined by those such that $0 \leqslant n_i < d_i$, $i = 1,\ldots,N$ (there are $\sqrt{\det Q} = \prod_{i=1}^{N} d_i$ of them). We will now show that for any choice of the coefficients, subject to the above restrictions, the series (5.4) converges absolutely on compact sets, together with all of its term-by-term derivatives. We may fix one multi-index $m_1,\ldots,m_N$ such that $0 \leqslant m_i < d_i$ and assume that

$$a_{m_1 \ldots m_N} = 1$$

$$a_{n_1 \ldots n_N} = 0 \quad \text{unless} \quad m_i \equiv n_i(d_i) \quad i = 1,\ldots,N$$

The dominant term in the right-hand side of formula 5.6 is $e^{\pi\sqrt{-1}\,\Sigma k_i d_i t_{ij} k_j}$, and to ensure convergence, it suffices to show that $\operatorname{Im} \Delta T$ is symmetric positive definite. To do so, notice that

$$d_j \delta_{jk} = E(\tilde{u}_j, u_k) = \operatorname{Im} H(\tilde{u}_j, u_k)$$

$$= \sum_i \operatorname{Im} t_{ij} h_{ik}$$

In other terms, if we write $H = (h_{ik})$

$${}^t\operatorname{Im} T \cdot H = \Delta$$

So,
$$\operatorname{Im}(\Delta T) = {}^t\operatorname{Im} T\, H\, \operatorname{Im} T$$

which is positive definite because H is (and $\operatorname{Im} T$ is invertible). In conclusion,

Proposition 5.7. The space of theta-functions relative to a polarization H, ρ has dimension

$$\sqrt{\det E} = d_1 \cdots d_N$$

where E is the imaginary part of H and the determinant is computed relative to any basis of Λ.

Example 5.8. Let $X = J(C)$ be the Jacobian of a curve C of genus g. Recall that

$$J(C) = H^1(C,\mathcal{O})/H^1(C,\mathbb{Z})$$

and that there is a positive definite hermitian form H on $H^1(C,\mathcal{O})$ whose imaginary part E restricts to minus the intersection pairing on $H^1(C,Z)$. Moreover, we have seen that we can choose a basis $u_1 \ldots u_g\ \tilde{u}_1 \ldots \tilde{u}_g$ of $H^1(C,\mathbb{Z})$ such that

$$E(u_i,u_j) = E(\tilde{u}_i,\tilde{u}_j) = 0$$

$$E(\tilde{u}_i,u_j) = \delta_{ij}$$

Choose the multiplicator ρ for H such that

$$\rho(u_i) = \rho(\tilde{u}_i) = 1 \quad , \quad i = 1,\ldots,g$$

Then Proposition 5.6 says that the space of sections of $L(H,\rho)$ is one-dimensional. Moreover, the discussion preceding (5.6) enables us to explicitly write down a theta-function θ on $\mathbb{C}^g$ which represents a generator of $H^0(X,L(H,\rho))$.

For any $z \in H^1(C,\mathbb{C})$ we write $z = \Sigma z_i u_i$. Then a generator for $H^0(X,L(H,\rho))$ is

$$\theta(z) = e^{\frac{\pi}{2}B(z,z)} \Sigma a_{n_1\ldots n_g} e^{2\pi\sqrt{-1}(\Sigma n_i z_i)}$$

where $a_{0\ldots 0} = 1$ and

$$a_{n_1\ldots n_g} = e^{\pi\sqrt{-1}\ \Sigma n_i t_{ij} n_j}$$

In conclusion

$$\theta(z) = e^{\frac{\pi}{2}\ \Sigma h_{ij} z_i z_j} \Big(\sum_{n_1\ldots n_g \in \mathbb{Z}^g} e^{2\pi\sqrt{-1}\ \Sigma n_i z_i + \pi\sqrt{-1}\ \Sigma n_i t_{ij} n_j} \Big)$$

$\theta(z)$, or rather its quotient by the nowhere vanishing factor $e^{\frac{\pi}{2}\Sigma h_{ij} z_i z_j}$, is called the <u>theta-function of C</u>. Its zero locus is a periodic divisor on $\mathbb{C}^N$, which is the pull-back of a divisor on

$J(C)$. Any one of these two equivalent objects is called the theta-divisor of $J(C)$. One of our goals in the next sections will be to explore the relationship between the geometry of the theta-divisor and the geometry of the curve C.

Now let us return to the problem of computing the dimension of the space of theta-functions relative to the given hermitian form H and a given multiplicator ρ.

Suppose that H is zero. Two cases are possible. Either ρ is trivial or it is not. If ρ is trivial the only theta-functions are the constants. If ρ is non-trivial, I claim that there are no non-zero theta-functions. In fact the functional equation reduces to

$$\theta(z + u) = \rho(u)\theta(z)$$

This implies that θ is bounded, hence constant, so θ must vanish identically if $\rho(u) \neq 1$ for some u. Now suppose that H is positive semi-definite. Let W be the complex vector subspace of $\mathbf{C}^N$ consisting of all the vectors z such that $H(z,w) = 0$ for any w. $W \cap \Lambda$ can be described as the set of all $u \in \Lambda$ such that $E(u,v) = 0$ for any $v \in \Lambda$, and is therefore a lattice in W. If ρ is trivial on $W \cap \Lambda$, any theta-function θ is constant on $W/W \cap \Lambda$. If ρ is non-trivial, θ vanishes on $W/W \cap \Lambda$. Moreover, if $w \in W$, $u \in W \cap \Lambda$ and $z \in \mathbf{C}^N$, the functional equation for θ implies that

$$\theta(w + u + z) = \rho(u)\theta(w + z)$$

so $w \to \theta(w+z)$ is a theta-function on $W/W \cap \Lambda$ and is constant or vanishes identically according to whether ρ is trivial on $W \cap \Lambda$ or not. In conclusion two cases are possible:

(a) ρ is non-trivial on $W \cap \Lambda$: no non-zero theta-functions.

(b) ρ is trivial on $W \cap \Lambda$. Then θ is constant on each coset $W+z$. Moreover, it is easily seen that H is the pullback of a positive definite hermitian H on $\mathbb{C}^N/W$ whose imaginary part is integral on $\Lambda/\Lambda \cap W$, and that ρ is the pullback of a multiplicator ρ' for H'. In this case any theta-function on X is the pullback of a theta-function on $(\mathbb{C}^N/W)/(\Lambda/\Lambda \cap W)$, relative to H' and ρ'. The dimension of the latter vector space is given by Proposition 5.7.

The last case to be considered is the one when H is not positive semidefinite. We claim that, if this is the case, where are no non-zero theta-functions relative to H and ρ. There is a subspace W of $\mathbb{C}^N$ on which H is negative definite. Suppose that there is a non-zero theta-function θ. As we showed in Section 3, the function

$$\psi(z) = e^{-\pi H(z,z)}|\theta(z)|^2$$

is bounded on $\mathbb{C}^N$. It is clear that the restriction of ψ to any one of the translates of W is plurisubharmonic (i.e. the matrix of second partials $\left(\frac{\partial^2\psi}{\partial w_i \partial \bar{w}_j}\right)$ is positive semidefinite, where the w_i's are coordinates on a translate of W), and strictly so (i.e. $\left(\frac{\partial^2\psi}{\partial w_i \partial \bar{w}_j}\right)$ is definite) where θ is not zero. The same is true of

$$\tilde{\psi}(z) = e^{-\varepsilon\Sigma|z_i|^2}\psi(z)$$

for small ε; $\tilde{\psi}$ vanishes at infinity, so it has a maximum. At this point $\left(\frac{\partial^2\psi}{\partial w_i \partial \bar{w}_j}\right)$ is positive, which is absurd.

Remark 5.9. We have shown that ΔT has symmetric imaginary part. Its real part is also symmetric. In fact, we know that

$$\text{(5.10)} \qquad H \operatorname{Im} T = \Delta$$

Moreover, if we write

$$\sqrt{-1}\, u_i = \sum_j a_{ji} u_j + \sum_j b_{ji} \tilde{u}_j$$

where $A = (a_{ij})$ and $B = (b_{ij})$ are real matrices, we conclude that

$$A + \mathrm{Re}\, T \cdot B = 0$$

$$\mathrm{Im}\, T \cdot B = I$$

Therefore, $$\mathrm{Re}\, T = -A\, \mathrm{Im}\, T.$$

On the other hand,

$$\Sigma\, \mathrm{Re}\, t_{ij} h_{ik} = \mathrm{Re}\, H(\tilde{u}_j, u_k) = -E(\tilde{u}_j, \sqrt{-1}\, u_k)$$

$$= -\sum_i a_{ik}\; E(\tilde{u}_j, u_i) = -\sum_j d_j a_{jk}$$

In other terms, $${}^t\mathrm{Re}\, T \cdot H = -\Delta A\ .$$

Therefore $$\Delta A\ \mathrm{Im}\, T = -{}^t\mathrm{Re}\, T\, H\, \mathrm{Im}\, T = -{}^t\mathrm{Re}\, T \Delta$$

or $$\Delta\ \mathrm{Re}\, T = {}^t\mathrm{Re}\, T\ \Delta$$

i.e. $\Delta\ \mathrm{Re}\, T$ is symmetric, and, in conclusion:

<u>ΔT is a symmetric matrix</u>

<u>with positive imaginary part</u>

This statement usually goes under the name of "Riemann bilinear relations". We may give a more intrinsic version of it as follows. Choose a basis $v_1, \dots, v_{2N}$ for Λ and a basis $b_1 \dots b_N$ for $\mathbb{C}^N$. Set

$$v_j = \Sigma \omega_{ij} b_i$$

$$\Omega = (\omega_{ij})$$

$$q_{ij} = E(v_i, v_j); \qquad Q = (q_{ij})$$

With our choice of bases

$$Q = \begin{pmatrix} 0 & -\Delta \\ \Delta & 0 \end{pmatrix}$$

$$\Omega = (I,T)$$

The symmetry of ΔT is equivalent to

(5.11) $$\Omega Q^{-1}\,{}^t\Omega = 0$$

and the positivity of $\operatorname{Im} \Delta T$ is equivalent to

(5.12) $$\sqrt{-1}\,\Omega Q^{-1}\,{}^t\bar{\Omega} > 0$$

The above statements are clearly invariant under change of bases for Λ and $\mathbb{C}^N$. They go under the names of first and second Riemann bilinear relations.

6. PROJECTIVE EMBEDDINGS

Let X be any compact complex variety, and L a line bundle on X that has at least one non-trivial section. Choose a basis $s_0 \ldots s_n$ for the vector space $H^0(X,L)$. Let x be a point on X where one among $s_0 \ldots s_n$ does not vanish, and ζ a fibre coordinate for L in a neighborhood of x. Then the point

$$[s_0(x):\ldots:s_n(x)] = [\zeta(s_0(x)):\ldots:\zeta(s_n(x))]$$

in projective n-space depends holomorphically on x and does not depend on the choice of ζ. Therefore if L has at least one non-zero section, the formula

$$\psi(x) = [s_0(x):\ldots:s_n(x)]$$

gives a mapping (defined everywhere except, possibly, on a proper subvariety of X)

$$\psi: X \to \mathbb{P}^n$$

Changing the basis $s_0 \ldots s_n$ has the effect of modifying ψ by a projective transformation of $\mathbb{P}^n$. Therefore ψ is essentially determined by L. ψ is everywhere defined if for any point x of X there is a section of L that does not vanish at x. ψ separates points if, given two points of X, x and y, there is a section of L that vanishes at x but not at y, and vice versa. ψ is a local embedding if for any point x of X and any tangent vector σ at x, there is a section s of L such that

$$\sigma(\zeta(s)) \neq 0 \quad \text{at} \quad x; \quad s(x) = 0$$

where ζ is a fixed fibre coordinate near x.

If ψ has the three properties described above, ψ is an isomorphism of X onto a smooth subvariety of projective n-space. We want to apply this to complex tori.

Proposition 6.1. *Let X be the complex torus $\mathbb{C}^N/\Lambda$. Suppose there is a line bundle $L(H,\rho)$ on X such that H is positive definite. Let L be $L(3H,\rho^3)$. Then ψ is an isomorphism of X onto a smooth subvariety of projective space.*

Proof: $L(H,\rho)$ has at least one section. Let θ be a non-zero theta function for H, ρ. Then, as we saw at the end of §3, for any choice of a, b in $\mathbb{C}^N$,

$$\theta(x+a)\theta(x+b)\theta(x-a-b) = \xi(x)$$

represents a section of L. If we choose a and b in such a way that θ does not vanish at $z+a$, $z+b$, $z-a-b$, then ξ does not vanish at z.

Now suppose there are two points a, b which are not separated by ψ and such that $a-b \notin \Lambda$. This implies that there is a non-zero constant K such that for any theta function $\hat{\theta}$ relative to $3H,\rho^3$,

$$\hat{\theta}(a) = K\hat{\theta}(b)$$

In particular, for any choice of x, y,

$$\theta(x+a)\theta(y+a)\theta(a-x-y) = K\theta(x+b)\theta(y+b)\theta(b-x-y)$$

Keep y fixed in the above identity and take logarithmic derivatives with respect to x. The result is

$$d \log \theta(x+a) - d \log \theta(a-x-y) = d \log \theta(x+b) - d \log \theta(b-x-y)$$

which in particular implies that

$$d \log \frac{\theta(x+a)}{\theta(x+b)} = d \log \theta(x+a) - d \log \theta(x+b)$$

is a translation-invariant differential $\Sigma\alpha_i dz_i$. The conclusion is that

$$\theta(x+a) = K\, e^{\Sigma\alpha_i x_i}\theta(x+b)$$

or

$$\theta(x+a-b) = K\, e^{-\Sigma\alpha_i b_i}\, e^{\Sigma\alpha_i x_i}\theta(x) \tag{6.2}$$

This implies that $T^*_{a-b}L(H,\rho) = L(H,\rho)$, i.e. that $E(a-b\,,u)$ is an integer for any u in Λ. In particular, the group generated by Λ and $a-b$ is a lattice Λ' since E is non-degenerate. There are a finite number of extensions ρ' of ρ by Λ' and each of them defines a line bundle $L(H,\rho')$ on $X' = \mathbb{C}^N/\Lambda'$ which pulls back to $L(H,\rho)$ on X. Moreover (6.2) shows that any section of $L(H,\rho)$ is the pullback of a section of $L(H,\rho')$ for some ρ'. But this is absurd since, by Riemann-Roch, the dimension of $H^0(X',L(H,\rho'))$ is strictly less than the dimension of $H^0(X, L(H,\rho))$.

There remains to show that ψ is a local embedding. Suppose this is not the case at some point $a \in \mathbb{C}^N$. Then there are a tangent vector $v = \Sigma c_i \frac{\partial}{\partial z_i}$ and a constant K such that

$$K\,\hat{\theta}(a) = v\,\hat{\theta}(a)\,, \quad \text{i.e.} \quad K = v \log \hat{\theta}(a)$$

for any theta function $\hat{\theta}$ relative to $3H,\rho^3$. In particular, for any $x, y \in \mathbb{C}^N$ and any theta-function θ relative to H, ρ,

$$f(a+x) + f(a+y) + f(a-x-y) = K \tag{6.3}$$

where $f = v(\log \theta)$. Then f is affine. In fact (6.3) implies that

$$f(\xi) + f(\eta) = K - f(3a-\xi-\eta)$$

Setting $\eta = 0$ gives

$$f(\xi) = K' - f(3a-\xi)$$

hence
$$f(\xi) + f(\eta) = f(\xi + \eta) + K''$$

and $f(\xi) - K''$ is linear. Therefore, if we write $c = (c_1,\dots,c_N)$, for a suitable constant k and any complex number w, we have

$$\theta(z+wc) = e^{wf(z) + kw^2}\theta(z)$$

As before, it follows that, for any w and any $u \in \Lambda$, $E(wc,u) = 0$, which is absurd since E is non-degenerate.

We have seen that if a complex torus admits a polarization, it is isomorphic to a smooth subvariety of projective space. These complex tori are called abelian varieties. The datum of an abelian variety and a polarization is called a polarized abelian variety.

What can we say of a complex torus X with a line bundle $L(H,\rho)$ such that H is positive semidefinite and has, say, p positive eigenvalues? We have seen in §5 that a quotient of X is a p-dimensional abelian variety; therefore there is a mapping of X onto a p-dimensional smooth subvariety of projective space whose differential has rank p everywhere.

Remark 6.4. With the notations of Proposition 6.1, it can also be proved that $\psi(X)$ is an algebraic subvariety of projective space, i.e. is the set of common zeros of a finite number of homogeneous polynomials, In fact, it is a general result (Chow's theorem) that a closed analytic subset S of projective space is algebraic. With the additional assumption that S be smooth, this will actually be proved in §7.

As an application we see that any compact Riemann surface C is algebraic; in fact the hypotheses of Proposition 6.1 are certainly satisfied for $J(C)$, and C is embedded in its Jacobian. When the genus of C is one, i.e. when C is a one-dimensional complex torus, this means that there is always a polarization $L(H,\rho)$ on C. Such an H is easy to produce. With the notations of Example 1.2, a good H is

$$H(z,w) = \frac{1}{\operatorname{Im} w} z \bar{w}$$

The situation is entirely different for two-dimensional (or higher-dimensional) tori, as we will see in the following section.

7. MEROMORPHIC FUNCTIONS

Let p, q be two holomorphic functions in a neighborhood of $0 \in \mathbb{C}^N$. We say that p, q are relatively prime at a point x if their power series at x are relatively prime in the ring of convergent power series. We will need the following:

Fact (See [1], for example): If p and q are relatively prime at 0, they are relatively prime at any point in a neighborhood of 0.

A meromorphic fraction on an open set U of $\mathbb{C}^N$ is a fraction p/q, where p and q are holomorphic and relatively prime at every point of U. A meromorphic fraction uniquely determines p and q up to multiplication by the same unit.

Let X be a connected complex manifold. A meromorphic function on X is the datum of a covering $\{U_i\}$ of X by coordinate patches and a collection $\{f_i\}$ of meromorphic fractions on the U_i's which agree on the overlaps $U_i \cap U_j$. Two meromorphic functions $(\{U_i\},\{f_i\})$, $(\{V_j\},\{g_j\})$ are regarded as the same if f_i and g_j agree on $U_i \cap V_j$ for every choice of i and j.

A divisor on X is a formal locally finite linear combination with integral coefficients $D = \Sigma m_i D_i$ of irreducible codimension 1 analytic subsets D_i. D is called effective if all the m_i are non-negative and at least one of them is positive.

The polar locus and zero locus of a meromorphic fraction p/q (i.e. the sets of points where p and q vanish, respectively) are not changed when p and q are multiplied by the same unit. Therefore it makes sense to speak about the zero locus and polar locus of a meromorphic function f; they are codimension 1 analytic subsets of X. Let Y be an irreducible component of, say, the zero locus of f. The multiplicity to which f vanishes along Y also has intrinsic meaning; the same is true for multiplicity of pole along components of the polar locus. Therefore we may define the polar divisor P_f and zero divisor Z_f of f as follows

$$P_f = \Sigma m_P P \ , \quad Z_f = \Sigma m_Z Z$$

where P(resp. Z) runs through the irreducible components of the polar locus (resp. zero locus) of f, and m_P (resp. m_Z) stands for the order of pole (resp. zero) of f along P (resp. Z). The divisor of f is

$$(f) = P_f - Z_f$$

Two divisors are said to be linearly equivalent if their difference is the divisor of a meromorphic function. Let D be an effective divisor. Cover X with sufficiently small open sets U_i and let f_i be a local defining equation for D in U_i and let f_i be a local defining equation for D in U_i, i.e. a holomorphic function on U_i that vanishes, with the appropriate multiplicities, exactly on $D \cap U_i$. The line bundle associated with D, denoted L_D, is defined as the line bundle whose transition functions, relative to the covering $\{U_i\}$, are f_i/f_j. The sections of L_D can be viewed as meromorphic functions having poles at most along D. If D is not effective, it may be decomposed as $D = E - F$, where E, F are effective, and L_D is defined as $L_E \otimes L_F^{-1}$. It is a straightforward consequence of the definitions that $L_{(f)}$ is trivial and that, conversely, two divisors are linearly equivalent if and only if the associated line bundles are isomorphic.

With this language, one of the consequences of Theorem 3.9 is that: On a complex torus X, any meromorphic function is the quotient of two theta-functions (relative to the same hermitian form and multiplier). In fact, let f be a meromorphic function on X; f can be viewed as a section of L_{P_f}. Another section of L_{P_f} "is" the function 1. f and 1, viewed as sections of L_{P_f} are represented by theta functions θ_1 and θ_2. Clearly $f = \theta_1/\theta_2$.

Meromorphic functions on X can be added and multiplied and are easily seen to form a field K(X), the so-called <u>function field</u> of X. The two main results that we wish to prove are:

<u>Theorem 7.1. Let X be a compact connected complex manifold of dimension N. Then K(X) is a finitely generated extension of</u> $\mathbb{C}$ <u>of transcendence degree at most N.</u>

<u>Proposition 7.2. Let</u> $X = \mathbb{C}^N/\Lambda$ <u>be a complex torus. The following are equivalent:</u>

(i) <u>The transcendence degree of K(X) over</u> $\mathbb{C}$ <u>is at least p;</u>

(ii) <u>There is a hermitian form H on</u> $\mathbb{C}^N$ <u>which is positive semidefinite with at least p positive eigenvalues and whose imaginary part is integral on</u> Λ.

We will prove (7.2) (using 7.1) before (7.1). We have seen that if (ii) is satisfied, then X maps onto a smooth algebraic variety S of dimension at least p in projective space $\mathbb{P}^m$. The homogeneous coordinate ring P of S is the quotient of the ring $\mathbb{C}[x_0,\dots,x_m]$ by the homogeneous ideal generated by the forms that vanish on S. The field of meromorphic functions on S is (we only need "contains" which is pretty obvious) the field of quotients of homogeneous elements of the same degree in P. K(S) has transcendence degree equal to the dimension of S; on the other hand, K(S) maps into K(X) by pull-back so (i) holds.

Conversely let $f_1,\dots,f_p$ be p algebraically independent meromorphic functions on X. For each i let D_i be the polar divisor of f_i. Set $L = L_{D_1+\dots+D_p}$. L is isomorphic to some

$L(H,\rho)$. H is positive semi-definite since L has sections. Suppose H has less than p positive eigenvalues. Then, as we showed in §6, there are an abelian variety Y of dimension less than p and a line bundle M on it which pulls back to L (we use again the fact that L has sections). On the other hand, as was shown before, f_i is a quotient $\theta_i/\hat{\theta}_i$ of theta-functions relative to H and ρ. As we showed in §6, θ_i and $\hat{\theta}_i$ are pullbacks of sections of M, and the f_i are pullbacks of algebraically independent meromorphic functions on Y. This contradicts Theorem 7.1.

We now turn to the proof of (7.1), which is entirely elementary and is based on the following form of the Schwarz lemma.

Lemma 7.3. *Let f be a function which is holomorphic in a neighborhood of the polycylinder* $P = \{(z_1,\dots,z_n)/\ |z_i| \leqslant 1,\ i = 1,\dots,n\}$ *and vanishes of order h at 0. Let M denote the maximum of $|f|$ on P. Then*

$$|f(z_1,\dots,z_n)| \leqslant M \max_i |z_i|^h,\quad |z_i| \leqslant 1,\quad i = 1,\dots,n.$$

Proof. It suffices to prove the statement on each line through the origin in $\mathbb{C}^n$, i.e. we may assume that $n = 1$. But then $f(z)/z^h$ is holomorphic and its modulus is bounded by M, by the maximum principle. The lemma follows.

Proof of 7.1. Let $f_1,\dots,f_{N+1}$ be holomorphic functions on X. Using the compactness of X we may find three finite coverings of X by coordinate open sets, $U_x \supset\supset V_x \supset\supset W_x$, where x runs through a finite set of r points in X, with the following properties:

(i) in U_x, $f_i = \dfrac{P_{i,x}}{Q_{i,x}}$ where $P_{i,x}$ and $Q_{i,x}$ are relatively prime at each point;

(ii) $x \in W_x$

(iii) Relative to a suitable system of coordinates $(z_{1,x},\ldots,z_{N,x})$ centered at x in U_x, V_x is the open polycylinder $|z_i| < 1$, $i = 1,\ldots,N$, and W_x is the open polycylinder $|z_i| < \frac{1}{2}$, $i = 1,\ldots,N$.

Condition (i) implies that

$$\varphi_{i,x,y} = Q_{i,x}/Q_{i,y}$$

is holomorphic and nowhere vanishing in $U_x \cap U_y$, and bounded in $V_x \cap V_y$. We set

$$\varphi_{x,y} = \prod_i \varphi_{i,x,y}$$

$$C = \max_{x,y} \max_{V_x \cap V_y} |\varphi_{x,y}|$$

C is not less than 1 since $\varphi_{xy}\varphi_{yx} = 1$. We must find a polynomial F of degree ℓ such that

$$F(f_1,\ldots,f_N,f_{N+1}) \equiv 0$$

Write $\quad F(f_1,\ldots,f_{N+1}) = \dfrac{R_x}{Q_x^\ell}$ in V_x where $Q_x = \prod_i Q_{i,x}$.

The proof will be in two steps. First we will see that R_x can be made to vanish to any order that we wish at x. Then we will show that if each R_x vanishes to a sufficiently high degree at x then all the R_x are identically zero.

To say that R_x vanishes to order h at x means that all the derivatives of R_x of order less than h vanish at x. These are all linear conditions on the coefficients of F and there are

$$\frac{h(h+1)\ldots(h+N-1)}{N!}$$

of them; in all $r\,\frac{h\ldots(h+N-1)}{N!}$ conditions.

On the other hand the number of polynomials of degree $\leqslant \ell$ in $N+1$ variables is

$$\frac{(\ell+1)\ \ldots\ (\ell+N+1)}{(N+1)!}$$

which is larger than $r\frac{h\ldots(h+N-1)}{N!}$ if ℓ is large enough. Set

$$M = \max_{x}\ \max_{V_x} |R_x|$$

Since $R_x = \varphi_{x,y}\, R_y$ we conclude, using Schwarz's lemma, that

$$M \leqslant \frac{C^{\ell}}{2^{h}}\, M$$

We can choose an ℓ such that $C^{\ell} < 2^{h}$ (i.e. $\ell \log_2 C < h$) and

$$\frac{(\ell+1)\ldots(\ell+N+1)}{(N+1)!} > r\,\frac{h\ldots(h+N-1)}{N!}$$

since the left-hand side has degree $N+1$ in ℓ and the right-hand side has degree N in h. So $M = 0$, i.e. $F(f_1,\ldots,f_{N+1}) = 0$.

Similar arguments would also show that if $\{f_1,\ldots,f_N\}$ contains a transcendence basis for $K(X)$ over $\mathbb{C}$, then the degree of F can be bounded independently of f_{N+1}, thus showing that $K(X)$ is finitely generated over $\mathbb{C}$.

Remark 7.4. One of the consequences of Theorem 7.1 is Chow's Theorem for projective manifolds. Let V be an n-dimensional connected sub-manifold of projective space. We want to show that V is algebraic. In fact, let Y be the smallest algebraic subset of projective space which contains V (i.e. the intersection of all the algebraic subsets of projective space that contain V). We must show that $V = Y$. Y is irreducible. In fact if the product PQ of two homogeneous polynomials vanishes on V, then either P or Q vanishes on an open subset of V, hence on all of V. Every rational function on Y restricts to a meromorphic function on V. By Theorem 7.1 the field of rational functions on Y cannot have transcendence degree greater than n over $\mathbb{C}$; therefore Y has dimension n. But now Y is an irreducible (algebraically, but it can be seen that this implies analytic irreducibility) complex space of dimension n containing an n-dimensional V and must therefore coincide with V.

Example 7.5. Let V be a real vector space of dimension $2N$ and Λ a lattice in V. By a complex structure on V we shall mean an automorphism J of V such that $J^2 = -I$. V becomes a complex vector space by setting: $(a + \sqrt{-1}\, b)v = av + bJv$, a, b real. This makes V/Λ into a complex torus.

Let E be a skew-symmetric bilinear form on V which is integral on Λ. Let J be a fixed complex structure on V. As we saw before, a necessary and sufficient condition for E to be the Chern class of a line bundle on V/Λ, i.e. for E to be the imaginary part of a hermitian H, is that

$$E(Ju,Jv) = E(u,v) \tag{7.6}$$

for every choice of u and v.

If $n \geqslant 2$ and $E \neq 0$, condition 7.6 is not satisfied for J in a dense open subset of the space of all complex structures on V. On the other hand, there is only a denumerable infinity of forms E

which are integral on Λ, so the Baire category theorem implies that on "most" complex tori of dimension two or more there are no line bundles with non-zero Chern class. It follows that on "most" complex tori of dimension two or more the only line bundle with sections is the trivial one, or, alternatively, there are no effective divisors, hence no non-constant meromorphic functions.

So far we have seen that for tori of dimension greater than one, at least two extreme cases can occur. $K(X)$ can have transcendence degree N or zero. It can be easily seen that all the intermediate cases can also occur. We will do this when $N = 2$.

Let Λ be the lattice in $\mathbb{C}^2$ generated by

$$(1,0), \quad (\sqrt{-1},0), \quad (0,1), \quad (\alpha,\beta)$$

where β is <u>not real</u>. There is a surjective mapping

$$\mathbb{C}^2/_\Lambda = X \xrightarrow{\ \varphi\ } \mathbb{C}/_{\mathbb{Z} + \beta\mathbb{Z}}$$

gotten by sending (z,w) onto w. Therefore $K(X)$ has transcendence degree at least one. We will see that, for a "generic" choice of α and β, $K(X)$ has transcendence degree exactly equal to one. We will do this by showing that, in general, the linear subspace $\{w = 0\}$ has no complementary subspace W in $\mathbb{C}^2$ such that $W \cap \Lambda$ is a lattice in W. This would be true in an abelian variety, where we could take as W the orthogonal complement of $\{w = 0\}$ with respect to a positive definite hermitian form whose imaginary part is integral on Λ.

If a W with the required property exists, then W is generated by a vector

$$a(1,0) + b(\sqrt{-1},0) + c(0,1) + d(\alpha,\beta) = (a + \sqrt{-1}\, b + d\alpha,\ c + d\beta) = v$$

where a, b, c, d are integers, and there is a non-real λ such that λv belongs to Λ. This implies that, first of all

$$\lambda \in \mathbb{Q}(\beta)$$

and secondly that

$$\alpha \in \mathbb{Q}(\sqrt{-1},\lambda) \subset \mathbb{Q}(\sqrt{-1},\beta)$$

This is usually not the case (e.g. $\beta = \sqrt{-1}$, $\alpha = \sqrt{2}$).

In the course of this discussion we have given an indication of the proof of the Poincaré complete reducibility theorem. Let $X = \mathbb{C}^N/\Lambda$ be an abelian variety, A an abelian subvariety. Then there is an abelian subvariety B of X such that the natural mapping $A \times B \to X$ is an isogeny (i.e. has finite kernel and is surjective).

This can be proved as follows. $A = V/V \cap \Lambda$ where V is a linear subspace of $\mathbb{C}^N$ such that $V \cap \Lambda$ is a sublattice. Choose any positive definite hermitian form on $\mathbb{C}^N$ which has integral imaginary part on Λ and set $B = V^{\perp}/V^{\perp} \cap \Lambda$ where $V^{\perp}$ is the orthogonal complement of V ($V^{\perp} \cap \Lambda$ is automatically a lattice in $V^{\perp}$).

8. GEOMETRY OF THE THETA-DIVISOR

Let C be a curve of genus $g \geqslant 2$. Let $X = J(C)$ be the Jacobian of C. H will be the standard polarization of X:

$$H(\omega,\omega') = \frac{2}{\sqrt{-1}} \int \omega \wedge \overline{\omega'}$$

As usual, E will denote the imaginary part of H. Choose a basis $u_1,\ldots,u_g, \tilde{u}_1,\ldots,\tilde{u}_g$ for $H^1(C,\mathbb{Z})$ such that

$$E(u_i,u_j) = E(\tilde{u}_i,\tilde{u}_j) = 0$$

$$E(\tilde{u}_i,u_j) = \delta_{ij}$$

We also write, as usual,

$$\tilde{u}_j = \sum_i t_{ij} u_i$$

The Riemann theta-function of C is

$$\theta(z) = \sum_{(n_1 \dots n_g) \in \mathbb{Z}^g} e^{2\pi\sqrt{-1}\,\Sigma n_i z_i + \pi\sqrt{-1}\,\Sigma n_i t_{ij} n_j}$$

Notice that θ is <u>even</u>, i.e. $\theta(-z) = \theta(z)$.

Choose a base point $q \in C$. Then there is defined a mapping $\psi_q: C \to X$. Let $C^{(r)}$ denote the r-fold symmetric product of C with itself (i.e. the quotient of C^r by the action of the full symmetric group, a smooth variety in our case). Points in $C^{(r)}$ will be written as formal sums $p_1+\dots+p_r$. ψ_q can be extended to a mapping

$$\psi_q^{(r)}: C^{(r)} \to X$$

defined as follows:

$$\psi_q^{(r)}(p_1+\dots+p_r) = \psi_q(p_1)+\dots+\psi_q(p_r)$$

ψ, ψ_q or $\psi^{(r)}$ will be used instead of $\psi_q^{(r)}$ when no confusion is likely to occur. We begin with two simple remarks.

<u>Lemma 8.1</u>. <u>Let $\omega_1,\dots,\omega_g$ be a basis for the holomorphic differentials on C. Let $p_1,\dots,p_r$ be distinct points of C, and $m_1,\dots,m_r$ positive integers. Let τ_i be a local coordinate in a neighborhood of p_i. Write $\omega_i = \alpha_{ij} d\tau_j$ near p_i. Then</u>

$$h^1(C, L_{\Sigma m_i p_i})$$

$$= g - \text{rank} \begin{pmatrix} \alpha_{11}(p_1), \ldots, \alpha_{11}^{(m_1-1)}(p_1), \alpha_{12}(p_2), \ldots, \alpha_{1r}^{(m_r-1)}(p_r) \\ \vdots \qquad\qquad\qquad \vdots \\ \alpha_{g1}(p_1), \ldots \qquad\qquad\qquad \alpha_{gr}^{(m_r-1)}(p_r) \end{pmatrix}$$

where $\alpha_{ij}^{(h)}$ *stands for the* h-th *derivative of* α_{ij} *with respect to* τ_j.

Proof. By duality:

$$h^1(C, L_{\Sigma m_i p_i}) = h^0(C, K \otimes L_{-\Sigma m_i p_i})$$

It follows from the exact sequence

$$0 \to H^0(C, K \otimes L_{-\Sigma m_i p_i}) \to H^0(C,K) \overset{\sigma}{\to} H^0(C, \sum_i \mathbb{C}_{p_i}^{m_i}) \to \ldots$$

that

$$g - \dim \sigma(H^0(C,K)) = g - \text{rank} \begin{pmatrix} \alpha_{11}(p_1) \ldots \\ \vdots \\ \vdots \end{pmatrix}$$

This concludes the proof.

Remark 8.2. Choose a basis $\omega_1, \ldots, \omega_g$ for the holomorphic differentials on C. Then the mapping $\psi_q^{(r)}$ is given by

$$\psi_q^{(r)}(p_1 + \ldots + p_2) = \text{class of} \left(\int_q^{p_1} \omega_1 + \ldots + \int_q^{p_r} \omega_1, \ldots, \int_q^{p_1} \omega_g + \ldots + \int_q^{p_r} \omega_g \right)$$

If $p_1, \ldots, p_r$ are all distinct and τ_i is a local coordinate near p_i, then $\tau_1, \ldots, \tau_r$ are local coordinates near $p_1 + \ldots + p_r$ on $C^{(r)}$.

Relative to these coordinates, the Jacobian matrix of $\psi^{(r)}$ is obviously $(\alpha_{ij}(p_j))$. There is a version of this statement that applies to the case when two or more of the p_i coincide, but we won't go into details.

A consequence of (8.1) is that for $r \leqslant g$, $\psi^{(r)}$ is generically 1-1 onto its image. In fact $\psi^{(g)}(C^{(g)}) = X$, so $\psi^{(r)}(C^{(r)})$ is r-dimensional and generically the Jacobian matrix of $\psi^{(r)}$ has maximal rank. Hence, for a generic $\zeta \in C^{(r)}$, $h^0(C,L_\zeta) = 1$ by (8.1) and Riemann-Roch. But $\psi(\zeta) = \psi(\zeta')$ if and only if ζ and ζ' are linearly equivalent. Since $h^0(C,L_\zeta) = 1$, this means that $\zeta' = \zeta$.

Now we return to the main goal of our investigation. Let Θ denote the theta-divisor of C, i.e. the zero divisor of $\theta(z)$. We want to show that Θ is a translate of $\psi(C^{(g-1)})$.

Consider a point $e = \psi(p_1+\ldots+p_g) \in X$. Let T_e denote translation by e. Intersect $T_e(\Theta)$ and $\psi(C)$ and assume that $\psi(C)$ is not contained in $T_e(\Theta)$. $T_e(\Theta)$ cuts out on C a divisor ζ_e. If this were the best of all possible worlds ζ_e would be linearly equivalent to $p_1+\ldots+p_g$. This is not quite true, but:

Lemma 8.3. *Either $T_e(\Theta) \supset \psi(C)$ or $\deg(\zeta_e) = g$ and the linear equivalence class of $\zeta_e - (p_1+\ldots+p_g)$ is independent of e (but depends on the base point q). In other terms*

$$\psi(\zeta_e) + \mathcal{K}_q = e \qquad (\mathcal{K}_q \text{ independent of } e)$$

If there is no danger of confusion we will usually write $\mathcal{K}$ for $\mathcal{K}_q$. Assume, for the time being, that (8.3) has been proved. Then choose distinct points $p_1,\ldots,p_g$ such that $h^0(C,L_\zeta) = 1$, where $\zeta = \Sigma p_i$. Set $e = \psi(\zeta) + \mathcal{K}$. If $T_e(\Theta) \supset \psi(C)$, then

$$\Theta \ni \psi(p_g) - e = \psi(\zeta) - e - \psi(p_1+\ldots+p_{g-1})$$

$$= -\mathcal{K} - \psi(p_1+\ldots+p_{g-1})$$

and
$$\psi(p_1+\ldots+p_{g-1}) + \mathcal{K} \in \Theta$$

because $\Theta = -\Theta$.

If, on the other hand, $T_e(\Theta)$ intersects $\psi(C)$ in a divisor of g points ζ_e , then $\zeta_e = \zeta$, by the choice of ζ, and it follows again that

$$\psi(p_1+\ldots+p_{g-1}) + \mathcal{K} \in \Theta$$

In conclusion, for a generic choice of $p_1,\ldots,p_{g-1}$, $\psi(p_1+\ldots+p_{g-1})$ belongs to $T_{-\mathcal{K}}(\Theta)$, hence

$$\psi(C^{(g-1)}) \subset T_{-\mathcal{K}}(\Theta)$$

To show that the converse inclusion is true, we first have to show that the intersection number of $\psi(C)$ and $T_{\psi(t)}(\psi(C^{(g-1)})$ is equal to g for generic $t \in C^{(g)}$.

We must find points $p, p_1,\ldots,p_{g-1}$ such that p is linearly equivalent to

$$p_1'+\ldots+p_{g-1}' + t - (2g-2)q = \Delta - p_1-\ldots-p_{g-1} + t - (2g-2)q$$

where $\Delta = p_1+\ldots+p_{g-1} + p_1'+\ldots+p_{g-1}'$ is the divisor of an abelian differential (a <u>canonical</u> divisor). The existence of Δ is guaranteed by the fact that

$$1 \leq h^0(C,L_{\Sigma p_i'}) = h^0(C,K \otimes L_{-\Sigma p_i'})$$

by Riemann-Roch. Write

$$\psi(\Delta + t) = \psi(r_1 + \ldots + r_g)$$

Then what we are looking for is points $p, p_1, \ldots, p_{g-1}$ such that $p + \Sigma p_i$ is linearly equivalent to Σr_i. For a generic choice of $r_1, \ldots, r_g$ (i.e. of t) this can happen only if p is one of the r's and the p_i's are the remaining ones. This proves that the intersection number of $\psi(C)$ and $\psi(C^{(g)})$ is g.

What we know so far is that $T_{-\mathcal{K}}(\Theta) = m\psi(C^{(g-1)}) + Y$. If we intersect both sides with $\psi(C)$, we can conclude that

$$m = 1, \qquad Y \cdot \psi(C) = 0$$

where the dot denotes intersection. All we must show is that $Y = 0$. Suppose not. Choose a smooth point p of Y. The tangent line to $\psi(C)$ at any point is parallel to the tangent space to Y at p (otherwise $\psi(C)$ and a translate of Y could be made to meet transversally, contradicting $Y \cdot \psi(C) = 0$). This is impossible since $\psi(C)$ generates X (every point of X is the sum of g points of $\psi(C)$). We have thus proved

<u>Proposition 8.4.</u> $\Theta = T_{\mathcal{K}}(\psi(C^{(g-1)}))$.

Now choose an effective divisor ζ of degree $g-1$ and set

$$e = \psi(\zeta) + \mathcal{K}$$

Then $\theta(e) = \theta(-e) = 0$, so

$$-e = \psi(\zeta') + \mathcal{K}$$

Adding the above equations gives

$$\psi(\zeta + \zeta') = -2\mathcal{K}$$

On the other hand $\zeta + \zeta'$ contains g-1 arbitrary points, so

$$h^0(C, L_{\zeta+\zeta'}) = g$$

i.e. $L_{\zeta+\zeta'} = K$, so

$$\psi(K) = -2\mathcal{K}$$

In particular we see that the symmetry of Θ corresponds, on $\psi(C^{(g-1)})$, to the operation of taking the residual divisor with respect to a canonical one, i.e. passing from $\zeta = p_1+\ldots+p_{g-1}$ to a divisor $\zeta' = p_1'+\ldots+p_{g-1}'$ (generically unique) such that $\zeta + \zeta'$ is the divisor of a holomorphic differential.

We now want to explore the relationship between singularities of Θ and special divisors, i.e. divisors $\zeta = p_1+\ldots+p_r$ such that

$$h^0(C, L_\zeta) > 1$$

The first result is

Proposition 8.5. *Let ζ be a special effective divisor of degree g-1. Then $e = \psi(\zeta) + \mathcal{K}$ is a singular point of Θ. More precisely, the multiplicity of e is at least equal to $h^0(C, L_\zeta)$.*

Proof. Let $d + 1$ be an integer not greater than $h^0(C, L_\zeta)$. We will prove, inductively on d, that the multiplicity of $\psi(\zeta) + \mathcal{K}$ is at least equal to $d + 1$, the case $d = 0$ being clear. The function:

$$(p_1, \ldots, p_{g-1}) \mapsto \theta(\psi(\Sigma p_i) + \mathcal{K})$$

vanishes identically. Differentiating with respect to $p_1,\dots,p_d$, we get the identity:

$$\sum_{i_1\dots i_d} \frac{\partial^d\theta}{\partial z_{i_1}\dots\partial z_{i_d}}(\psi(\Sigma p_i) + \mathcal{K})\omega_{i_1}(p_1)\dots\omega_{i_d}(p_d) +$$

$$+ Q(p_1,\dots,p_{g-1}) = 0$$

where Q involves derivatives of θ of order less than d. By induction hypothesis:

$$\sum_{i_1\dots i_d} \frac{\partial^d\theta}{\partial z_{i_1}\dots\partial z_{i_d}}(\psi(\zeta) + \mathcal{K})\omega_{i_1}(p_1)\dots\omega_{i_d}(p_d) = 0$$

if Σp_i is linearly equivalent to ζ. The assumption that

$$h^0(C,L_\zeta) \geqslant d + 1$$

implies that $p_1,\dots,p_d$ can be chosen arbitrarily without changing $\psi(\zeta)$, so

$$\frac{\partial^d\theta}{\partial z_{i_1}\dots\partial z_{i_d}}(\psi(\zeta) + \mathcal{K}) = 0$$

for every choice of $i_1,\dots,i_d$. This concludes the proof.

We now prove a partial converse of (8.5).

Lemma 8.6. *The singular set of* Θ *consists precisely of the points* $\psi(\zeta) + \mathcal{K}$, *where* ζ *is an effective special divisor of degree* $g-1$.

Proof. Suppose $\psi(\zeta) + \mathcal{K}$ is a singular point of Θ. We must show that $\zeta = \sum_{i=1}^{g-1} p_i$ is special. These two assertions clearly do not depend on the choice of the base point; we can thus assume that $q \neq p_i$, $i = 1,\dots,g-1$. Consider the curve

$$\Gamma = \{\psi(\zeta) - \psi(p) + \mathcal{K} \mid p \in C\}$$

If ζ is not special, Γ does not lie in Θ, and meets it at $\psi(\zeta) + \mathcal{K} = \psi(\zeta) - \psi(q) + \mathcal{K}$ with multiplicity one, showing that $\psi(\zeta) + \mathcal{K}$ is not a singular point of Θ.

Remark 8.7. Another proof of (8.6) can be obtained by combining (8.1) and (8.2). Taken together, (8.1) and (8.2) say that, when $p_1,\dots,p_{g-1}$ are distinct points of C such that $\zeta = \Sigma p_i$ is non-special, the Jacobian matrix of ψ at ζ has rank $g-1$, and therefore $\psi(\zeta) + \mathcal{K}$ is a smooth point of Θ (notice that if $\psi(\zeta') = \psi(\zeta)$, then $\zeta = \zeta'$ because ζ is not special). With a little more work, this method applies also to the case when $p_1,\dots,p_{g-1}$ are not all distinct.

We are now in a position to prove the full converse of (8.5). This result is essentially due to Riemann, and can be formally stated as follows:

Theorem 8.8. Let ζ be an effective divisor of degree $g-1$ on C. The multiplicity of $\psi(\zeta) + \mathcal{K}$ in Θ is precisely equal to $h^0(C,L_\zeta)$.

Proof. Set $d+1 = h^0(C,L_\zeta)$. All we have to do is show that the multiplicity of $\psi(\zeta) + \mathcal{K}$ is not greater than $d+1$. We will use induction on d. The case $d = 0$ is Lemma 8.6. For any choice of $\bar{p}_1,\dots,\bar{p}_d$ we may write

$$\psi(\zeta) = \psi(q_1+\dots+q_{g-d-1} + \bar{p}_1+\dots+\bar{p}_d)$$

where $q_1,\ldots,q_{g-d-1}$ depend on $\bar{p}_1,\ldots,\bar{p}_d$. For generic $\bar{p} = (\bar{p}_1,\ldots,\bar{p}_d)$

$$h^0(C,L_{\Sigma q_i}) = 1$$

Now, for fixed $\bar{p}$, set

$$\mu(p_1,\ldots,p_d,v) = \psi(\Sigma p_i + \Sigma q_j) - \psi(v) + \mathcal{K}$$

$p_1,\ldots,p_d$, $v \in C$. Note that

$$\mu(\bar{p},v) = \psi(\zeta) - \psi(v) + \mathcal{K} \in \Theta$$

and moreover if $\zeta + q$ is linearly equivalent to $\zeta' + v$, and q (the base point) and v are generic,

(8.9)
$$h^0(C,L_{\zeta+q}) = d+1$$
$$h^0(C,L_{\zeta}) = d$$

This means, by induction hypothesis, that for generic v there is a erivative of order d of θ that does not vanish at $\mu(\bar{p},v) = \psi(\zeta') + \mathcal{K}$. In general $h^0(C,L_{\zeta'}) \geqslant d$. Now we claim that $\theta(\mu(p_1,\ldots,p_d,v))$, as a function of $p_1,\ldots,p_d$ alone, vanishes at $\bar{p}$ of order at most $d-1$, for generic $\bar{p},v$. If this were not the ase, in fact, we would have:

$$0 = \sum \frac{\partial^d\theta}{\partial z_{i_1}\ldots\partial z_{i_d}} (\psi(\zeta') + \mathcal{K})\, \omega_{i_1}(\bar{p}_1)\ldots\omega_{i_d}(\bar{p}_d)$$

or generic $\bar{p}$ and v, so that θ vanishes of order at least d at $\psi(\zeta') + \mathcal{K}$, contradicting (8.9) and the induction hypothesis.

We can thus find a smooth curve $(p_1(t),\ldots,p_d(t))$ such that

$$p_i(0) = \bar{p}_i \quad , \quad i = 1,\ldots,d$$

$$\Sigma q_j + \Sigma p_i(t) + q \text{ is non-special for generic } q \text{ and } t \neq 0 ,$$

$$\frac{\partial^d}{\partial t^d}\theta(\mu(p_1(t),\ldots,p_d(t),v))|_{t=0} \not\equiv 0$$

This implies that, for each fixed t,

$$t^{-d}\theta(\mu(p_1(t),\ldots,p_d(t),v))$$

is not identically zero, in fact vanishes precisely at $q,q_1,\ldots,q_{g-d-1}, p_1(t),\ldots,p_d(t)$. In particular, for generic q,

$$\frac{\partial^d}{\partial t^d}\theta(\mu(p_1(t),\ldots,p_d(t),v))|_{t=0}$$

has a simple zero at q, corresponding to $\mu(\bar{p},q) = \psi(\zeta) + \mathcal{K}$. Hence θ cannot vanish at $\psi(\zeta) + \mathcal{K}$ of order greater than d. This concludes the proof of Theorem 8.8.

The last missing piece is

Proof of Lemma 8.3. We first establish a few notations. Consider a polygon S :

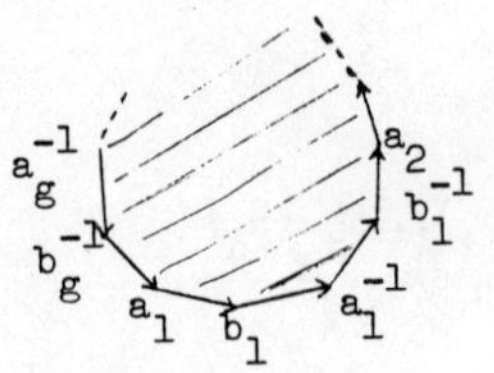

If we identify a_i and a_i^{-1} , b_i and b_i^{-1}, with opposite orientation (meaning that the initial point of a_i will be identified with the end point of a_i^{-1}, and vice versa), the result is an orientable compact two-manifold of genus g. Moreover the 1-cycles corresponding to $a_1 \ldots a_g, b_1 \ldots b_g$ generate the first homology of this manifold and, relative to a suitable orientation the intersection numbers of the a_i's and the b_j's are

$$I(a_i, a_j) = I(b_i, b_j) = 0$$

$$I(a_i, b_j) = \delta_{ij}$$

Therefore we can put a complex structure on S such that C is obtained from the Riemann surface with boundary S by identifying sides. Choose a basis $\omega_1 \ldots \omega_g$ for the holomorphic differentials on C such that

$$\int_{a_j} \omega_i = \delta_{ij}$$

Set $u_i = i^{th}$ standard basis vector for $\mathbb{C}^g$. We also set

$$\int_{b_j} \omega_i = t_{ij}$$

$$\tilde{u}_j = \Sigma t_{ij} u_i$$

The Jacobian of C is the quotient of $\mathbb{C}^g$ by the lattice Λ generated by the u_i and the $\tilde{u}_i$. Relative to the standard polarization of $J(C)$:

$$E(u_i, u_j) = E(\tilde{u}_i, \tilde{u}_j) = 0$$

$$E(\tilde{u}_i, u_j) = \delta_{ij}$$

so the notations that we have just established are consistent with those introduced at the beginning of the section.

The theta-function of C satisfies the functional equations

$$\theta(z+u_i) = \theta(z)$$

$$\theta(z+\tilde{u}_i) = e^{-2\pi\sqrt{-1}\, z_i - \pi\sqrt{-1}\, t_{ii}}\theta(z)$$

If we set $\varphi(z) = \theta(z-e)$, then θ satisfies functional equations:

$$\varphi(z+u_i) = \varphi(z)$$

$$\varphi(z+\tilde{u}_i) = e^{2\pi\sqrt{-1}\, e_i}\, e^{-2\pi\sqrt{-1}\, z_i - \pi\sqrt{-1}\, t_{ii}}\varphi(z)$$

Let f be a multivalued function on C. Choose a single-valued determination of f on S (in the application it won't matter which) and denote by f^+ the value of f on a_k, b_k and by f^- the value of f on a_k^{-1}, b_k^{-1}, $k = 1,\ldots,g$. A typical example will be the ith coordinate ψ_i of ψ viewed as a multivalued mapping $C \to \mathbb{C}^g$. On a_k

$$\psi_i^- = \psi_i^+ + t_{ik}$$

whereas on b_k

$$\psi_i^+ = \psi_i^- + \delta_{ik}$$

Another example is $\varphi\circ\psi$, or φ, for short. One has

$$\varphi^+ = \varphi^- \quad \text{on } b_k$$

$$\varphi^- = e^{2\pi\sqrt{-1}\, e_k}\, e^{-2\pi\sqrt{-1}\, \psi_k^+ - \pi\sqrt{-1}\, t_{kk}}\varphi^+ \quad \text{on } a_k$$

It follows that

$$d \log \varphi^+ = d \log \varphi^- \quad \text{on } b_k$$

$$d \log \varphi^- = d \log \varphi^+ - 2\pi\sqrt{-1}\ \omega_k \quad \text{on } a_k$$

Assume that $\theta(z-e)$ does not vanish identically on $\psi(C)$. We want to show that φ has exactly g zeros. We may assume that all the zeros of φ occur in the interior of S, in which case the number of zeros of φ is

$$\frac{1}{2\pi\sqrt{-1}} \sum_k \int_{a_k} + \int_{b_k} (d \log \varphi^+ - d \log \varphi^-)$$

$$= \frac{1}{2\pi\sqrt{-1}} \sum \int_{a_k} 2\pi\sqrt{-1}\ \omega_k = g$$

This proves the first part of Lemma 8.3. To prove the second part of (8.3), let $p_1, \ldots, p_g$ be the zeros of φ in S. We must evaluate $\sum_i \psi(p_i)$. We will do this component by component. Using the residue theorem again, we get:

$$2\pi\sqrt{-1} \sum_i \psi_h(p_i) = \sum_k \int_{a_k} + \int_{b_k} (\psi_h^+\ d \log \varphi^+ - \psi_h^-\ d \log \varphi^-)$$

$$\int_{b_k} \cdots = \delta_{hk} \int_{b_k} d \log \varphi^+$$

$$\int_{a_k} \cdots = 2\pi\sqrt{-1} \int_{a_k} \psi_h^+ \omega_k + 2\pi\sqrt{-1}\ t_{hk} - t_{hk} \int_{a_k} d \log \varphi^+$$

Denote by $\mathcal{I}_k, \mathcal{E}_k, \mathcal{I}_k', \mathcal{E}_k'$ the initial and end points of a_k and b_k, respectively. The functional equation for φ implies that

$$\int_{a_k} d \log \varphi^{+} = \log \varphi^{+}(\mathcal{E}_k) - \log \varphi^{+}(\mathcal{J}_k') = 2\pi\sqrt{-1}\, \nu_k$$

$$\int_{b_k} d \log \varphi^{+} = \log \varphi^{+}(\mathcal{E}_k') - \log \varphi^{+}(\mathcal{J}_k') =$$

$$= 2\pi\sqrt{-1}\, e_k - 2\pi\sqrt{-1}\, \psi_k(\mathcal{J}_k') - \pi\sqrt{-1}\, t_{kk} + 2\pi\sqrt{-1}\, \mu_k$$

where ν_k, μ_k are integers. Putting everything together, we obtain

$$\sum_i \psi(p_i) \equiv e - \mathcal{K} \qquad (\mathrm{mod}\ \Lambda)$$

where

$$\mathcal{K}_h = \sum_k \int_{a_k} \psi_h^{+} \omega_k - \psi_h(\mathcal{J}_h') - \frac{1}{2} t_{hh}$$

$\mathcal{K}$ clearly does not depend on e. With this the proof of (8.3), and hence of Theorem 8.8 is complete.

REFERENCES

[1] GUNNING, R., ROSSI, H., Analytic Functions of Several Complex Variables, Prentice-Hall (1965).
[2] LEWITTES, J., Riemann surfaces and the theta function, Acta Math. **111** (1964) 37–61.
[3] MUMFORD, D., Abelian Varieties, Oxford Univ. Press (1970).
[4] WEIL, A., Variétés Kähleriennes, Hermann, Paris (1958).

IAEA-SMR-18/13

INTRODUCTION TO COMPLEX TORI

P. DE LA HARPE
Institut de mathématiques,
Université de Lausanne,
Dorigny,
Switzerland

Abstract

INTRODUCTION TO COMPLEX TORI.

I. Real tori: closed subgroups and lattices in $\underline{R}^n$; Connected abelian real Lie groups and real tori; A complex example; Vocabulary of Lie groups. II. Complex tori – generalities and abelian curves: Isomorphisms of complex tori; Classification of complex tori of dimension 1; One-dimensional tori as non-singular cubics. III. The group of periods of a meromorphic function: Meromorphic functions on $\underline{C}^n$; Groups of periods; Statement of a fundamental reduction theorem. IV. Period relations: Lattices in $\underline{C}^n$ and matrices; Frobenius relations; A complex torus without non-constant meromorphic functions.

"Machines can write theorems and proofs, and read them. The purpose of mathematical exposition for people is to communicate ideas, not theorems and proofs. Experience shows that almost always the best way to communicate a mathematical idea is to talk about concrete examples and unsolved problems".
(P.R. Halmos in "Ten problems in Hilbert space", Bull. Am. Math. Soc. **76** (1970) 887–933)

INTRODUCTION AND REFERENCES

Real and complex tori are nice examples of two kinds of objects: Lie groups and (in some cases) algebraic manifolds. We shall emphasize here the first aspect, and actually hope that these examples could motivate a more comprehensive study. The few proofs where general facts about Lie groups are used (mostly in Section I.2 and II.1) can safely be jumped over in a first reading. For a possible introduction to Lie groups, read first Chapter I in

[1] CHEVALLEY, C., Theory of Lie Groups, Princeton Univ. Press (1946).

Then Chapters 1 and 2 in

[2] ADAMS, J.F., Lectures on Lie Groups, Benjamin (1969).

And then one of the many classical books on this subject, such as (besides the two above)

[3] HOCHSCHILD, G., The Structure of Lie Groups, Holden-Day (1965).

Part of the very fundamental work of E. Cartan on Lie groups is contained in

[4] HELGASON, S., Differential Geometry and Symmetric Spaces, Academic Press (1962).

Later, enrich your collection of examples by practising the exercises in (you may start with those of Chapter 3)

[5] BOURBAKI, N., Groupes et algèbres de Lie; so far published by Hermann, Ch.1 (1960), Ch.2 and 3 (1972 Ch. 4, 5 and 6 (1968), Ch. 7 and 8 (1975).

Some of the words most often used are listed in our Section I.4.

Chapter I is a discussion of real tori and other abelian real Lie groups, which have a particula simple structure. An easy example given in Section I.3 will show that this simplicity does not car over to the complex case. The material of this chapter can be found scattered among numerous textbooks on various subjects; we have used those of N. Bourbaki and J. Dieudonné.

There are many exercises in this chapter and in the next one, usually given with an almost complete solution. The reader will probably be able to solve many of them in his head. Other exercises may, on the other hand, take much effort; they are intended to indicate developments or connections with other areas of study.

In Chapter II we first look at some general properties of complex tori of dimension n (but read n = 1 all through if you wish) and then consider in more detail the classification problem in complex dimension one. The third and last section should connect both with a traditional chapter of many function-theory lecture courses ("elliptic functions") and with elementary algebraic geometry (in particular, Exercise 35 may be taken as a justification for the expressions "complex torus of dimension one" and "abelian curve" to be synonymous). Our main sources have been the two following:

[6] ROBERT, A., Elliptic Curves, Springer Lecture Notes in Mathematics **326** (1973).
[7] SERRE, J.P., Cours d'arithmétique, Presses Universitaires de France (1970) (an English translation has been published by Springer).

Material on elliptic functions may be found in (among many others):

[8] CARTAN, H., Théorie élémentaire des fonctions analytiques d'une ou plusieurs variables complexes, Hermann (1961) (exists in English).
[9] MARKUSHEVICH, A.I., Theory of Functions of a Complex Variable, 3 vols, Prentice-Hall (1965 and 196
[10] SAKS, S., ZYGMUND, A., Analytic Functions, 3rd Edn, PWN, Warsaw (1971).
[11] SWINNERTON-DYER, H.P.F., Analytic Theory of Abelian Varieties, London Math. Soc. Lecture Note Series **14**, Cambridge Univ. Press (1974).
[12] WHITTAKER, E.T., WATSON, G.N., A Course of Modern Analysis, 4th Edn, Cambridge Univ. Press (192

There are a few flippant mentions of some books mostly concerned with algebraic geometry – the following:

[13] SHAFAREVICH, I.R., Basic Algebraic Geometry, Springer (1974).
[14] MUMFORD, D., Abelian Varieties, Oxford Univ. Press (1970).

Chapter III contains important statements without proofs. They describe a convenient way for our purpose to write down a meromorphic function on a torus, namely as a quotient of two theta functions. Besides the often-quoted paper by M. Field (these Proceedings), we call on one the basic references:

[15] GUNNING, R.C., ROSSI, H., Analytic Functions of Several Complex Variables, Prentice-Hall (1965).

Three very classical textbooks on complex tori that we have used frequently are:

[16] CONFORTO, F., Abelsche Funktionen und algebraische Geometrie, Springer (1956).
[17] SIEGEL, C.L., Analytic Functions of Several Complex Variables, Inst. Adv. Study, Princeton (1950).
[18] SIEGEL, C.L., Topics in Complex Function Theory, 3 Vols, Wiley-Interscience (1969, 1971 and 1973).

One way to approach the voluminous [18] is to start with:

[19] BAILY, W.L., Jr., Review of C.L. Siegel's Topics in Complex Function Theory, Bull. Am. Math. Soc. **81** (1975) 528–536.

The final (and long) exercise in Chapter III is in fact an alternative to Chapter IV and sketches a section in:

[20] WEIL, A., Variétés kählériennes, Hermann (1971).

Chapter IV is in fact a small part of [16] and [17]. There are strong conditions for the existence of "many" meromorphic functions on a complex torus of complex dimension at least two, which are known as **Frobenius relations.** Modulo the important statements of Chapter III (unproved there), we show in Section IV.2 that these conditions are necessary. This is good enough to write down in Section IV.3 a numerical example of a torus without any non-constant meromorphic function.

Chapter V of the lectures given during the Summer Seminar Course (not included in these Proceedings) was essentially the beginning of:

[21] SIEGEL, C.L., Symplectic geometry, Am. J. Math. **65** (1943) 1–86.

with some motivations from the classification problem for algebraic complex tori that the reader will find in [17]. But as we have not introduced the notion of **polarization**, we rather give references only.

We have not mentioned examples of either current research or unsolved problems.

The reader interested in the historical importance of complex tori for algebraic geometry should read Ch. VII, § 7, in:

[22] DIEUDONNE, J., "Cours de géométrie algébrique, 1: aperçu historique sur le développement de la géométrie algébrique", Presses Universitaires de France (1974).

See also the historical sketch in [13].

Though not mentioned in this text, we finally cite two books studying the analogues of complex tori, in some sense, for other ground fields than $\underline{C}$:

[23] WEIL, A., Variétés abéliennes et courbes algébriques, Hermann (1948).
[24] LANG, S., Abelian Varieties, Interscience (1959).

Out of these many references, we would advise from personal taste to start further reading in Robert [6], Serre [7], Swinnerton-Dyer [11] and Siegel [17].

Digressions and exercises refer also to the following books and papers:

[25] (BOREL, A., MOSTOW, G.D., Eds),"Algebraic groups and discontinuous subgroups", Proc. Symp. Pure Mathematics IX, Am. Math. Soc.,1966.
[26] AUSLANDER, L., MACKENZIE, R.E., Introduction to Differentiable Manifolds, McGraw Hill (1963).
[27] BERGER, M., GAUDUCHON, P., MAZET, E., Le spectre d'une variété riemannienne, Springer Lecture Notes in Mathematics **194** (1971).
[28] BOREL, A., "Compact Clifford-Klein forms of symmetric space", Topology **2** (1963) 111 – 122.
[29] BOREL, A., Introduction aux groupes arithmétiques, Hermann (1969).
[30] BOREL, A., CHOWLA, S., HERZ, C.S., IWASAWA, K., SERRE, J.P., Seminar on Complex Multiplication, Springer Lecture Notes in Mathematics **21** (1966).
[31] BORN, M., HUANG, K., Dynamical Theory of Crystal Lattices, Oxford Univ. Press (1968).

[32] COXETER, H.S.M., Introduction to Geometry, Wiley (1961). See also, by the same author: "L'œuvre d'Escher et les mathématiques", in Le monde de M.C. Escher (LOCHER, J.L., Ed.), Editions du Chêne, Paris (1972).

[33] DIEUDONNE, J., Fondements de l'analyse moderne, Gauthier-Villars (1963). This is the first volume of "Eléments d'analyse", with six volumes published up to July 1975.

[34] The Graphic Work of M.C. Escher (new, revised and expanded edition) Oldbourne (1967).

[35] GODEMENT, R., Cours d'algèbre, Hermann (1963).

[36] HIRZEBRUCH, F., NEUMANN, W.D., KOH, S.S., Differentiable Manifolds and Quadratic Forms, Dekker (

[37] KOBAYASHI, S., NOMIZU, K., Foundations of Differential Geometry **II**, Interscience (1969).

[38] KUROSH, A.G., The Theory of Groups **2**, 2nd English Edn, Chelsea (1960).

[39] MOSTOW, G.D., Strong Rigidity of Locally Symmetric Spaces, Princeton Univ. Press (1973).

[40] NARASIMHAN, R., Several Complex Variables, Chicago Univ. Press (1971).

[41] Séminaire Henri Cartan, 10e année, 1957/1958: Fonctions automorphes, Secrétariat mathématique, 11 rue Pierre Curie, Paris 5e (1958).

[42] SERRE, J.P., Cohomologie des groupes discrets, in Annals of Mathematics Studies **70**, Princeton Univ. Press (1971) 77–169.

[43] SERRE, J.P., Arbres, amalgames et SL_2 , to appear in Springer Lecture Notes in Mathematics.

[44] DU VAL, P., Elliptic Functions and Elliptic Curves, London Math. Soc. Lecture Note Series **9**, Cambridge Univ. Press (1973).

[45] WANG, H.C., Complex parallelizable manifolds, Proc. Am. Math. Soc. **5** (1954) 771–776.

[46] WHITNEY, H., Complex Analytic Varieties, Addison-Wesley (1972).

SOME NOTATION

$\underline{\mathrm{N}}$ $= \{0, 1, 2, ...\}$ is the set of natural integers, zero included.

$\mathbb{Z}$ is the group of rational integers.

$\underline{\mathrm{Q}}$ is the field of rational numbers.

$\underline{\mathrm{R}}$ is the field of real numbers.

$\underline{\mathrm{R}}_+$ is the set of real numbers $\geqslant 0$.

$\underline{\mathrm{R}}^*$ is the set of non-zero real numbers.

$\underline{\mathrm{C}}$ is the field of complex numbers.

If $z = x + iy$ with $x, y \in R$, $\mathrm{Re}(z)$ is x and $\mathrm{Im}(z)$ is y.

$\underline{\mathrm{C}}^*$ is the set of non-zero complex numbers.

$\mathbb{S}^1$ $= \{ z \in C \mid |z| = 1 \}$ is the unit circle in $\underline{\mathrm{C}}$.

The sets $\underline{\mathrm{R}}$, $\underline{\mathrm{R}}_+$, $\underline{\mathrm{R}}^*$, $\underline{\mathrm{C}}$, $\underline{\mathrm{C}}^*$ and $\mathbb{S}^1$ are always furnished with their usual topology.

π denotes sometimes a **canonical projection**, sometimes the real number 3, 14 ... .

Chapter I
REAL TORI

1. CLOSED SUBGROUPS AND LATTICES IN R^n

We start with two easy exercises. The first should be kept in mind in the whole of the present section, and the second provides the idea of the proof of Propositions 1 and 2.

Exercise 1. Check that the map from $\underline{R}$ to $\underline{C}$ defined by $x \mapsto \exp(i2\pi x)$ induces an isomorphism of topological groups $\underline{R}/\mathbb{Z} \to \mathbb{S}^1$. State and check the analogous fact for $\underline{R}^2/\mathbb{Z}^2 \to \mathbb{S}^1 \times \mathbb{S}^1$.

Exercise 2. Classify all closed subgroups of the additive group $\underline{R}$.
[Hint: if Γ is such a subgroup, Γ-$\{0\}$ is either empty or has elements of minimal absolute value, or has elements of arbitrarily small absolute value.] Give a few examples of other subgroups.

Let now n, ν, m be three natural integers with $1 \leqslant n + \nu \leqslant m$. We shall identify the abelian group $\mathbb{Z}^n \oplus \underline{R}^\nu$ with the subgroup consisting of those vectors $(\lambda_1, ..., \lambda_n, x_1, ..., x_\nu, 0, ..., 0)$ for which the λ_j's are rational integers and the x_k's are real numbers. In particular, $\mathbb{Z}^n$ is a discrete subgroup of $\underline{R}^m$ and $Z^n \oplus \underline{R}^\nu$ is a closed subgroup of $\underline{R}^m$; we want to show that there are essentially no others. (Of course $\underline{R}^m$ is furnished with its standard topology.)

Proposition 1. Let Γ be a discrete subgroup of $\underline{R}^m$, with $m \geqslant 1$ and $\Gamma \neq \{0\}$. Then there exist an integer n with $1 \leqslant n \leqslant m$ and an invertible linear transformation ϕ: $\underline{R}^m \to \underline{R}^m$ such that $\phi(\Gamma) = \mathbb{Z}^n$.

Proof. Let $\mathscr{P}$ be the set of all subspaces P of $\underline{R}^m$ such that $P \cap \Gamma$ is a free $\mathbb{Z}$-module generated by a basis of P over $\underline{R}$. Choose a maximal element P_0 in $\mathscr{P}$(which is ordered by inclusion) and let $\{e_1, ..., e_n\}$ be a basis of P_0 over $\underline{R}$ which generates $P_0 \cap \Gamma$. We want to show that $\Gamma \subset P_0$.

Suppose it is not true, and choose $a \in \Gamma$ with $a \notin P_0$. Let C be the subset of $\underline{R}^m$ consisting of points of the form

$$ra + \sum_{k=1}^{n} r_k e_k$$

where r and the r_k's are in the closed interval [0,1] of $\underline{R}$. Then C is compact and $C \cap \Gamma$ is finite because it is both compact and discrete. Let now $D = \{c \in C \cap \Gamma | c \notin P_0\}$, which is a finite non-empty set ($a \in D$), and choose a point $d_0 = s_0 a + \Sigma s_k e_k$ in D with the component s_0 positive and smallest possible ($0 < s_0 \leqslant 1$). (In this proof, summation signs without indicated limits are summations over k from 1 to n.)

We claim that $(\underline{R}d_0 \oplus P_0) \cap \Gamma = \mathbb{Z}d_0 \oplus (P_0 \cap \Gamma)$; the non-trivial part of the claim is that the left-hand side is included in the right-hand side, and we check this now. Let $x = t_0 a + \Sigma t_k e_k \in (Rd_0 \oplus P_0) \cap \Gamma$. Subtracting from x first a convenient integral multiple of d_0 and then convenient integral multiples of the e_k's, we can obtain a vector $y = u_0 a + \Sigma u_k e_k$ in Γ with $0 \leqslant u_0, u_1, ..., u_n \leqslant 1$, namely a vector $y \in C \cap \Gamma$. Now u_0 is an integral multiple of s_0 [otherwise there would exist $q \in N$ with $qs_0 < u_0 < (q + 1)s_0$; by subtracting from y-qd_0 convenient integral multiples of the e_k's, one would find $z = (u_0 - qs_0) a + \Sigma v_k e_k \in C \cap \Gamma$ with $0 < u_0 - qs_0 < s_0$, and that would contradict our choice of d_0]. Hence t_0 is also an integral multiple of s_0. It follows that x has integral components $t_0, t_1, ..., t_n$, so that the claim is proved.

But now the claim contradicts the fact that P_0 is maximal in $\mathscr{P}$, so that we were not allowed to choose a as above; consequently $\Gamma = P_0 \cap \Gamma = \{x = \Sigma t_k e_k \in P_0 \mid t_1, ..., t_n \in \mathbb{Z}\}$. Finally, let ϕ be an invertible linear transformation of $\underline{R}^m$ such that $\{\phi(e_1), \phi(e_2), ..., \phi(e_n)\}$ are the first n vectors of the canonical basis in $\underline{R}^m$. Then $\phi(\Gamma) = \mathbb{Z}^n$. ■

Remark. Clearly, there are no **maximal discrete subgroups** of $\underline{R}^m$. But it follows from Proposition that any discrete subgroup of $\underline{R}^m$ is contained in a (non-unique) subgroup isomorphic to $\mathbb{Z}^m$, i.e. in a subgroup of **maximal rank.**

Exercise 3. Classify all discrete subgroups of the direct product $\underline{R}^m \times A$ where A is a finite abelian group. [A n s w e r: Let Γ be a discrete subgroup of $\underline{R}^m \times A$; then there exists a subgroup Δ of $\underline{R}^m \times A$ such that the projection $\underline{R}^m \times A \to \underline{R}^m$ induces an isomorphism of Δ onto a discrete subgroup of $\underline{R}^m$, and there exists a subgroup B of A, such that $\Gamma = \Delta \cdot B$ is isomorphic to the direct product $\Delta \times B$, hence to $\mathbb{Z}^n \times B$ for some n.
[C o m m e n t: Let G be a connected compact real Lie group; then the centre of the universal covering $\tilde{G}$ of G is of the form $\underline{R}^m \times A$, and any real Lie group which is locally isomorphic to G is of the form $\tilde{G}/\Gamma$, where Γ is a discrete subgroup of $\underline{R}^m \times A$.]

Corollary. Let Γ be a subgroup of a real vector space V of dimension n. Then the following conditions are equivalent:

(i) Γ is discrete and V/Γ is compact;
(ii) Γ is discrete and generates V as a real vector space;
(iii) there exists a basis $\{e_1, ..., e_n\}$ of V over $\underline{R}$ which is a basis of Γ over $\mathbb{Z}$.

Definition. A **lattice** in V is a subgroup of V which satisfies the conditions of the corollary.

Exercise 4. Let Γ be a lattice in $\underline{R}^n$ and let $\| \ \|$ denote a norm on $\underline{R}^n$. Show that the series $\Sigma' \|\omega\|^{-s}$ converges absolutely for all complex numbers s with $\mathrm{Re}(s) > n$; in this exercise, the sign Σ' indicates a summation over all elements in Γ, except the origin.
[S k e t c h: Choose a basis $\{e_1, ..., e_n\}$ of $\underline{R}^n$ over $\underline{R}$ which is a basis of Γ over $\mathbb{Z}$. All norms being equivalent on $\underline{R}^n$, there is no loss of generality if one chooses that defined by

$$\left\| \sum_{j=1}^{n} x_j e_j \right\|^2 = \sum_{j=1}^{n} |x_j|^2$$

Then

$$\sum \|\omega\|^{-s} \leqslant 2^n \sum \|\lambda_1 e_1 + ... + \lambda_n e_n\|^{-s}$$

where the right-hand side summation is over all n-tuples $(\lambda_1, ..., \lambda_n)$ with $\lambda_1, ..., \lambda_n \in N$ and $\lambda_1 + ... + \lambda_n \geqslant 1$. Group those terms for which $\lambda_1 + ... + \lambda_n$ is a given strictly positive integer λ. Show by induction on n that there are

$$\frac{(\lambda + n - 1)!}{\lambda! \ (n-1)!} \leqslant c\lambda^{n-1}$$

of them, where c is a constant independent of λ.

(Or even better: get the estimate without computing the actual number, knowing that the surface of a sphere of radius λ in $\underline{R}^n$ is λ^{n-1} up to a multiplicative constant.) It follows that the terms for which $\lambda_1 + \dots + \lambda_n = \lambda$ contribute to the series by a factor like λ^{n-1-s}. According to classical criteria, the series converges if $\mathrm{Re}(-n+1+s) > 1$.]

Digression. More generally, if G is any connected real Lie group, a lattice in G is a discrete subgroup Γ of G such that G/Γ is compact (some authors consider as lattices in G more general objects than these). For example, it is known that any connected semi-simple Lie group has lattices (see Borel [28]). The **deformation theory** of lattices has important applications in geometry (see the papers by Garland in [25], or Mostow [39]). For a Lie group without lattice, see Bourbaki [5], Ch. III, §9, Exercise 30.

Proposition 2. Let S be a closed subgroup of $\underline{R}^m$, with $m \geqslant 1$ and $S \neq \{0\}$. Then there exist two positive integers n, ν with $1 \leqslant n + \nu \leqslant m$ and an invertible linear transformation $\phi: \underline{R}^m \to \underline{R}^m$ such that $\phi(S) = \mathbb{Z}^n \oplus \underline{R}^\nu$.

Proof

Step one. Let us first show that if S is not discrete then S contains a real line.

We furnish $\underline{R}^m$ with some norm $\| \ \|$. By hypothesis, there exists a sequence $(a_j)_{j\in N}$ of non-zero vectors in S which converges towards the origin. Define $b_j = a_j/\|a_j\|$ for each $j \in N$. As the unit sphere in $\underline{R}^m$ is compact, there is a subsequence $(b_{j_k})_{k\in N}$ of $(b_j)_{j\in N}$ which converges towards some point b of the unit sphere of $\underline{R}^m$.

Let $x \geqslant 0$ be a real number, fixed for a while. For each k in N, let x_k be the integer ($\geqslant 0$) defined by

$$x_k \|a_{j_k}\| \leqslant x < (x_k+1)\|a_{j_k}\|$$

Then $x_k a_{j_k} \in S$ and

$$\|x_k a_{j_k} - xb\| = \|x_k \|a_{j_k}\|(b_{j_k} - b) + (x_k\|a_{j_k}\| - x)b\|$$

$$\leqslant x_k\|a_{j_k}\| \, \|b_{j_k} - b\| + \|a_{j_k}\| \, \|b\| \leqslant x\|b_{j_k} - b\| + \|a_{j_k}\| \, \|b\|$$

for all $k \in N$, so that

$$\lim_{k\to\infty} \|x_k a_{j_k} - xb\| = 0$$

As S is closed, it follows that $xb \in S$.

As this holds for any $x \in \underline{R}_+$, one has $\underline{R}b \subset S$ and step one is proved.

Step two. Let V be the subspace of $\underline{R}^m$ which is the union of all the lines in S; then V is the largest subgroup of S which is a subspace of $\underline{R}^m$. Let W be any supplement of V in $\underline{R}^m$: one has then $S = V \oplus (W \cap S)$ with $W \cap S$ discrete. [Otherwise $W \cap S$ contains a line by step one, which contradicts our choice for V.] We know from Proposition 1 what $W \cap S$ can be like, and Proposition 2 follows. ■

Exercise 5. Describe closed subgroups of closed subgroups of $\underline{R}^m$.

We give one solution. Let S and T be closed subgroups of $\underline{R}^m$ with $S \subset T$. Let U and V be the maximal vector subgroup of S and T, respectively; then U is a subspace of V. Choose a supplement X of U in V. By Proposition 2, one has $V \cap S = U \oplus (X \cap S)$.

For any supplement W of V in $\underline{R}^m$, let $p_W : \underline{R}^m \to W$ be the parallel projection along the direction of V. Then $p_W(S)$ is a discrete subgroup of W; indeed $p_W(S) \subset p_W(T) = W \cap T$, which is discrete. It follows from Proposition 1 that there are $\underline{R}$-linearly independent vectors $w_1, \ldots, w_k$ in W which generate $p_W(S)$ over $\mathbb{Z}$. Choose $y_1, \ldots, y_k$ in S such that $p_W(y_1) = w_1, \ldots, p_W(y_k) = w_k$ and let Y be a supplement of V in $\underline{R}^m$ which contains the y_j's. Clearly $S = (V \cap S) \oplus (Y \cap S)$ and one has:

$$\begin{array}{llll} S = U \oplus (X \cap S) \oplus (Y \cap S) & & & U \oplus X = V \\ \cap & & & \\ T = \quad V \quad \oplus (Y \cap T) & & & V \oplus Y = \underline{R}^m \end{array}$$

The two following exercises go beyond our subject, but they should make it clear that lattices show up in numerous contexts; see also the historical note after Ch. VII of Bourbaki's "Topologie générale" (Les groupes additifs $\underline{R}^n$).

Exercise 6. Let $\{e_1, \ldots, e_n\}$ be the canonical basis of $\underline{R}^n$ where $n \geqslant 2$, let $(|)$ denote the standard scalar product on $\underline{R}^n$, and let Γ_0 be the lattice in $\underline{R}^n$ generated by the e_j's.

(i) Check that $\{\alpha \in \Gamma \mid (\alpha|\alpha) = 1\} = \{\pm e_1, \ldots, \pm e_n\}$;

(ii) Check that $\{\alpha \in \Gamma \mid (\alpha|\alpha) = 2\} = \{\pm e_j \pm e_k \in \Gamma \mid j < k\}$.

It is known that $\{\alpha \in \Gamma \mid (\alpha|\alpha)$ is 1 or 2$\}$ is a **root system of type** B_n; see Bourbaki [5] Ch.VI, §4, No.5.

Exercise 7. The notation being as in Exercise 6, let $\Gamma_1 = \{\alpha \in \Gamma_0 \mid (\alpha|\alpha)$ is even$\}$ and let Γ be the subgroup of $\underline{R}^n$ generated by Γ_1 and $\frac{1}{2}(e_1 + \ldots + e_n)$.

(i) Check that

$$\Gamma_1 = \{(x_1, \ldots, x_n) \in \Gamma_0 \mid \sum_{j=1}^{n} x_j \text{ is even}\}$$ is a lattice in $\underline{R}^n$ and that the quotient group Γ_0/Γ_1 has two elements. [**Hint:** $x^2 \equiv x$ (mod.2) for all $x \in \mathbb{Z}$.]

(ii) Suppose $n \equiv 0$ (mod.4). Check that Γ is a lattice in $\underline{R}^n$, that the quotient group Γ/Γ_1 has two elements, and that

$$\Gamma = \left\{ (x_1, \ldots, x_n) \in \underline{R}^n \;\middle|\; \begin{array}{ll} 2x_j \in \mathbb{Z} & j = 1, \ldots, n \\ x_j - x_k \in \mathbb{Z} & j, k = 1, \ldots, n \\ \sum_{j=1}^{n} x_j \in 2\mathbb{Z} & \end{array} \right\}$$

[Observe that $2e \in \Gamma_1$ and that $e \notin \Gamma_1$.]

(iii) If $n \equiv 0$ (mod.4), show that $(\alpha|\beta) \in \mathbb{Z}$ for all $\alpha, \beta \in \Gamma$. [Consider the three cases: α and β in Γ_1; α in Γ_1 and $\beta = e$; $\alpha = \beta = e$.]

(iv) If $n \equiv 0$ (mod.8), $(\alpha|\alpha)$ is even for all α in Γ.

Comment: If n = 8, the set $\{\alpha \in \Gamma \mid (\alpha|\alpha) = 2\}$ is known to contain 240 elements and to underline **a root system of type** E_8; the form

$$\begin{cases} \Gamma \times \Gamma \to \mathbb{Z} \\ (\alpha, \beta) \mapsto (\alpha|\beta) \end{cases}$$

happens to play a crucial role in the theory of integral quadratic forms (see Serre [7], Ch.V), and hence in algebraic topology (see Hirzebruch, Neumann and Koh [36]).

Lattices occur naturally in physics, and in particular in crystallography: open any book on classical solid-state physics. As a more specific example, let us mention that the evaluation of sums like those of Exercise 4 is important for knowledge of the potential energy in various crystal systems; indications of computational methods are given, e.g., in Appendix III to Born and Huang [31].

2. CONNECTED ABELIAN REAL LIE GROUPS AND REAL TORI

Let V be a real vector space (of finite dimension n) and let Γ be a lattice in V. The quotient group $T = V/\Gamma$ has a unique structure of a real analytic manifold which makes the canonical projection $\pi: V \to T$ a covering. Furnished with this, T is a real Lie group known as a **real torus.** Notice that T is connected (because V is connected and π is continuous) and compact (see the definition of a lattice). From the point of view of Lie group theory, V is the Lie algebra of T and π is its exponential map. The next result will allow us to speak about *the* real n-torus.

Proposition 3. Between any two n-dimensional real tori, there exists a group isomorphism which is also an analytic diffeomorphism.

Proof. Let V_j be an n-dimensional real vector space, let Γ_j be a lattice in V_j, let $T_j = V_j/\Gamma_j$ be the associated torus and let $\{e_1^j, \ldots, e_n^j\}$ be a basis of V_j over $\underline{R}$ which is a basis of Γ_j over $\mathbb{Z}$ $(j = 1, 2)$. Let $\phi: V_1 \to V_2$ be the isomorphism of vector spaces defined by $\phi(e_k^1) = e_k^2$ for $k = 1, \ldots, n$. Then ϕ induces a quotient map $\varphi: T_1 \to T_2$ which has the desired properties. ∎

Proposition 3 expresses the fact that two real n-tori are isomorphic as real Lie groups. In this sense, any n-torus can be looked at as $\mathbb{S}^1 \times \mathbb{S}^1 \times \ldots \times \mathbb{S}^1$ (n times), see Exercise 1.

Exercise 8. Check that any continuous group-homomorphism $\varphi: \underline{R}^m \to T^n$ can be written as $\pi \circ \phi$, where ϕ is a linear map from $\underline{R}^m$ to $\underline{R}^n$ and where $\pi: \underline{R}^n \to T^n$ is the canonical projection. [**Solution**: φ lifts to $\underline{R}^m \to \underline{R}^n$ because $\underline{R}^m$ is simply connected.] What can you then say about continuous homomorphisms from T^m to T^n?

Exercise 9. Let $T^n = \underline{R}^n/\mathbb{Z}^n$ and let

$$\left\{ \begin{array}{c} \underline{R}^n \longrightarrow T^n \\ (x_1, \ldots, x_n) \mapsto [x_1, \ldots, x_n] \end{array} \right.$$

be the canonical projection. For any n-tuple $(\lambda_1, \ldots, \lambda_n)$ of rational integers, let $\rho(\lambda_1, \ldots, \lambda_n)$ be the map which sends $[x_1, \ldots, x_n]$ in T^n onto $\exp\{i2\pi(\lambda_1 x_1 + \ldots + \lambda_n x_n)\}$ in $\mathbb{S}^1$. Show that these maps are pairwise distinct continuous homomorphisms from T^n to $\mathbb{S}^1$, and that any continuous homomorphism from T^n to $\mathbb{S}^1$ is one of them.
Comment: these maps are the **irreducible complex representations** of T^n (see Adams [2], Ch.3, for more details).

Proposition 4. Let G be a connected abelian real Lie group of dimension m. Then there exists a natural integer $n \leqslant m$ such that G is isomorphic to the direct product $T^n \times \underline{R}^{m-n}$. In particular, any connected compact abelian real Lie group is a torus, and any simply-connected abelian real Lie group is a vector group.

Proof. We use some general facts about Lie groups. Let V be the underlying vector space of the Lie algebra of G. As G is abelian, $\exp: V \to G$ is a group homomorphism. As exp is a local diffeomorphism, its kernel Γ is a discrete subgroup of V. The proposition follows as $G = V/\Gamma$. ∎

Exercise 10. Show that there exist non-abelian real Lie groups which are diffeomorphic to manifolds of the form $T^n \times \underline{R}^{m-n}$.
Comment: It is known, e.g., that any connected real Lie group G which is solvable is diffeomorphic to some $T^n \times \underline{R}^{m-n}$; but if G is moreover compact, then G must be a torus. See the proof in Hochschild [3], Theorem 2.2 of Ch. XII and Theorem 1.3 of Ch. XIII.

Corollary. Let S be a connected closed subgroup of a torus T. Then S and T/S are both real tori and T is isomorphic to the direct product $S \times (T/S)$.

Proof. Both S and T/S are connected compact abelian real Lie groups, hence they are tori. As T and $S \times (T/S)$ are tori of the same dimension, they are isomorphic. ∎

Exercise 11. Show that there are subgroups of tori which are not closed, or not connected, or both.

Exercise 12. Let P be the polynomial function

$$\begin{cases} \underline{R}^{2n} \longrightarrow \underline{R} \\ x = (x_1, \ldots, x_{2n}) \mapsto \displaystyle\sum_{k=1}^{n} (x_{2k-1}{}^2 + x_{2k}{}^2 - 1)^2 \end{cases}$$

and let $\mathscr{V}$ be the submanifold $\{x \in \underline{R}^{2n} | P(x) = 0\}$ of $\underline{R}^{2n}$. In sophisticated terms, $\mathscr{V}$ is an **affine algebraic submanifold** of $\underline{R}^{2n}$. Let

$$\mu \begin{cases} \mathscr{V} \times \mathscr{V} \to \mathscr{V} \\ (x,y) \mapsto z \end{cases}$$

be the **algebraic map** defined by

$$\begin{array}{ll} z_1 = x_1 y_1 - x_2 y_2 & z_2 = x_1 y_2 + x_2 y_1 \\ \vdots & \vdots \\ z_{2n-1} = x_{2n-1} y_{2n-1} - x_{2n} y_{2n} & z_{2n} = x_{2n-1} y_{2n} + x_{2n} y_{2n-1} \end{array}$$

Then μ endows $\mathscr{V}$ with a structure of real Lie group. Check that $\mathscr{V}$ is isomorphic to T^n, so that any real torus is **algebraic** in this sense.
[**Hint:** if $x_1 = \cos\theta_1$; $y_1 = \cos\varphi_1$; $x_2 = \sin\theta_1$; $y_2 = \sin\varphi_1$, then $z_1 = \cos(\theta_1 + \varphi_1)$ and $z_2 = \sin(\theta_1 + \varphi_1)$.]

The last proposition of this section is a generalization of the corollary above.

Proposition 5. Let S be a connected closed subgroup of a connected abelian real Lie group A. Then A is isomorphic to the direct product $S \times (A/S)$.

Proof. We identify the underlying subspace of the Lie algebra of A with $\underline{R}^m$ and we denote by $\pi:\underline{R}^m \to A$ the natural projection (in other words $\pi = \exp_A$). Let Γ be the kernel of π and let $T = \pi^{-1}(S)$, which is a closed subgroup of $\underline{R}^m$ containing Γ. Let V be the maximal vector subgroup of T and let Y be a supplement of V in $\underline{R}^m$ such that (see Exercise 5)

$$\Gamma = (V \cap \Gamma) \oplus (Y \cap \Gamma)$$
$$T = V \oplus (Y \cap T)$$

As Γ is also the kernel of the restriction of π to T, one has $S \cong T/\Gamma \cong \{V/(V \cap \Gamma)\} \oplus \{(Y \cap T)/(Y \cap \Gamma)\}$. But S is connected; this implies $Y \cap T = Y \cap \Gamma$ and so $S \cong V/(V \cap \Gamma)$. On the other hand, $A/S \cong (\underline{R}^m/\Gamma)/(T/\Gamma) \cong \underline{R}^m/T = (V \oplus Y)/T = Y/(Y \cap T) = Y/(Y \cap \Gamma)$. It follows that $S \times (A/S) \cong \{V/(V \cap \Gamma)\} \oplus \{Y/(Y \cap \Gamma)\} = (V \oplus Y)/\Gamma \cong A$. ■

Exercise 13 (Alternative proof of Proposition 5, of which we keep the notation). By Proposition 4, there exist natural integers m, n, p, a, b, c such that $A = T^m \times \underline{R}^a$, $S = T^n \times \underline{R}^b$ and $A/S = T^p \times \underline{R}^c$. Show that $m = n + p$ and $a = b + c$. [**Sketch**: Counting dimensions makes it sufficient to check only one of these equalities. Now m (or n, p, n + p) is the dimension of a (in fact the) maximal compact subgroup of A (or S, A/S, S × (A/S), respectively). Consider maximal compact subgroups in the middle term of the sequence $S \to A \to A/S$ and check that $m = n + p$.]

3. A COMPLEX EXAMPLE

The statements of Section I.2 do *not* carry over to the complex case. The aim of this section is to show in particular that Proposition 5 does *not* hold for connected abelian complex Lie groups. (Neither does the corollary to Proposition 4, but examples to show this are more complicated; we shall give one in Chapter IV.)

Let $\mathbb{Z}^2$ be canonically embedded in $\underline{C}^2$, and let $A = \underline{C}^2/\mathbb{Z}^2 = (\underline{C}/\mathbb{Z}) \times (\underline{C}/\mathbb{Z})$ be considered as a **complex** Lie group; we denote by $\pi:\underline{C}^2 \to A$ the canonical projection. Let S be the image of $\underline{C}$ by the injective homomorphism

$$\varphi \begin{cases} \underline{C} \to A \\ z \mapsto (z, iz) \end{cases}$$

Then S is a closed subgroup of A, and also a complex submanifold of A; in other words, S is a sub complex Lie group of A. Let us show that A and S × (A/S) are *not* isomorphic as complex manifolds (*a fortiori* not isomorphic as complex Lie groups).

We claim that A/S is compact. Indeed, let $\psi:(\underline{R}/\mathbb{Z}) \times (\underline{R}/\mathbb{Z}) \to A/S$ be the continuous map obtained by composition of the inclusion $(\underline{R}/\mathbb{Z}) \times (\underline{R}/\mathbb{Z}) \to A$ and of the canonical projection $A \to A/S$. As the domain of ψ is compact, it is sufficient to prove that ψ is onto. But let a be an arbitrary point in A. Choose $(u,v) \in \underline{C}^2$ with $\pi(u,v) = a$, and define $z = \mathrm{Im}(v) + i\mathrm{Im}(u) \in \underline{C}$. Then $(x,y) = (u,v) - (z,iz)$ is clearly in $\underline{R}^2$, and so defines an element $[x,y]$ in $(\underline{R}/\mathbb{Z}) \times (\underline{R}/\mathbb{Z})$. In other words $[x,y] \equiv a(\mathrm{mod}.S)$, so that ψ is onto.

Then we claim that any holomorphic map $f:S \times (A/S) \to A$ is constant on the "fibres" $\{s\} \times (A/S)$. Indeed, the map

$$\begin{cases} \underline{C} \to \underline{C}^* \\ z \mapsto \exp(i2\pi z) \end{cases}$$

defines an isomorphism $\underline{C}/\mathbb{Z} \to \underline{C}^*$, so that A and $\underline{C}^* \times \underline{C}^*$ are holomorphically isomorphic. Now for each $s \in S$, the map f induces two holomorphic maps:

$$\{s\} \times (A/S) \to A \overset{\approx}{\to} \underline{C}^* \times \underline{C}^* \to \underline{C}^2 \overset{pr_j}{\to} \underline{C} \qquad (j = 1, 2)$$

where pr_1 and pr_2 are the co-ordinate projections; these two maps are constant by the maximum principle, and the claim is proved.

It follows that $S \times (A/S)$ and A are not isomorphic as complex manifolds, as stated above.

4. VOCABULARY OF LIE GROUPS

This section is not intended to be a crash introduction to Lie groups, but merely a recall of some words used here and there in these notes. The letter K denotes either the real field $\underline{R}$ or the complex field $\underline{C}$.

A **Lie algebra** over K is a (finite-dimensional) vector space $\underline{g}$ over K furnished with a bilinear mapping $(X, Y) \mapsto [X, Y]$ from $\underline{g} \times \underline{g}$ to $\underline{g}$ such that $[X, X] = 0$ for all $X \in \underline{g}$ and such that $[X, [Y, Z]] = [[X, Y], Z] + [Y, [X, Z]]$ for all $X, Y, Z \in \underline{g}$. An **abelian** Lie algebra over K is a vector space V over K furnished with the **bracket** (or product) defined by $[X, Y] = 0$ for all $X, Y \in V$.

A **Lie group** over K is a group furnished with a structure of analytic manifold over K which makes the group operations analytic. (The implicit function theorem makes it sufficient to require that the multiplication $G \times G \to G$ of the group is analytic.) If a Lie group is denoted by a capital roman letter such as G, its **Lie algebra** will often be denoted by the corresponding underlined small letter $\underline{g}$; it is by definition the tangent space to G at the identity, which is identified with the space of left-invariant vector fields on G, and so is furnished with a bracket which makes it a Lie algebra (see Robertson's paper in these Proceedings). The **exponential map** is written $\exp: \underline{g} \to G$ and assigns to $X \in \underline{g}$ the element $\varphi_X(1) \in G$, where φ_X is the integral curve $\underline{R} \to G$ defined by X and by the initial value $\varphi_X(0) = e =$ identity in G. (The map φ_X is a group homomorphism from $\underline{R}$ to G.)

The group of all invertible linear transformations of a vector space V over K, known as the **general linear group of** V, is a Lie group over K which is denoted by GL(V). Its Lie algebra $\underline{g\ell}(V)$ is the space of all linear endomorphisms of V together with the bracket defined by $[X, Y] = XY - YX$, where XY is the usual composition of the maps Y and X. Lie subgroups of GL(V) are called **linear Lie groups** and provide many examples of Lie groups; however, not any group is isomorphic to a linear one (see Hochschild [3], Ch. XVIII).

Exercise 14. Write down a subset of the space of complex (n × n)-matrices which is a complex Lie group for the usual topology and the usual product for matrices, and which contains *no* invertible matrix. [**Answer**: the group G of matrices $\begin{pmatrix} z & z \\ z & z \end{pmatrix}$ where z is a non-zero complex numbe which is isomorphic to $\underline{C}^*$ by

$$\begin{cases} \underline{C}^* \to G \\ z \mapsto \begin{pmatrix} \frac{1}{2}z & \frac{1}{2}z \\ \frac{1}{2}z & \frac{1}{2}z \end{pmatrix} \end{cases} \quad]$$

Let G be a Lie group over K and let $g \in G$. The **inner automorphism**

$$\begin{cases} G \to G \\ h \mapsto ghg^{-1} \end{cases}$$

is denoted by Int(g), and its derivative at the identity of G by Ad(g). The **adjoint representation of** G is the analytic homomorphism

$$\mathrm{Ad}: \begin{cases} G \to GL(\underline{g}) \\ g \mapsto \mathrm{Ad}(g) \end{cases}$$

Its derivative at the identity of G is the **adjoint representation of** $\underline{g}$ and is known to be given by

$$\mathrm{ad}: \begin{cases} \underline{g} \to \underline{g\ell}(\underline{g}) \\ X \mapsto (Y \to [X, Y]) \end{cases}$$

Then the formula $\exp(\mathrm{ad}(X)) = \mathrm{Ad}(\exp(X))$ holds for all $X \in \underline{g}$, where the left-hand exp is given by the traditional power series and where the right-hand exp is that defined for G. It follows among other things that $\underline{g}$ is an abelian Lie algebra if and only if the connected component G_0 of G is an abelian group.

Let G be a complex Lie group; the real Lie group obtained by **restricting the scalars** is denoted by $(G)_{\underline{R}}$. Similarly, if $\underline{g}$ is a complex Lie algebra, the real Lie algebra obtained by restricting the scalars is denoted by $(\underline{g})_{\underline{R}}$. If, moreover, $\underline{g}$ is the Lie algebra of G, then $(\underline{g})_{\underline{R}}$ is that of $(G)_{\underline{R}}$. Clearly, if G and H are isomorphic as complex Lie groups, then $(G)_{\underline{R}}$ and $(H)_{\underline{R}}$ are isomorphic as real Lie groups; Section I.3 shows that the converse does *not* hold.

If G is a real Lie group, if there exists a complex Lie algebra $\underline{h}$ such that $(\underline{h})_{\underline{R}} = \underline{g}$, and if G is **connected**, then there exists a complex Lie group H such that $(H)_{\underline{R}} = G$. Giving H or $\underline{h}$ is by definition giving a **complex structure** on G or $\underline{g}$, respectively.

Exercise 15. It is known that any continuous homomorphism of a real Lie group into another is real analytic (see e.g. Hochschild [3], Ch.VII, Theorem 4.2). Show that the analogous statement for complex Lie groups does *not* hold. [**Solution**: consider complex conjugation in $\underline{C}$.]

Exercise 16. It is known that any closed subgroup of a real Lie group is a sub Lie group (Hochschild [3], Ch. VIII, §1). Show that the analogous statement for complex Lie groups does *not* hold. [**Solution**: consider $\underline{R} \subset \underline{C}$.]

Exercise 17. Let G_0 be the group of all real (2×2)-matrices of the form $\begin{pmatrix} x & y \\ -y & x \end{pmatrix}$, with $x^2 + y^2 \neq 0$. Let j be the matrix $\begin{pmatrix} 1 & 0 \\ 0 & -1 \end{pmatrix}$ and let G_1 be the set of all real (2×2)-matrices of the form $\begin{pmatrix} x & y \\ y & -x \end{pmatrix}$, with $x^2 + y^2 \neq 0$. Show that $G = G_0 \cup G_1$ is a linear real Lie group with two connected components

G_0 and G_1. Show that G_0 (hence $\underline{g}$) has a complex structure, but that G has *no* complex structure [**Hint:** G_0 is isomorphic to $\underline{C}^*$, and the isomorphism carries

$$\begin{cases} G_0 \to G_0 \\ g \mapsto jgj^{-1} \end{cases} \quad \text{to} \quad \begin{cases} \underline{C}^* \to \underline{C}^* \\ z \mapsto \bar{z} \end{cases}$$

Comment: As pointed out by K. Honda and S. Yamaguchi (A note on non-connected complex Lie groups, Mem. Faculty Sci. Kyushu Univ. Ser. A, 27 (1973) 65–67), this implies that there is a misprint in many standard text books, such as Helgason [4], p. 323.]

Let G be a real Lie group with identity e and let M be a smooth manifold. A smooth **action** of G on M (say from the left) is a smooth map $\varphi: G \times M \to M$ such that $\varphi(e,m) = m$ for all $m \in M$ and $\varphi(g,\varphi(h,m)) = \varphi(gh,m)$ for all $g, h \in G$ and $m \in M$; one often writes gm instead of $\varphi(g,m)$; G is said to **act** on M (in these notes, if a Lie group acts on a smooth manifold, the action will always be smooth). The action is said to be:

effective	if $gm = m$ for all $m \in M$ implies $g = e$;
almost effective	if there is a discrete subgroup D in G such that $gm = m$ for all $m \in M$ implies $g \in D$;
transitive	if for any pair (m_1, m_2) of points in M there exists $g \in G$ with $gm_1 = m_2$.

If G acts on M, the **isotropy subgroup** of G at a point $m \in M$ is $Is(m) = \{g \in G | gm = m\}$ and is a closed subgroup of G. If the action is transitive and if m_1 and m_2 are points in M, it is easy to check that $Is(m_1) = g^{-1} Is(m_2) g$ for any g in G such that $gm_1 = m_2$; otherwise said, the isotropy subgroups of G are **conjugate** by inner automorphisms.

A real Lie group G acting on a complex manifold M (i.e. acting in the above sense on the real smooth manifold underlying M) acts by **holomorphic transformation** if the map $m \mapsto gm$ from M to M is holomorphic for each $g \in G$.

If Γ is an abstract group acting on a set S, the **orbit** of a point s in S under Γ is the subset $\{t \in S \mid \text{there exists } \gamma \in \Gamma \text{ with } t = \gamma s\}$.

The set of orbits is then the **quotient set** of S by the action of Γ, and is denoted either by $\Gamma \backslash S$ or by S/Γ.

Chapter II

COMPLEX TORI – GENERALITIES AND ABELIAN CURVES

1. ISOMORPHISMS OF COMPLEX TORI

Now let V be a **complex** vector space of finite complex dimension, say n. By definition, a **lattice** in V is a lattice in the underlying real vector space $(V)_{\underline{R}} \cong \underline{R}^{2n}$ of V. The quotient group $T = V/\Gamma$ of V by the lattice Γ has a unique structure of a complex analytic manifold which makes the canonical projection $\pi: V \to T$ a holomorphic covering. Furnished with this, T is a complex Lie group known as a **complex torus**. Notice that T is a connected compact complex Lie group. Keeping in mind Proposition 4, we now state the converse (it will not be used in what follows, and you may jump over the proof if you feel uncomfortable with the matter of Section I.4).

Proposition 6. Any connected compact complex Lie group is a complex torus and, in particular, is an abelian group.

Proof. Let G be a connected compact complex Lie group, let $\underline{g}$ be its Lie algebra, and consider the adjoint representation $\mathrm{Ad}: G \to GL(\underline{g})$. As this is a holomorphic map and as G is compact, Ad is locally constant (maximum principle); as G is connected, Ad is constant. It follows that its derivative $\mathrm{ad}: \underline{g} \to \underline{g\ell}(\underline{g})$ is the zero map, i.e. that $\underline{g}$ is abelian. Hence G is abelian (see Section I.4) and the result follows from Proposition 4. ■

Exercise 18. Show that any complex torus can be given a Kähler structure.
[**Sketch:** Furnish $\underline{C}^n$ with the standard Kähler structure defined by the form

$$\tilde{\omega} = -2i \sum_{k=1}^{n} dz_k \wedge d\bar{z}_k$$

Show that $\tilde{\omega}$ is invariant by any translation. If Γ is any lattice in $\underline{C}^n$, it follows that $\tilde{\omega}$ induces a Kähler structure defined by a form ω_Γ on the torus $\underline{C}^n/\Gamma$ (see Robertson's paper, Section 10, these Proceedings).]

Digression. A complex manifold M of complex dimension n is said to be **complex parallelizable** if there exist n holomorphic vector fields on M which are $\underline{C}$-linearly independent at each point of M. Observe that, for any complex Lie group G and any discrete subgroup D of G, the quotient space G/D is complex parallelizable. It is a result due to Wang [45] that all connected compact complex parallelizable manifolds are of this type; among these, complex tori are the only ones which admit Kähler metrics.

Proposition 7. Let Γ_1 and Γ_2 be two lattices in $\underline{C}^n$, and let $\varphi: \underline{C}^n/\Gamma_1 \to \underline{C}^n/\Gamma_2$ be a holomorphic map. Then φ is induced by an affine transformation, i.e. there exist a linear map $A: \underline{C}^n \to \underline{C}^n$ and a vector $b \in \underline{C}^n$ such that the map

$$\phi: \begin{cases} \underline{C}^n \to \underline{C}^n \\ z \mapsto Az + b \end{cases} \qquad \text{makes the diagram} \qquad \begin{array}{ccc} \underline{C}^n & \xrightarrow{\ \phi\ } & \underline{C}^n \\ \downarrow & \varphi & \downarrow \\ \underline{C}^n/\Gamma_1 & \longrightarrow & \underline{C}^n/\Gamma_2 \end{array}$$

commutative.

In particular, when $n = 1$, φ is a diffeomorphism if and only if it is not a constant; and φ is a group homomorphism if and only if it maps the identity in $\underline{C}^n/\Gamma_1$ onto the identity in $\underline{C}^n/\Gamma_2$.

Proof (written for $n=1$): The map φ being given, let $\phi: \underline{C} \to \underline{C}$ be any lifting of φ to the universal coverings $(\widetilde{\underline{C}/\Gamma_1}) = \underline{C}$ and $(\widetilde{\underline{C}/\Gamma_2}) = \underline{C}$. Let $\omega_1, \omega_2 \in \underline{C}$ be such that $\Gamma_1 = \mathbb{Z}\omega_1 \oplus \mathbb{Z}\omega_2$ (ω_1 and ω_2 are linearly independent over $\underline{R}$, i.e. $\mathrm{Im}(\omega_1/\omega_2) \neq 0$). The images of the maps $z \mapsto \phi(z + \omega_1) - \phi(z)$ and $z \mapsto \phi(z + \omega_2) - \phi(z)$ must be in Γ_2, hence these maps are constant by continuity. Taking derivatives:

$$\phi'(z + \omega_1) - \phi'(z) = \phi'(z + \omega_2) - \phi'(z) = 0$$

for all $z \in \underline{C}$. It follows that ϕ' is a bounded holomorphic map from $\underline{C}$ to $\underline{C}$, hence is a constant, so that ϕ must be of the form $z \mapsto az + b$, where a and b are complex constants.

The same proof works for $n > 1$: consider then the ϕ_j's, where $\phi:(z_1, ..., z_n) \mapsto (\phi_1(z_1, ..., z_n), ..., \phi_n(z_1, ..., z_n))$. One may also consider left-invariant holomorphic 1-forms on $\underline{C}^n/\Gamma_2$ and their pull-back by φ. ■

Corollary. (i) If two complex tori are isomorphic as complex manifolds, then they are also isomorphic as complex Lie groups. (ii) Two complex tori $T_1 = \underline{C}^n/\Gamma_1$ and $T_2 = \underline{C}^n/\Gamma_2$ are isomorphic if and only if the lattices Γ_1 and Γ_2 are homothetic. (A homothety is an invertible linear map from $\underline{C}^n$ to $\underline{C}^n$.)

Exercise 19. More generally, let $T = \underline{C}^n/\Gamma$ be a complex torus, let G be a complex Lie group, and let $\varphi: T \to G$ be a holomorphic map; prove that

$$\psi: \begin{cases} T \to G \\ t \mapsto \varphi(0)^{-1}\varphi(t) \end{cases}$$

is a homomorphism.

[**Sketch:** Assume first that $\varphi(0) = e$ = the identity in G and let

$$F: \begin{cases} T \times T \to G \\ (t,u) \mapsto \varphi(t)\varphi(u)\varphi(tu)^{-1} \end{cases} \qquad \text{so that } F(T,\{0\}) = F(\{0\}, T) = \{e\}$$

If u is near enough 0, then $F(T,\{u\})$ is in a co-ordinate patch of G around e and so is constant by the maximum principle. It follows that $F(T,T) = \{e\}$. **Comment:** both the corollary and the statement of this exercise hold in algebraic geometry (see Shafarevich [13], Ch.III, §3, Section 4.

Exercise 20. With the notation of the proof of Proposition 7, check that a is well defined by φ, and that b is well defined modulo the lattice Γ_2.

Comment. It can be shown that the set of complex structures on the vector space $\underline{R}^{2n}$ is in bijective correspondence with the homogeneous space $J_n = GL_{2n}(\underline{R})/GL_n(\underline{C})$ (see e.g. Kobayashi and Nomizu [37], Ch.IX, Prop. 1.3). One can deduce from this that the set of (isomorphism classes of) complex tori of dimension n is in bijective correspondence with the set of orbits of J_n under the natural action of $GL_{2n}(\mathbb{Z})$; in particular, this set is uncountable. The following is striking to us: the classification of complex n-tori, essentially carried out by Proposition 7 (though one can do better: see Section II.2 for the case n = 1), is easy; but to know which tori are "algebraic" is hard (see Chapter IV).

Exercise 21. Let Γ be a lattice in $\underline{C}$ and let $\omega \in \Gamma - \{0\}$ be a point as near as possible from the origin (for the usual metric on $\underline{C}$). Show that there exists a basis of Γ over $\mathbb{Z}$ which contains ω. [**Sketch:** Let $\omega' \in \Gamma - \mathbb{Z}\omega$ be as near as possible from the origin. Then $\{\omega,\omega'\}$ is a basis of Γ; otherwise, there is a point in Γ which is either in the open triangle defined by 0, ω and ω', or in the open triangle defined by 0, $-\omega$ and $-\omega'$, and in both cases this contradicts our choices for ω and ω'.]

Exercise 22. If Γ is a lattice in $\underline{C}$, denote by $\bar{\Gamma}$ the lattice $\{z \in \underline{C} \mid \bar{z} \in \underline{C}\}$ and say Γ is **real** if $\Gamma = \bar{\Gamma}$. Show that Γ is real if and only if it has either a basis consisting of two complex conjugate numbers or a basis consisting of a real number and a pure imaginary number; in the first case Γ is "rhombic", and in the second case Γ is "rectangular". [**Hint:** use Exercise 21.]

Exercise 23. Let Γ be a lattice in $\underline{C}$ which is invariant by a rotation ρ of the plane of angle θ. Show that one may take $\theta = 2\pi/n$ where $n \in \{1, 2, 3, 4, 6\}$. [**Sketch**: If $P \in \Gamma - \{0\}$, $(\rho^j(P))_{j \in N}$ must be discrete, so that $\theta = 2\pi/q$ for some rational number q. Let then $Q = \rho^k(P)$ be as near P as possible, and let σ be the rotation ρ^k; show that σ is a rotation of angle $2\pi/n$ for some $n \in \mathbb{Z} - \{0\}$. Let $P' \in \Gamma$ be obtained by a rotation of angle $2\pi/n$ round Q, and let $Q' \in \Gamma$ be obtained by a rotation of angle $2\pi/n$ round P'. If $P = Q'$ then $n = 6$. If $P \neq Q'$, then the distance PQ' is larger than or equal to PQ; this implies $n \leqslant 4$. You may read this, and much more about plane lattices, in Coxeter [32]. If there exist such a σ with $n = 4$ (or $n = 3$ or/hence $n = 6$), the lattice Γ is said to be **square**, and is then both rectangular and rhombic (or, respectively, **triangular**, and is then rhombic in many ways).]

Exercise 24. Let τ be one of the numbers $\sqrt{3}i$, i, $\exp(i\pi/3)$. Then the triangle defined by the origin, the number 1 and the number τ has angles $(\pi/2, \pi/3, \pi/6)$, $(\pi/2, \pi/4, \pi/4)$ and $(\pi/3, \pi/3, \pi/3)$ respectively. Show that, up to similarity, these are the only planar triangles having angles π/m, π/n, π/p where m, n and p are integers larger than or equal to 2. [**Comment**: These triangles are connected with regular tessellations of the plane and with the so-called Coxeter groups (see Bourbaki [5], Ch.V, §4, Exercises 4 and others; see also Serre [42], Section 1.9).]

Exercise 25. Let Γ be a lattice in $\underline{C}$ spanned by 1 and $\tau \in \underline{C}$ $(\tau \notin \underline{R})$. Let φ be a continuous endomorphism of T/Γ; then there exists $a \in \underline{C}$ with $a\Gamma \subset \Gamma$ and such that φ is induced by

$$\begin{cases} \underline{C} \to \underline{C} \\ z \mapsto az \end{cases}$$

If $a \notin \mathbb{Z}$, show that there exist $m, n, p, q \in \mathbb{Z}$ with $n \neq 0$ and $n\tau^2 + (m-q)\tau - p = 0$. [**Hint**: write $a = m + n\tau$, $a\tau = p + q\tau$. Given Γ, observe that $\{a \in \underline{C} | a\Gamma \subset \Gamma\}$ is a sub-ring of $\underline{C}$; if you like field theory, read more about that in Borel et al. [30].]

Exercise 26. A **fundamental domain** for a lattice Γ in $\underline{C}$ is a "reasonable" subset D of $\underline{C}$ (e.g. a connected Borel subset) such that the restriction of the canonical projection is a bijection from D to $\underline{C}/\Gamma$. The favourite fundamental domains in classical textbooks are parallelograms; draw others. [**Comment**: Escher can make D look like a beetle, or like two knights on horseback; see [34], or [32] if the first is not available to you.]

CLASSIFICATION OF COMPLEX TORI OF DIMENSION 1

We follow in this section Serre [7], Ch.VII, §2.

Let $\mathscr{M}$ be the set of pairs (ω_1, ω_2) of non-zero complex numbers with $\mathrm{Im}(\omega_1/\omega_2) > 0$, and let $\mathscr{L}$ be the set of all lattices in $\underline{C}$. There is an onto map from $\mathscr{M}$ to $\mathscr{L}$ which assigns to $(\omega_1, \omega_2) \in \mathscr{M}$ the lattice $\Gamma(\omega_1, \omega_2) = \mathbb{Z}\omega_1 \oplus \mathbb{Z}\omega_2$.

Let us denote by $SL_2(\mathbb{Z})$ the group of (2×2) matrices with entries in $\mathbb{Z}$ and with determinant +1. Then $SL_2(\mathbb{Z})$ acts (from the left) on $\mathscr{M}$ by

$$(*)\quad \begin{cases} SL_2(\mathbb{Z}) \times \mathscr{M} \to \mathscr{M} \\ \left(\begin{pmatrix} a & b \\ c & d \end{pmatrix}, (\omega_1, \omega_2)\right) \mapsto (a\omega_1 + b\omega_2, c\omega_1 + d\omega_2) \end{cases}$$

Indeed, if $(\omega_1, \omega_2) \in \mathscr{M}$ and if τ denotes ω_1/ω_2, then

$$\operatorname{Im}\left(\frac{a\omega_1 + b\omega_2}{c\omega_1 + d\omega_2}\right) = \operatorname{Im}\left(\frac{a\tau + b}{c\tau + d}\,\frac{c\bar{\tau} + d}{c\bar{\tau} + d}\right) = |c\tau + d|^{-2}\,(ad - bc)\operatorname{Im}(\tau)$$
$$= |c\tau + d|^{-2}\operatorname{Im}(\tau) > 0$$

so that the map (*) is well defined. We leave it to the reader to check that it is an action.

Lemma. Two elements (ω_1, ω_2) and (ω_1', ω_2') of $\mathscr{M}$ define the same lattice if and only if they are on the same orbit of $SL_2(\mathbb{Z})$.

Proof. If there exists $\begin{pmatrix} a & b \\ c & d \end{pmatrix} \in SL_2(\mathbb{Z})$ with $\omega_1' = a\omega_1 + b\omega_2$ and $\omega_2' = c\omega_1 + d\omega_2$, then clearl[y] $\Gamma(\omega_1, \omega_2) = \Gamma(\omega_1', \omega_2')$. Conversely, if the two lattices are the same, then there exists a matrix $\begin{pmatrix} a & b \\ c & d \end{pmatrix}$ with entries in $\mathbb{Z}$ such that $\omega_1' = a\omega_1 + b\omega_2$ and $\omega_2' = c\omega_1 + d\omega_2$. As the matrix must b[e] invertible over $\mathbb{Z}$, its determinant must be invertible in $\mathbb{Z}$ and so is either $+1$ or -1. The comput[ation] just above the lemma shows that the signs of

$\operatorname{Im}(\omega_1'/\omega_2')$ and of $\det\begin{pmatrix} a & b \\ c & d \end{pmatrix} \operatorname{Im}(\omega_1/\omega_2)$ are the same; the lemma follows. ■

The lemma allows us to identify $\mathscr{L}$ and $SL_2(\mathbb{Z})\backslash\mathscr{M}$.

Exercise 27. Let $\mathbb{Z}^2 = \mathbb{Z} \oplus \mathbb{Z}i \in \underline{C}$ and let $\omega_1 = m + ni \in \mathbb{Z}^2$.
Show that the following are equivalent:

(i) m and n are coprime;
(ii) There exists $\omega_2 \in \mathbb{Z}^2$ such that $\{\omega_1, \omega_2\}$ is a basis of the lattice $\mathbb{Z}^2$;
(iii) The open segment from the origin to ω_1 does not meet $\mathbb{Z}^2$;
(iv) There exist $p, q \in \mathbb{Z}$ with $\begin{pmatrix} m & n \\ p & q \end{pmatrix} \in SL_2(\mathbb{Z})$.

Generalize to $\mathbb{Z}^n \subset \mathbb{R}^n$ for $n \geqslant 2$. [**Hint**: use Bezout's theorem. **Comment**: the main point her[e] is that $\mathbb{Z}$ is a principal ring; see Godement [35], Exercise 2 in §18.]

Now let $\underline{C}^*$ operate on $\mathscr{M}$ by

$$\begin{cases} \underline{C}^* \times \mathscr{M} \to \mathscr{M} \\ (\lambda, (\omega_1, \omega_2)) \mapsto (\lambda\omega_1, \lambda\omega_2) \end{cases}$$

and on $\mathscr{L}$ by

$$\begin{cases} \underline{C}^* \times \mathscr{L} \to \mathscr{L} \\ (\lambda, \Gamma) \mapsto \lambda\Gamma \end{cases}$$

We shall identify the quotient set of $\mathscr{M}$ by the action of $\underline{C}^*$ to the **Poincaré half-plane** $P_1 = \{\tau \in \underline{C} \mid \operatorname{Im}(\tau) > 0\}$; the identification is of course that induced by

$$\begin{cases} \mathscr{M} \to P_1 \\ (\omega_1, \omega_2) \mapsto \omega_1/\omega_2 \end{cases}$$

For all $g \in SL_2(\mathbb{Z})$, $\lambda \in \underline{C}^*$ and $(\omega_1, \omega_2) \in \mathcal{M}$, one has $g(\lambda(\omega_1, \omega_2)) = \lambda(g(\omega_1, \omega_2))$; it follows that the action $(*)$ induces an action of $SL_2(\mathbb{Z})$ on $C^* \backslash \mathcal{M} = P_1$, which is clearly given by the standard expression

$$\begin{cases} SL_2(\mathbb{Z}) \times P_1 \to P_1 \\ \left(\begin{pmatrix} a & b \\ c & d \end{pmatrix}, \tau\right) \mapsto \dfrac{a\tau + b}{c\tau + d} \end{cases}$$

As the matrix $-I = \begin{pmatrix} -1 & 0 \\ 0 & -1 \end{pmatrix}$ acts as the identity on P_1, one introduces the **modular group** $PSL_2(\mathbb{Z})$, which is the quotient of $SL_2(\mathbb{Z})$ by the central subgroup with two elements $\{I, -I\}$. It is easy to check that the quotient action of $PSL_2(\mathbb{Z})$ on P_1 is effective.

We shall call the quotient set of $\mathcal{L}$ by the action of $\underline{C}^*$ the set of **isomorphism classes of lattices** in $\underline{C}$, two lattices being isomorphic if they are homothetic (this terminology is justified by the Corollary to Proposition 7). The onto map from $\mathcal{M}$ to $\mathcal{L}$ defines an onto map from P_1 to this set $\underline{C}^* \backslash \mathcal{L}$. And the lemma above says precisely that two elements in P_1 define the same isomorphism class of lattices if and only if they are on the same orbit of $PSL_2(\mathbb{Z})$.

Proposition 8. For each τ in the Poincaré half-plane P_1, let Γ_τ be the lattice $\mathbb{Z} \oplus \mathbb{Z}\tau$ in $\underline{C}$. Then

(i) Any complex torus of dimension one is isomorphic to $\underline{C}/\Gamma_\tau$ for some $\tau \in P_1$.

(ii) Two tori $\underline{C}/\Gamma_\tau$ and $\underline{C}/\Gamma_{\tau'}$ are isomorphic if and only if there exists an element in the modular group $PSL_2(\mathbb{Z})$ which transforms τ into τ'.

In other words, complex tori of dimension one are classified by the quotient set $PSL_2(\mathbb{Z}) \backslash P_1$.

Proof: follows from the Corollary to Proposition 7 and from the beginning of the present section. ■

As remarked after Exercise 20, Proposition 8 implies in particular that there are uncountably many isomorphism classes of complex tori of dimension one (indeed, $PSL_2(\mathbb{Z})$ is countable and P_1 is not). Compare with Proposition 3.

It is intentional that the word "isomorphic" in Proposition 8 is not precise: you can choose "isomorphic as complex Lie groups" or "isomorphic as complex manifolds", again because of the Corollary to Proposition 7.

Exercise 28. Show that $\{+I, -I\}$ is exactly the centre of $SL_2(\mathbb{Z})$.

Exercise 29. Let $SL_2(\underline{R})$ be the group of (2×2) matrices with entries in $\underline{R}$ and with determinant $+1$. It is a closed subset – indeed a closed submanifold – of the vector space $\underline{R}^4$ of (2×2) real matrices, and in fact a linear Lie group. [$GL_2(\underline{R})$ is an open submanifold in $\underline{R}^4$ and det: $\begin{pmatrix} a & b \\ c & d \end{pmatrix} \mapsto ad\text{-}bc$ is of constant rank on $SL_2(\underline{R})$.]

(i) Show that

$$\begin{cases} SL_2(\underline{R}) \times P_1 \to P_1 \\ \left(\begin{pmatrix} a & b \\ c & d \end{pmatrix}, \tau\right) \mapsto \dfrac{a\tau + b}{c\tau + d} \end{cases}$$

defines a smooth action by holomorphic transformations of P_1 which is almost effective and transitive.

(ii) Let B be the closed subgroup

$$\left\{\begin{pmatrix} p & q \\ 0 & 1/p \end{pmatrix} \in SL_2(\underline{R}) \,\middle|\, p > 0, q \in \underline{R}\right\} \quad \text{of } SL_2(R),$$

let $N = \left\{\begin{pmatrix} 1 & q \\ 0 & 1 \end{pmatrix} \,\middle|\, q \in \underline{R}\right\}$, let $A = \left\{\begin{pmatrix} a & 0 \\ 0 & 1/a \end{pmatrix} \,\middle|\, a > 0\right\}$,

and let $SO(2) = \left\{\begin{pmatrix} a & b \\ -b & a \end{pmatrix} \,\middle|\, a^2 + b^2 = 1\right\}$

be the special orthogonal group. Show that the multiplication

$$\begin{cases} SO(2) \times B \to SL_2(\underline{R}) \\ (k, s) \mapsto ks \end{cases}$$

is a smooth diffeomorphism. [**Hint**: To show it is onto, apply the Gram-Schmidt orthonormalizati process to the columns of a matrix in $SL_2(R)$:

$$\begin{pmatrix} a & b \\ c & d \end{pmatrix} = \begin{pmatrix} p & q \\ r & s \end{pmatrix} \begin{pmatrix} (a^2 + c^2)^{\frac{1}{2}} & * \\ 0 & (a^2 + c^2)^{-\frac{1}{2}} \end{pmatrix}$$

To show it is one-to-one: $ks = k's' \Rightarrow k'^{-1}k = s's^{-1} \in SO(2) \cap B = \{I\}$.]

(iii) Show that the multiplication map $\begin{cases} SO(2) \times A \times N \to SL_2(\underline{R}) \\ (k, a, n) \mapsto kan \end{cases}$

is a smooth diffeomorphism ("Iwasawa decomposition" for $SL_2(\underline{R})$). State and prove an analogou decomposition for $SL_n(\underline{R})$, $n > 2$.

(iv) Show that for the action defined in (i), one has $Is(i) = SO(2)$, so that P_1 can be identifie to the homogeneous space $SL_2(\underline{R})/SO(2)$. Check that $SO(2)$ is a maximal compact subgroup and also a maximal torus in $SL_2(\underline{R})$, hence that the same is true of $Is(\tau)$ for any $\tau \in P_1$.

(v) Observe that $\sigma : \begin{cases} SL_2(\underline{R}) \to SL_2(\underline{R}) \\ g \mapsto {}^t(g^{-1}) \end{cases}$, also given by $\begin{pmatrix} a & b \\ c & d \end{pmatrix} \mapsto \begin{pmatrix} 0 & 1 \\ -1 & 0 \end{pmatrix} \begin{pmatrix} a & b \\ c & d \end{pmatrix} \begin{pmatrix} 0 & 1 \\ -1 & 0 \end{pmatrix}^{-1}$,

is an automorphism of $SL_2(\underline{R})$ with $\sigma^2 = id$; what are the fixed points of σ? what is the transformation of P_1 induced by σ?

Comment: It can be shown that $SL_2(\underline{R})$ – or more precisely the quotient $PSL_2(\underline{R}) = SL_2(\underline{R})/\{+I, -I\}$ – is the group of **all** holomorphic diffeomorphisms of P_1. In particular, the latter is a real Lie group (see e.g. Cartan [8], Section VI.2.6). There are many other complex manifolds (but not all: see $\underline{C}^2$) for which the group of holomorphic diffeomorphisms is a Lie group; this is, e.g., the case of E. Cartan's "bounded domains", by a famous theorem due to H. Cartan; for an expository proof, see Narasimhan [40].

Exercise 30

(i) Let S and T be the two elements of $PSL_2(\mathbb{Z})$ defined by $\begin{pmatrix} 0 & -1 \\ 1 & 0 \end{pmatrix}$ and $\begin{pmatrix} 1 & 1 \\ 0 & 1 \end{pmatrix}$, respective Check that $S^2 = 1$ and $(ST)^3 = 1$.

(ii) Let $D = \{\tau \in P_1 \mid |\tau| \geqslant 1 \text{ and } |Re(\tau)| \leqslant \frac{1}{2}\}$. Draw a picture of the sets D, TD, $T^{-1}D$, $T^{\pm 2}D$, ..., SD, TSD, $T^{-1}SD$, $T^{\pm 2}SD$, ..., STD, $ST^{-1}D$.

(iii) For any fixed τ in P_1, show that the function

$$\begin{cases} PSL_2(\mathbb{Z}) \to \underline{R}_+ \\ g \mapsto \mathrm{Im}(g\tau) \end{cases} \quad \text{has a maximum.}$$

(iv) Let G′ be the subgroup of $PSL_2(\mathbb{Z})$ generated by S and T and let τ be an arbitrary point in P_1. Show that there exists $g \in G'$ such that $g\tau \in D$.

(v) What is the set $\{\tau \in D \mid \text{there exists } g \in PSL_2(\mathbb{Z}) \text{ with } g \neq 1 \text{ and } g\tau \in D\}$?

(vi) For each $\tau \in D$, what is the subgroup $G_\tau = \{g \in PSL_2(\mathbb{Z}) | g\tau = \tau\}$?

(vii) Show that $PSL_2(\mathbb{Z}) = G'$.

The **solution** is e.g. in Serre [7], Ch.VII, § 1. It is known that $PSL_2(\mathbb{Z})$ is the free product $\mathbb{Z}_2 * \mathbb{Z}_3$ where $\mathbb{Z}_2$ is generated by S and $\mathbb{Z}_3$ by ST; see Serre [43] for a geometric proof, or Appendix B to Kurosh [38] for an algebraic proof.

We have shown (Propositions 7 and 8, Exercise 29 (iv)) that the double coset space $PSL_2(\mathbb{Z})\backslash SL_2(\underline{R})/SO(2)$ is in natural bijection with the following sets:

1. Lattices in $\underline{C}$ (identified when they are homothetic).
2. Compact complex connected Lie groups of dimension one, i.e. complex tori of dimension one (identified when they are isomorphic as complex Lie groups).
3. Riemann surfaces of genus one, with base point (identified when they are holomorphically equivalent).

We shall make it plausible in the next section that this list can be completed by:

4. Non-singular plane cubics of equation $y^2 = x^3 + ax + b$. For more details, see Ch. I in Robert [6].
5. One could also add to this list an important item on positive definite binary quadratic forms with real coefficients. See A. Weil: exposé 1 in [41] and/or Borel [29], No.2.3.
6. A **flat 2-torus** is a 2-dimensional real torus T together with the Riemannian structure induced by a scalar product on the universal covering $\underline{R}^2$ of T. The study of flat 2-tori, including their classification, has much in common with that of complex 1-tori. See Berger et al. [27].

3. ONE-DIMENSIONAL TORI AS NON-SINGULAR CUBICS

In this section, we translate into our context some standard material from the theory of elliptic functions. Proofs can be found in numerous classical textbooks, from which we quote:

H. Cartan [8], Sections V.2.5 and VI.5.3
A.I. Markushevich [9], Ch.5 of Vol. III
A. Saks and A. Zygmund [10], Ch. VIII
E.T. Whittaker and G.N. Watson [12], Ch. XX

or in more modern language, two Lecture Notes which are among our main sources:

A. Robert [6], Ch.I
H.P.F. Swinnerton-Dyer [11], Section I.2

Let τ be a point in the Poincaré half-plane P_1; let $L_\tau = \mathbb{Z} \oplus \mathbb{Z}_\tau$ and $T_\tau = \underline{C}/L_\tau$ be the corresponding lattice and torus respectively. A meromorphic function on T_τ can be considered (or actually defined) as a L_τ-**elliptic function,** i.e. as a meromorphic function f on $\underline{C}$ such that $f(z + \omega) = f(z)$ for all z in $\underline{C}$ and ω in L_τ.

It is shown that the expression

$$\mathscr{P}_\tau(z) = \frac{1}{z^2} + \sum_{\omega \in L_\tau - \{0\}} \left(\frac{1}{(z-\omega)^2} - \frac{1}{\omega^2} \right)$$

[For technical reasons $\mathscr{P}$ is used instead of a gothic p.]

defines a L_τ-elliptic function known as the **Weierstrass function** corresponding to L_τ. (The proof that $\mathscr{P}_\tau$ is meromorphic is an easy corollary of Exercise 4, where it is sufficient to consider $n = 2$ and $s = 3$; the proof that $\mathscr{P}_\tau$ is L_τ-invariant is best seen by first looking at the derivative $\mathscr{P}'_\tau$ of $\mathscr{P}_\tau$.) If the point τ in P_1 is understood from the context and fixed, we also write L instead of L_τ and $\mathscr{P}$ instead of $\mathscr{P}_\tau$.

Exercise 31. Let Γ be a lattice in $\underline{C}$ with basis $\{\omega_1, \omega_2\}$. A **theta-function** for Γ is a function θ (holomorphic here, as in Ref.[11] or [14], but notice it may be meromorphic in Ref.[6]) such that

$$(*) \quad \begin{aligned} \theta(z + \omega_1) &= \theta(z)\exp(a_1 z + b_1) \\ \theta(z + \omega_2) &= \theta(z)\exp(a_2 z + b_2) \end{aligned} \qquad \text{for all } z \in \underline{C}$$

where a_1, b_1, a_2, b_2 are complex numbers.

(i) Show that $a_2\omega_1 - a_1\omega_2 = -i2\pi n$ for some $n \in \mathbb{Z}$.

(ii) Check that $z \mapsto \exp(az^2 + bz + c)$ is a theta-function for any triple (a, b, c) of complex numbers. Show that any theta-function without zeros is of this form.

(iii) If θ satisfies (*), show that

$$\psi(z) = \theta(z) \exp\left\{ -\frac{a_1 z(z-\omega_1)}{2\omega_1} - \frac{b_1 z}{\omega_1} \right\}$$

satisfies

$$\begin{aligned} (**) \quad & \psi(z + \omega_1) = \psi(z) \\ (***) \quad & \psi(z + \omega_2) = \psi(z)\exp(-i2\pi n \frac{z}{\omega_1} + b) \end{aligned} \qquad \text{for all } z \in \underline{C}$$

where n is as in (i) and where b is some complex number.

(iv) From (**), write

$$\psi(z) = \sum_{k \in \mathbb{Z}} c_k \exp(i2\pi k \frac{z}{\omega_1})$$

and show formally that (***) implies $c_{k+n} = c_k \exp(i2\pi k(\omega_2/\omega_1) - b)$; if $\text{Im}(\omega_2/\omega_1) > 0$ and if $n > 0$, observe that the sequence $c_k, c_{k+n}, c_{k+2n}, \ldots$ is fast decreasing.

Comment: Of course, the quotient of two theta-functions satisfying (*) with the same a's and b's is an elliptic function. Starting from the formal identity of (iv), one can prove the existence of non-trivial elliptic functions without using $\wp$. In higher dimensions, there are analogues of theta-functions, but not of the Weierstrass function.

Exercise 32. Notations being as in Exercise 31, suppose $\mathrm{Im}(\omega_2/\omega_1) > 0$, and consider $\underline{C}$ as $\underline{R}^2$ with the basis $\{\omega_1, \omega_2\}$. Define

$$A: \begin{cases} \underline{R}^2 \times \underline{R}^2 \longrightarrow \underline{R} \\ (x_1\omega_1 + x_2\omega_2, y_1\omega_1 + y_2\omega_2) \mapsto \det\begin{pmatrix} x_1 & x_2 \\ y_1 & y_2 \end{pmatrix} \end{cases}$$

Check that $A(\omega_1', \omega_2') = \pm 1$ if and only if $\{\omega_1', \omega_2'\}$ is a basis of Γ, and that, moreover, the sign + holds if and only if $\mathrm{Im}(\omega_2'/\omega_1') > 0$; in other words, A depends on Γ in $\underline{C}$ but not on the choice of any particular basis. Now both A and

$$\begin{cases} \underline{C} \times \underline{C} \to \underline{R} \\ (x, y) \mapsto \mathrm{Im}(\bar{x}y) \end{cases}$$

are $\underline{R}$-bilinear alternate non-zero forms on $\underline{R}^2$, so that there exists $\lambda \in \underline{R}^*$ with $A(x,y) = (1/\lambda)\mathrm{Im}(\bar{x}y)$ for all $x,y \in \underline{C}$; give a geometrical meaning to the number λ. [**Hint:** $\lambda = \mathrm{Im}(\bar{\omega}_1\omega_2) = |\omega_1|^2\,\mathrm{Im}(\omega_2/\omega_1)$ = Lebesgue area of a domain in $\underline{R}^2$.] Finally, show that $(x, y) \mapsto A(x,iy)$ defines on $\underline{C}$ an $\underline{R}$-bilinear form which is symmetric positive definite.

Comment: To sum up the outcome of Exercise 32:

(i) $A(x,y) = -A(y,x)$ for all $x,y \in \underline{C}$

(ii) $A(\Gamma, \Gamma) \subset \mathbb{Z}$

(iii) $(x,y) \mapsto A(x,iy)$ is a scalar product on $\underline{R}^2$.

It is hoped that, after Chapter IV, the reason will be clear for placing this exercise just after asserting the existence of non-constant elliptic functions, and not with other exercises on lattices.

We recall now some facts about the Weierstrass function.

List of some properties of $\wp$ and of its derivative $\wp'$

(i) $\wp$ and $\wp'$ are meromorphic functions on $\underline{C}$.

(ii) $\wp$ has a double pole at each point of L, and no other pole.

(iii) $\wp$ is an even function: $\wp(-z) = \wp(z)$ for all $z \in \underline{C}$ (this means in particular: $-z$ is a pole if and only if z is a pole).

(iv) $\wp'$ has a triple pole at each point of L, and no other pole.

(v) $\wp'$ is an odd function: $\wp'(-z) = -\wp'(z)$ for all $z \in \underline{C}$.

(vi) $\wp$ satisfies the differential equation $(\wp')^2 = 4\wp^3 - g_2\wp - g_3$ where $g_2 = 60\,\Sigma'\omega^{-4}$ and $g_3 = 140\,\Sigma'\omega^{-6}$ (see Exercise 4).

(vii) $(g_2)^3 - 27(g_3)^2 \neq 0$.

(viii) $\wp'$ has simple zeros at points of $\underline{C}$ which are congruent modulo L to $\frac{1}{2}, \frac{1}{2}\tau$ and $\frac{1}{2}(1+\tau)$, and no other zero.

(ix) For each $A \in \underline{C} \cup \{\infty\}$, the equation $\wp(z) = A$ has exactly two solutions (counting multiplicities) in any "period parallelogram", e.g. in

$$D = \{z \in \underline{C} \mid z = a + b\tau \text{ with a,b real and } -\tfrac{1}{2} \leqslant a,b < \tfrac{1}{2}\}$$

(x) The two solutions of $\wp(z) = A$ coincide only if either $A = \infty$ ($z \in L$ is a double pole) or $A \in \{\wp(\frac{1}{2}), \wp(\frac{1}{2}\tau), \wp(\frac{1}{2}(1+\tau))\}$ (such z's are zeros of $\wp'$).

Exercise 33. Show that $z \mapsto \exp(\mathscr{P}_\tau(z))$ defines a L_τ-periodic function which is *not* L_τ-elliptic, because it is not meromorphic. [Sketch: Near $z = 0$ one has $\exp(\mathscr{P}_\tau(z)) = \exp(z^{-2} + f(z)) = F(z)\exp(z^{-2})$, where F is holomorphic but where $\exp(z^{-2})$ has an essential singularity.]

Define E_0 to be the complex curve

$$\{(x, y) \in \underline{C}^2 \mid y^2 = 4x^3 - g_2 x - g_3\}$$

and E to be

$$\{[X, Y, Z] \in P_2(\underline{C}) \mid Y^2 Z = 4X^3 - g_2 XZ^2 - g_3 Z^3\}$$

Lemma. E is a complex submanifold of $P_2(\underline{C})$ (i.e. an algebraic subvariety of $P_2(\underline{C})$ *without* singularity).

Proof. Notation is as in Robertson's lectures (these Proceedings):

$$U_1 = \{[X, Y, Z] \in P_2(\underline{C}) \mid X \neq 0\}$$

$\varphi_1: U_1 \to \underline{C}^2$ is given by $[X, Y, Z] \mapsto \left(\frac{Y}{X}, \frac{Z}{X}\right)$; idem for φ_2 and φ_3.

$$\text{Let } f: \begin{cases} \underline{C}^2 \to \underline{C} \\ (x, y) \mapsto y^2 - 4x^3 + g_2 x + g_3 \end{cases}$$

Then clearly,

$$\varphi_3(U_3 \cap E) = E_0 = \{(x, y) \in \underline{C}^2 \mid f(x, y) = 0\}$$

Let us first check that E_0 is a submanifold of $\underline{C}^2$ by showing that $\partial f/\partial x$ and $\partial f/\partial y$ have no common zero on E_0.

Suppose $\partial f/\partial x = -12x^2 + g_2 = 0$ and $\partial f/\partial y = 2y = 0$. Let s be one of the square roots of g_2. Then $x = \pm s/2\sqrt{3}$ and $y = 0$. Suppose this point is on E_0: then $f(\pm s/2\sqrt{3}, 0) = (-4x^2 + g_2)x + g_3 = 2g_2/3\,(\pm s/2\sqrt{3}) + g_3 = 0$ so that $\pm g_2 s = -3\sqrt{3}\, g_3$, which contradicts property (vii) above. This proves our claim about E_0.

Now the "points at ∞ on E" are points [X, Y, 0] satisfying the equation of E; the latter gives $0 = 4X^3$, so that E has only one point at ∞: [0, 1, 0]. Call it P, so that $E = E_0 \cup \{P\}$. The chart

$$\varphi_2: \begin{cases} U_2 \to \underline{C}^2 \\ [X, Y, Z] \mapsto \left(x' = \frac{X}{Y}, \quad z' = \frac{Z}{Y}\right) \end{cases}$$

maps $E \cap U_2$ onto $\{(x', z') \in \underline{C}^2 \mid f'(x', z') = 0\}$ where $f'(x', z') = -z' + 4x'^3 - g_2 x' z'^2 - g_3 z'^3$. It is easily seen that $f'(\varphi_2(P)) = f'(0, 0) = 0$ and that $(\partial f'/\partial z')(0, 0) = 1 \neq 0$. By the implicit function theorem (see e.g. Dieudonné [33], Nos. 10.2.1 and 10.2.4), this implies that there is a ma

$$\begin{cases} \underline{C} \to \underline{C} \\ x' \mapsto z' = \psi(x') \end{cases}$$

holomorphic in a neighbourhood of zero and with $f'(x', \psi(x')) = 0$ in this neighbourhood. It follows that $x' \mapsto (x', \psi(x'))$ defines a local co-ordinate on $\varphi(E \cap U_2)$ around $\varphi(P)$, i.e. a local co-ordinate on E around P.

We leave it to the reader to check compatibility conditions. Up to this, we have proved the lemma. ■

Exercise 34 = Second proof of the lemma (actually the same, but looked at differently). Let F be the function from $\underline{C}^3$ to $\underline{C}$ given by $(X, Y, Z) \mapsto Y^2Z - 4X^3 + g_2XZ^2 + g_3Z^3$ and let $\mathscr{E} = \{(X, Y, Z) \in \underline{C}^3 - \{0\} \mid F(X, Y, Z) = 0\}$. Show that $\partial F/\partial X$, $\partial F/\partial Y$ and $\partial F/\partial Z$ have no common zero on $\underline{C}^3$ outside (0, 0, 0). Hence $\mathscr{E}$ is a conical complex submanifold of $\underline{C}^3 - \{0\}$. If $\pi: \underline{C}^3 - \{0\} \to P_2(\underline{C})$ is the canonical projection, this implies that $E = \pi(\mathscr{E})$ is an algebraic subvariety of $P_2(\underline{C})$ without singularity. (This last implication is proved, e.g. in Auslander and Mackenzie [26], §3.4.)

Now let

$$\tilde{\varphi}_0: \begin{cases} \underline{C} - L \to \underline{C}^2 \\ z \mapsto (\mathscr{P}(z), \mathscr{P}'(z)) \end{cases} \quad \text{and let}$$

$\varphi_0: T - \text{id} \to \underline{C}^2$ be the quotient map (recall it is understood that a point τ has been chosen in the Poincaré half-plane P_1). Then φ_0 has the following properties:

- It is **injective**, by properties (ix), (iii), (viii) and (x) above. Indeed: Either $z \not\equiv -z(\text{mod.}L)$; then $\mathscr{P}(w) = \mathscr{P}(z)$ and $w \not\equiv z(\text{mod.}L)$ imply $w \equiv -z(\text{mod.}L)$, so that $\tilde{\varphi}_0(-z) \neq \tilde{\varphi}_0(z)$ by (viii). Or $z \equiv -z(\text{mod.}L)$; then $\mathscr{P}(w) = \mathscr{P}(z)$ implies $w \equiv z(\text{mod.}L)$ by (x).
- It maps $\underline{C} - L$ to E_0, because of (vi) and the equation of E_0.
- It is **onto** E_0, by (ix).
- It is **holomorphic** by (i), (ii) and (iv).
- Its derivative **never vanishes**, by (viii).

It follows that φ_0 has a holomorphic inverse, so that $T - \text{id}$ and E_0 are isomorphic as complex manifolds.

Extend $\tilde{\varphi}_0$ to $\tilde{\varphi}: \underline{C} \to P_2(\underline{C})$ by defining $\tilde{\varphi}(L) = P$ (that point P defined in the proof of the above lemma; this is coherent with $\mathscr{P}(z) \sim z^{-2}$ and $\mathscr{P}'(z) \sim z^{-3}$ if $z \sim 0$). Let $\varphi: T \to P_2(\underline{C})$ be the quotient map. Then φ is holomorphic at $\text{id} \in T$ (this is merely to say that $\mathscr{P}$ and $\mathscr{P}'$ are meromorphic), and φ defines a holomorphic bijection from T to E which clearly has a holomorphic inverse.

Proposition 9. Let T be a complex torus of dimension one. Then there exist an algebraic submanifold E of $P_2(\underline{C})$ of degree 3 and an analytic isomorphism $\varphi: T \to E$. Moreover, any holomorphic map $T \to T$ is algebraic. In other words, T is an algebraic group in a natural way.

Proof. The first statement was shown above, and the second follows from Proposition 7. ■

Corollary. Meromorphic functions separate points on one-dimensional complex tori.

Proof. They do on $P_2(\underline{C})$. ■

Comments

(i) The second statement in Proposition 9 holds for any complex submanifold of $P_2(\underline{C})$; see Shafarevich [13], Ch. VIII, §3, Theorem 2.

(ii) A holomorphic map $T \to P_1(\underline{C})$ cannot be injective because there is no elliptic function of order 1. In this sense, Proposition 9 is the best possible.

(iii) It is *not* true that Proposition 9 carries over to higher dimensions; see Chapter IV.

(iv) It is known that any non-singular cubic in $P_2(\underline{C})$ is analytically isomorphic to a complex torus; see e.g. Robert [6].

(v) The terms "abelian curve" and "complex torus of dimension 1" may be used one for the other.

Exercise 35. Notation is as before Proposition 9. Let u, $v \in \underline{C}^2$ with $u \not\equiv v (\text{mod.} L)$. Then $\widetilde{\varphi}(u) = [u^3 \wp(u), u^3 \wp'(u), u^3] \in E$ and $\widetilde{\varphi}(v) = [v^3 \wp(v), v^3 \wp'(v), v^3] \in E$ span a line in $P_2(\underline{C})$, say $l(u,v)$. Let $w \in \underline{C}^2$ be the point defined by $u + v + w = 0$. A standard property of the Weierstrass function is that

$$\det \begin{pmatrix} u^3 \wp(u) & u^3 \wp'(u) & u^3 \\ v^3 \wp(v) & v^3 \wp'(v) & v^3 \\ w^3 \wp(w) & w^3 \wp'(w) & w^3 \end{pmatrix} = 0$$

Show that this is equivalent to the statement that $\widetilde{\varphi}(w) \in l(u,v)$. [Conversely, this may be used to define an abelian multiplication on any non-singular cubic; see comment (iv) above; see also Point 4 at the end of Section II.2.]

Exercise 36. Let Γ be a lattice in $\underline{C}$ and for each integer $n \geqslant 3$ let $S_n = \Sigma' \omega^{-n}$ (see Exercise 4).

(i) Check that $S_n = 0$ if n is odd.

(ii) When Γ is a square lattice (Exercise 23), check that $S_n = 0$ if $n \not\equiv 0$ (mod 4); in particular $g_3 = 0$ in this case.

(iii) When Γ is a triangular lattice, check that $S_n = 0$ if $n \not\equiv 0$ (mod.6); in particular $g_2 = 0$ in this case.

(iv) Show that Γ is real (Exercise 22) if and only if g_2 and g_3 are both real.

(v) Conversely to (ii) and (iii), show that Γ is square [rectangular] as soon as $g_3 = 0$ [$g_2 = 0$].

[**Hint:** $S_n(a\Gamma) = a^n S(\Gamma)$ and $S_n(\overline{\Gamma}) = \overline{S_n(\Gamma)}$ for any lattice. The "if" part in (iv) and question (v) are the only non-trivial points in this exercise; see e.g. du Val [44], Section 22, and Robert [6], Nos. 3.11 and 3.16 of Ch.I. **Question:** can you characterize as in (ii) and (iii) the third isomorphis class of lattices singled out in Exercise 24?

Note: We are grateful to A. Robert, who has indicated to us the good reference for the question of Exercise 36. Let $j = 2^6\ 3^3\ g_2^3 (g_2^3 - 27 g_3^2)^{-1}$, which is a function of τ (or of Γ). If Γ is square, $j = 2^6 3^3$. If Γ is triangular $j = 0$. If Γ is the "third class", $j = 2^4 3^3 5^3$. See J.P. Serre, "Complex multiplication", pp 292–296 of "Algebraic Number Theory", edited by J.W.S. Cassels and A. Fröhlich, Academic Press (1967).

Another point is worth mentioning: let Γ be a lattice, let $\text{End}(\Gamma) = \{a \in | a\Gamma \subset \Gamma\}$ and let $\text{Aut}(\Gamma) = \{a \in \text{End}(\Gamma) | a^{-1} \in \text{End}(\Gamma)\}$ (see Exercise 25). If Γ is square or triangular it is easy to check that $\text{Aut}(\Gamma) \supsetneqq \{+1, -1\}$. If $\Gamma = \Gamma_\tau$ with $\tau = i\sqrt{3}$, then $\text{Aut}(\Gamma) = \{+1, -1\} \subsetneqq \text{End}(\Gamma)$.]

Chapter III
THE GROUP OF PERIODS OF A MEROMORPHIC FUNCTION

1. MEROMORPHIC FUNCTIONS ON $\underline{C}^n$

In this section, by [F] we refer to Field's paper in these Proceedings.

Let n be a fixed integer ($n \geqslant 1$); a point in $\underline{C}^n$ will be denoted either by z or by $(z_1, ..., z_n)$. We denote by $A(\underline{C}^n)$ the ring of analytic functions on $\underline{C}^n$. It is a commutative ring with identity, written 1; it is also an integral domain (see [F], Prop. 2.2), so that the notion of a **unit** in $A(\underline{C}^n)$ makes sense.

Exercise 37. Let $f \in A(\underline{C}^n)$. Show that the following properties are equivalent:
  (i) f is a unit (it means: there exists $g \in A(\underline{C}^n)$ such that $f(z)g(z) = 1$ for all $z \in \underline{C}^n$, i.e. such that $fg = 1$);
  (ii) f has no zero on $\underline{C}^n$;
  (iii) There exists $h \in A(\underline{C}^n)$ with $f(z) = \exp(h(z))$ for all $z \in C^n$.
[Hint for (ii) $\Rightarrow$ (iii): observe that $\underline{C}^n$ is simply connected, so that the lifting problem

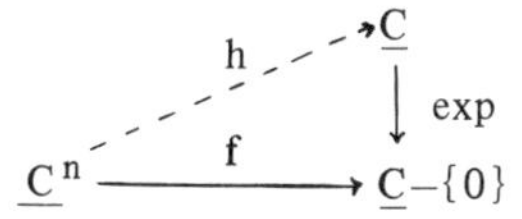

has always a solution.]

Given f and g in $A(\underline{C}^n)$, then g is said to **divide** f if there exists $h \in A(\underline{C}^n)$ with $f = gh$. Given f in $A(\underline{C}^n)$, not zero and not a unit, it is clear that any unit u in $A(\underline{C}^n)$ divides f, and so does any product uf with u a unit; if these are the only elements in $A(\underline{C}^n)$ which divide f, then f is **irreducible**.

Remark on "irreducible" and "prime". Let A be an integral domain with unit $1 \neq 0$. Let $f \in A$ and let (f) be the principal ideal generated by f.
  If (f) is prime, then f is irreducible. Indeed, suppose $f = gh$; then either $g \in (f)$, or $h \in (f)$ (or both); choose the notation such that $g \in (f)$, i.e. such that $g = fk$ for some $k \in A$. Then $f = fkh$; it follows that $kh = 1$, i.e. that h is a unit.
  Suppose, moreover, that A is a unique factorization domain. If f is irreducible, then (f) is clearly prime. This is why "irreducible" and "prime" have the same meaning in [F] Ch.II, which deals with rings like $\mathcal{O}_a$.
  Now the ring $A(\underline{C}^n)$ is *not* a unique factorization domain. Indeed, let $n = 1$ and let f be the function $z \mapsto \sin(2\pi z)$. For any finite subset K of the integers $\mathbb{Z}$, the polynomial

$$\prod_{a \in K} (z - a)$$

divides f. The assertion follows now from the definition (see [F], Def. 2.6).
  It is however a fact that, if f in $A(\underline{C}^n)$ is irreducible, then (f) is prime. But there exists a connected open subset U of $\underline{C}^3$ for which there exists $f \in A(U)$, f irreducible and (f) not prime; see Whitney [46], Section I.13.
  Given f and g in $A(\underline{C}^n)$, it is clear that any unit in $A(\underline{C}^n)$ divides both f and g; if these are the only elements in $A(\underline{C}^n)$ with this property, then f and g are **coprime**, written $(f,g) = 1$. (Though $f,g) = 1$ is not defined here as after Proposition 2.13 of [F], the meaning is the same because of the "nice" properties of $\underline{C}^n$.)

Theorem. Let $\widetilde{f}, \widetilde{g} \in A(\underline{C}^n)$. Then there exist f, g, h $\in A(\underline{C}^n)$ with $\widetilde{f} = hf$, $\widetilde{g} = hg$ and $(f, g) = 1$.

The proof goes beyond our project; see e.g. Gunning and Rossi [15], Section VIII.B, Prop.13. ∎

  Now one could define a **meromorphic function** on $\underline{C}^n$ to be an element of the quotient field of $A(\underline{C}^n)$. The theorem above would then imply that any meromorphic function on $\underline{C}^n$ can be written as a quotient g/h with g, $h \in A(\underline{C}^n)$ and $(g, h) = 1$. But this gives little geometrical idea and, worse, cannot carry over to many other manifolds.
  Meromorphic functions on $\underline{C}^n$ are rather defined by **local** requirements: as an assignment to each $z \in \underline{C}^n$ of a germ in the quotient field M_z of the ring $\mathcal{O}_z$, the various germs satisfying

compatibility conditions discussed in [F] Ch.II. We shall call set of **regular points** of f and denote by $\delta(f)$ the set of those points z in $\underline{C}^n$ to which f assigns a germ in $\mathcal{O}_z$; otherwise express and with the notation of [F], the space $\underline{C}^n$ is the disjoint union of $\delta(f)$, the pole set P(f) and the indeterminancy set T(f). The geometrical meaning of $\delta(f)$ is that it is the largest subset of $\underline{C}^n$ on which f defines a complex-valued continuous function.

After the **local** definition, which could be repeated for any complex manifold, comes the **global** result, which applies only to some of them.

Theorem. Let f be a meromorphic function on $\underline{C}^n$; then there exist:

(i) A function $q \in A(\underline{C}^n)$ with $\delta(f) = \{z \in \underline{C}^n | q(z) \neq 0\}$.

(ii) A function $p \in A(\underline{C}^n)$ with $(p,q) = 1$ and $f(z) = p(z)/q(z)$ for all $z \in \delta(f)$.

Moreover, if q' and p' have the same properties, then there exists a unit $u \in A(\underline{C}^n)$ with q'= up and p'= up.

See Ref. [15] again for the proof. ∎

We shall denote by $M(\underline{C}^n)$ the field of meromorphic functions on $\underline{C}^n$. The above theorem says in particular that $M(\underline{C}^n)$ is indeed the quotient field of $A(\underline{C}^n)$.

If f and g are in $M(\underline{C}^n)$, we shall write freely "f(z) = g(z) for all $z \in \underline{C}^n$" instead of "f = g". This can be read as:

(1) $\delta(f) = \delta(g)$

(2) $f(z) = g(z)$ for all $z \in \delta(f)$

(3) $P(f) = P(g)$

(4) $T(f) = T(g)$

and there is no trouble coming from expressions such as "$\infty = \infty$" or "0/0 = 0/0".

Exercise 38. Write down two meromorphic functions f and g for which $\delta(f + g) \supsetneq \delta(f) \cup \delta(g)$; same question with $\delta(fg) \supsetneq \delta(f) \cup \delta(g)$.

Exercise 39. What are the zeros, the poles and the indeterminancy set of $f \in M(\underline{C}^2)$ defined by $f(z) = z_1/z_2$? if $\gamma:\underline{R} \to \underline{C}^2$ is a continuous map with $\gamma(0) = (0,0)$, what can you say about limit(of $f(\gamma(t))$ when t tends towards zero? Same questions if f is defined by $f(z) = (z_1^3 + z_2^2)^{-1}$. Draw pictures.

2.   GROUPS OF PERIODS

Let f be a meromorphic function on $\underline{C}^n$. A **period** of f is a vector $\omega \in \underline{C}^n$ such that $f(z + \omega) = f(z)$ for all $\underline{z} \in \underline{C}^n$. The **period group** of f is the set G(f) of all periods of f, which is clearly a subgroup of the additive group $\underline{C}^n \approx \underline{R}^{2n}$.

Lemma. For any $f \in M(\underline{C}^n)$, the group G(f) is closed in $\underline{C}^n$.

Proof. Let $(\omega_j)_{j \in N}$ be a Cauchy sequence in G(f); suppose it has a limit in the space C^n, say ω. Choose $z_0 \in \underline{C}^n$ with $z_0 + \omega$ a regular point for f. By continuity, $z_0 + \omega_j$ is also a regular point for f, for large enough j. Then

$$f(z_0 + \omega) = f(z_0 + \lim_{j \to \infty} \omega_j) = \lim_{j \to \infty} f(z_0 + \omega_j) = \lim_{j \to \infty} f(z_0) = f(z_0)$$

by continuity of f at $z_0 + \omega$. Similarly $f(z + \omega) = f(z)$ for z near enough z_0, which implies that it is true for all z in $\underline{C}^n$ by uniqueness of analytic continuation. The lemma follows. ∎

Notation. Let Q be an invertible $\underline{C}$-linear map from $\underline{C}^n$ to $\underline{C}^n$ and let $f \in M(\underline{C}^n)$. We denote by ^Qf the function defined by $^Qf(z) = f(Q^{-1}z)$. If Q' and Q'' are two such maps, one checks that

$({}^{Q'Q''})f = {}^{Q'}({}^{Q''}f)$. [In other words, $(Q,f) \mapsto {}^{Q}f$ defines a linear representation of $GL_n(\underline{C})$ in the vector space $M(\underline{C}^n)$.]

Proposition 10. Let $f \in M(\underline{C}^n)$. Then the following are equivalent:

(1) There exist an integer $q < n$, a $\underline{C}$-linear map $\pi: \underline{C}^n \to \underline{C}^q$ and a meromorphic function $\tilde{f} \in M(C^q)$ such that $f = \tilde{f} \cdot \pi$.

(2) There exists $Q \in GL_n(\underline{C})$ such that ${}^{Q}f$ depends on $z_1, ..., z_{n-1}$ only and not on z_n.

(3) $G(f)$ is not discrete.

(4) f is solution of a partial differential equation

$$\sum_{j=1}^{n} a_j \frac{\partial f}{\partial z_j} = 0$$

where the a_j's are constants, not all zero.

Proof. (1) ⇔ (2) is an elementary exercise of linear algebra. (2) ⇒ (3): If (2) holds, then $V = \{(0, ..., 0, z) \in \underline{C}^n | z \in \underline{C}\}$ is in $G({}^{Q}f)$, so that $Q^{-1}(V)$ is a non-trivial subspace of $\underline{C}^n$ which is in $G(f)$.

(3) ⇒ (4): If (3) holds, it follows from the lemma above and from Proposition 2 that there exists $a = (a_1, ..., a_n) \in \underline{C}^n - \{0\}$ with $\underline{R}a \subset G(f)$. If z_0 is a regular point of f, there is a neighbourhood U of the origin in $\underline{C}$ such that the function

$$\begin{cases} U \to \underline{C} \\ z \mapsto f(z_0 + za) - f(z_0) \end{cases}$$

is holomorphic and vanishes on $U \cap \underline{R}$, hence on all of U. It follows that $\underline{C}a \subset G(f)$ and that the derivative of f in the direction of a vanishes, so that (4) holds.

(4) ⇒ (2): If (4) holds, there exists a matrix $Q \in GL_n(\underline{C})$ with

$$Q^{-1} = \begin{pmatrix} * & \cdots & * & a_1 \\ \vdots & & \vdots & \vdots \\ * & \cdots & * & a_n \end{pmatrix}$$

Write $w = Q^{-1}z$, so that ${}^{Q}f(z) = f(w)$. Then

$$\frac{\partial\, {}^{Q}f}{\partial z_n} = \sum_{j=1}^{n} \frac{\partial f}{\partial w_j} \frac{\partial w_j}{\partial z_n} = \sum_{j=1}^{n} \frac{\partial f}{\partial w_j} a_j = 0$$

and ${}^{Q}f$ does not depend on z_n. ■

Definition. A meromorphic function satisfying the equivalent conditions of Proposition 10 is said to be **degenerate**. Other functions in $M(\underline{C}^n)$ are **non-degenerate**.

Corollary 1. Let f be in $M(\underline{C}^n)$ and suppose f depends indeed on n variables, i.e. satisfies the negation of (2) above. Then:

(i) $G(f)$ is a discrete subgroup of $\underline{C}^n$.

(ii) If $\omega_1, ..., \omega_k$ are in $G(f)$, then they are linearly independent over $\underline{R}$ [= **reel unabhängig** in Conforto [16] = independent vectors in Siegel [17]] if and only if they are linearly independent over $\mathbb{Z}$ [= unabhängig and strongly independent, respectively].

(iii) $G(f)$ is isomorphic to $\mathbb{Z}^k$ for some integer k with $0 \leqslant k \leqslant 2n$. In particular, there exist $\omega_1, ..., \omega_k$ in $G(f)$ which generate $G(f)$.

Proof. Clear from Propositions 1 and 10. ∎

Definition. An **abelian function** on $\underline{C}^n$ is a non-degenerate meromorphic function on $\underline{C}^n$ which has 2n linearly independent periods. If f is an abelian function, we write $\Gamma(f)$ instead of $G(f)$ and call it the **lattice of periods** of f; a primitive system of periods of f is a basis of $\Gamma(f)$ over $\mathbb{Z}$. An abelian function f can be thought of as a non-degenerate meromorphic function on the comple torus $\underline{C}^n/\Gamma(f)$.

Example. Let $\tau_1, ..., \tau_n$ be points in the Poincaré half-plane P_1. Define

$$\begin{array}{ll} \omega_1 = (1, 0, ..., 0) & \omega_{n+1} = (\tau_1, 0, ..., 0) \\ \omega_2 = (0, 1, 0, ..., 0) & \omega_{n+2} = (0, \tau_2, 0, ..., 0) \\ \vdots & \vdots \\ \omega_n = (0, ..., 0, 1) & \omega_{2n} = (0, ..., 0, \tau_n) \end{array}$$

and let Γ be the lattice in $\underline{C}^n$ generated by these. With the notation of Section II.3, the assignmen

$$(z_1, z_2, ..., z_n) \mapsto \mathscr{P}_{\tau_1}(z_1)\, \mathscr{P}_{\tau_2}(z_2) \dots \mathscr{P}_{\tau_n}(z_n)$$

defines an abelian function on $\underline{C}^n$ with $\Gamma(f) = \Gamma$. (Check it is not degenerate.)

This is a very simple case, because it shows "separation of variables". If Γ is an arbitrary lattice in $\underline{C}^n$, we shall see in Chapter IV that there may not exist any non-constant meromorphic function f with $\Gamma \subset G(f)$.

Corollary 2. Let f be a meromorphic function on $\underline{C}^2$; then $G(f)$ is isomorphic to one of the following groups:

$\{0\}, \mathbb{Z}, \mathbb{Z}^2, \mathbb{Z}^3, \mathbb{Z}^4$	(when f is non-degenerate)
$\underline{C}, \underline{C} \oplus \mathbb{Z}, \underline{C} \oplus \mathbb{Z}^2$	(when f is degenerate non-constant)
$\underline{C}^2$	(when f is a constant).

Proof. Clear from Propositions 2 and 10. ∎

Exercise 40. Write down nine meromorphic functions on $\underline{C}^2$ so that the nine corresponding perio groups are pairwise not isomorphic.

Caution: some authors do not ask for an abelian function to be non-degenerate.

3. STATEMENT OF A FUNDAMENTAL REDUCTION THEOREM

Let Γ be a discrete subgroup of $\underline{C}^n$ and let $f \in M(\underline{C}^n)$ be such that $\Gamma \subset G(f)$ (we allow degenerate functions). From the existence result quoted in the second theorem of Section III.1, there exist two coprime functions $p,q \in A(\underline{C}^n)$ such that $f = p/q$. For each $\omega \in \Gamma$, write now p_ω and q_ω, the functions defined by

$$\left.\begin{array}{l} p_\omega(z) = p(z + \omega) \\ q_\omega(z) = q(z + \omega) \end{array}\right\} \quad \text{for all } z \in \underline{C}^n$$

For all $z \in \delta(f)$, one has $p(z)/q(z) = f(z) = f(z+\omega) = p_\omega(z)/q_\omega(z)$. It is easy to check that p_ω and q_ω are again coprime; from the unicity result quoted in the same theorem, it follows that there exist $v_\omega \in A(\underline{C}^n)$ such that

$$\begin{aligned} p(z+\omega) &= p(z)\exp(v_\omega(z)) \\ q(z+\omega) &= q(z)\exp(v_\omega(z)) \end{aligned} \qquad \text{for all } z \in \underline{C}^n$$

Of course, the v_ω's have to satisfy **compatibility conditions:**

$$\exp(v_{\omega+\omega'}(z)) = \frac{p(z+\omega+\omega')}{p(z+\omega)}\,\frac{p(z+\omega)}{p(z)} = \exp(v_{\omega'}(z+\omega))\exp(v_\omega(z))$$

for all $\omega, \omega' \in \Gamma$ and for all $z \in \underline{C}^n$. If we had multiplied both p and q by the same unit in $A(\underline{C}^n)$ to get $f = p'/q'$, we could have obtained other functions v'_ω. The natural question to ask is: how can one use this freedom to get functions v_ω as simple as possible? Here is the answer.

Theorem. Let Γ be a lattice in $\underline{C}^n$ and let f be a meromorphic function on $\underline{C}^n$ with $\Gamma \subset G(f)$. Then there exist $p, q \in A(\underline{C}^n)$ with $(p,q) = 1$ and $f = p/q$, and there exist **linear** polynomials v_ω $(\omega \in \Gamma)$ such that

$$\begin{aligned} p(z+\omega) &= p(z)\exp(v_\omega(z)) \\ q(z+\omega) &= q(z)\exp(v_\omega(z)) \end{aligned} \qquad \underline{\text{for all}}\ z \in \underline{C}^n\ \underline{\text{and}}\ \omega \in \Gamma.$$

Proof. See Siegel [17], §14–16, or Siegel [18], Ch.5, §7–8, or Conforto [16], Sections 14–20 for a "classical" proof; see also Swinnerton-Dyer [11], §4 for a more "modern" proof (which, to our mind, is easier). ■

Definition. Let Γ be a lattice in $\underline{C}^n$. A **theta-function** for Γ is a holomorphic function $\theta \in A(\underline{C}^n)$ such that

$$\theta(z+\omega) = \theta(z)\exp(v_\omega(z)) \qquad \text{for all } z \in \underline{C}^n$$

and where the v_ω's are (not necessarily homogeneous) linear polynomials on $\underline{C}^n$ (see Exercise 31). The collection of the v_ω's is called an **exponent system** for the function θ.

The theorem above now reads: any Γ-invariant meromorphic function on $\underline{C}^n$ is a quotient of two coprime theta-functions with the same exponent system.

Exercise 41. Open your favourite textbook with a chapter on elliptic functions of one complex variable, and look for the proof of the theorem above when $n = 1$. [**Hint**: Look out for "Weierstrass sigma function".]

Exercise 42. (This is quite a long one, which is very much parallel to Section IV.2 below. It is given here for readers who do not like matrices at the dose of Chapter IV (see the paper by Cornalba in these Proceedings). The notation is almost the same as in Weil [20], Ch. VI, No.3.)

First part. Let θ be a theta function for a lattice Γ in $\underline{C}^n$ with exponent system $(v_\omega)_{\omega\in\Gamma}$. For each $\omega \in \Gamma$, write

$$(1)\quad v_\omega(z) = i2\pi\,\{L_\omega(z) + c(\omega)\}$$

where L_ω is a $\underline{C}$-linear form on $\underline{C}^n$. Check that the compatibility conditions imply

$$(2)\quad L_{\omega+\omega'} = L_\omega + L_{\omega'}$$

$$(3)\quad c(\omega+\omega') \equiv L_{\omega'}(\omega) + c(\omega') + c(\omega) \quad (\text{mod.}\mathbb{Z}) \qquad \text{for all } \omega, \omega' \in \Gamma$$

It follows from (2) that

$$\begin{cases} \Gamma \times \underline{C}^n \to \underline{C} \\ (\omega, z) \mapsto L_\omega(z) \end{cases}$$

extends to a $\underline{R}$-bilinear form $F: \underline{C}^n \times \underline{C}^n \to \underline{C}$ which is $\underline{C}$-linear in the second variable. Define now $A(w,z) = F(w,z) - F(z,w)$ for all $w,z \in \underline{C}^n$. Deduce from (3) that A takes integral values on $\Gamma \times \Gamma$, hence real values on $\underline{C}^n \times \underline{C}^n$, so that $A: \underline{C}^n \times \underline{C}^n \to \underline{R}$ is $\underline{R}$-bilinear, alternate and integral on $\Gamma \times \Gamma$.

Second part. Using the $\underline{C}$-linearity of F in the second variable, check that $A(w,z) - A(iw,iz) = -i\{A(w,iz) - A(z,iw)\}$, so that $A(w,z) = A(iw,iz)$ for all $w,z \in \underline{C}^n$. Deduce from this that

$$(4)\quad H: \begin{cases} \underline{C}^n \times \underline{C}^n \to \underline{C} \\ (z,w) \mapsto A(z,iw) + iA(z,w) \end{cases}$$

is a hermitian form ($\underline{C}$-linear in the second variable) and that $\phi = 2iF - H$ is a $\underline{C}$-bilinear symmetric form on $\underline{C}^n$.

Third part. Define now for each $\omega \in \Gamma$

$$(5)\quad d(\omega) = c(\omega) - \tfrac{1}{2}F(\omega,\omega)$$

$$(6)\quad g(\omega) = \operatorname{Im}(d(\omega))$$

$$(7)\quad L(\omega) = g(i\omega) + ig(\omega)$$

Deduce from (3) and from the fact that A is integral on $\Gamma \times \Gamma$ that $d(\omega+\omega') - d(\omega) - d(\omega')$ is an integer for all $\omega, \omega' \in \Gamma$, so that g extends to a $\underline{R}$-linear form on $\underline{C}^n$ and so that L extends to a $\underline{C}$-linear form on $\underline{C}^n$. Check that $d(\omega) - L(\omega)$ is real for all $\omega \in \Gamma$.

Fourth part. Define

$$\psi: \begin{cases} \Gamma \to S^1 \\ \omega \mapsto \exp\{i2\pi(d(\omega) - L(\omega))\} \end{cases}$$

and

$$T: \begin{cases} \underline{C}^n \to \underline{C} \\ z \mapsto \dfrac{1}{4i}\,\phi(z,z) + L(z) \end{cases}$$

If $\theta_r \in A(\underline{C}^n)$ is defined by $\theta_r(z) = \exp\{-i2\pi T(z)\}\,\theta(z)$, check that

$$(8)\quad \theta_r(z+\omega) = \theta_r(z)\,\psi(\omega)\exp\left\{\pi H(\omega,z) + \frac{\pi}{2}H(\omega,\omega)\right\} \qquad \text{for all } z \in \underline{C}^n \text{ and for all } \omega \in \Gamma$$

Fifth part. Now let $f = p/q$ be as in the theorem above. Define as above a hermitian form H and two "reduced theta-functions" p_r and q_r, so that one has again $f = p_r/q_r$. Deduce from the analogue of (8) that the function

$$\phi: \begin{cases} \underline{C}^n \to \underline{R}_+ \\ z \mapsto \left|p_r(z)\exp\left\{-\dfrac{\pi}{2}H(z,z)\right\}\right| \end{cases}$$

is Γ-periodic, hence bounded, so that

$$(9p)\qquad |p_r(z)| \leqslant \mu\exp\left\{\frac{\pi}{2}H(z,z)\right\} \qquad \text{for all } z \in \underline{C}^n$$

for some positive real constant μ, and similarly

$$(9q)\qquad |q_r(z)| \leqslant \mu\exp\left\{\frac{\pi}{2}H(z,z)\right\} \qquad \text{for all } z \in \underline{C}^n$$

uppose that there exist $z_0 \in \underline{C}^n$ with $H(z_0, z_0) \leqslant 0$. Deduce from Liouville's theorem and from 9) that the functions from $\underline{C}$ to $\underline{C}$ given by $\lambda \mapsto p_r(\lambda z_0)$ and $\lambda \mapsto q_r(\lambda z_0)$ are constant, so that f s degenerate.

ummary of Exercise 42. Let Γ be a lattice in $\underline{C}^n$. Suppose that there exists an abelian function or Γ. Then there exists a positive definite $\underline{C}$-valued hermitian form H on $\underline{C}^n$ such that $I(\omega, \omega') - H(\omega', \omega) \in 2i\mathbb{Z}$ for all $\omega, \omega' \in \Gamma$. Equivalently there exists a skew-symmetric $\underline{R}$-alued $\underline{R}$-bilinear form A on $\underline{C}^n$ such that $A(\Gamma, \Gamma) \subset \mathbb{Z}$ and such that

$$\begin{cases} \underline{C}^n \times \underline{C}^n \to \underline{R} \\ (z, w) \mapsto A(z, iw) \end{cases}$$

s symmetric and positive definite, i.e. is a scalar product on $\underline{R}^{2n} = (\underline{C}^n)_{\underline{R}}$. [Hint for the equivalence: H is given, define $A(w, z) = \mathrm{Im}(H(w, z))$; if A is given, define H as in the second part above.] ee Exercise 32 for the case $n = 1$.

Chapter IV
PERIOD RELATIONS

1. LATTICES IN $\underline{C}^n$ AND MATRICES

For any pair (m, n) of strictly positive integers and for any ring $\mathscr{R}$ (hereafter one of $\mathbb{Z}$, Q, $\underline{R}$, $\underline{C}$), we denote by $M_{m,n}(\mathscr{R})$ the free $\mathscr{R}$-module of all matrices with m lines, n columns, and entries in $\mathscr{R}$. If X is in $M_{m,n}(\mathscr{R})$, its **transpose** is denoted by ${}^tX \in M_{n,m}(\mathscr{R})$. When $\mathscr{R} = \underline{C}$, its **conjugate** is denoted by $\overline{X} \in M_{m,n}(\underline{C})$. We shall often write matrices "in blocks"; for example, if $X \in M_{p,n}(\mathscr{R})$ and $Y \in M_{q,n}(\mathscr{R})$, then $\binom{X}{Y}$ is in $M_{p+q,n}(\mathscr{R})$. The product of two matrices $X \in M_{m,\ell}(\mathscr{R})$ and $Y \in M_{\ell,n}(\mathscr{R})$ is just written XY and lives of course in $M_{m,n}(\mathscr{R}$

When m = n, we write $M_n(\mathscr{R})$ instead of $M_{n,n}(\mathscr{R})$. This is in a natural way a (non-abelian ring with identity, with group of invertible elements denoted by $GL_n(\mathscr{R})$.

In this chapter (not until now, for typographical reasons), we think of vectors in $\underline{C}^n$ as matrices in $M_{n,1}(\underline{C})$, i.e. as "column-vectors". We introduce the C-bilinear form:

$$\langle | \rangle \begin{cases} \underline{C}^n \times \underline{C}^n \to \underline{C} \\ \left(z = \begin{pmatrix} z_1 \\ \vdots \\ z_n \end{pmatrix}, \quad w = \begin{pmatrix} w_1 \\ \vdots \\ w_n \end{pmatrix} \right) \mapsto \sum_{j=1}^{n} z_j w_j \end{cases}$$

so that $\langle z|w\rangle = {}^tzw = {}^twz = \langle w|z\rangle$ for all $z,w \in \underline{C}^n$. We introduce in the same way the $\underline{R}$-bilinear form $\underline{R}^n \times \underline{R}^n \to \underline{R}$, again denoted by $\langle | \rangle$.

Let Γ be a lattice in $\underline{C}^n$ and let $\{\omega_1, ..., \omega_{2n}\}$ be a basis of Γ over $\mathbb{Z}$. The **period matrix** associated with this situation is the matrix:

$$\Omega = (\omega_1 \ \omega_2 \ \dots \ \omega_{2n}) = \begin{pmatrix} \omega_{1,1} & \dots & \omega_{1,2n} \\ \vdots & & \vdots \\ \omega_{n,1} & & \omega_{n,2n} \end{pmatrix} \in M_{n,2n}(\underline{C})$$

Any "period" $\omega \in \Gamma$ can be written as $\omega = \Omega m$ for some $m \in M_{2n,1}(\mathbb{Z})$. Geometrically speaking the matrix Ω describes a map γ: $\underline{R}^{2n} \to \underline{C}^n$ which sends $\mathbb{Z}^{2n}$ (see just before Proposition 1) onto Γ, and which is clearly $\underline{R}$-linear and bijective.

As in the lemma of Section II.2, any other basis $\omega_1^*, ..., \omega_{2n}^*$ of Γ would be given by

$$\omega_1^* = \sum_{j=1}^{2n} \omega_j M_{j,1} \ \dots \ \omega_{2n}^* = \sum_{j=1}^{2n} \omega_j M_{j,2n}$$

for some matrix $M = (M_{j,k}) \in GL_{2n}(\mathbb{Z})$. (As we have not introduced oriented basis for lattices in $\underline{C}^n$, we cannot impose here that M has determinant +1.) The corresponding period matrix would then be

$$\Omega^* = (\omega_1^* \ \omega_2^* \ \dots \ \omega_{2n}^*) = \Omega M$$

As in the Corollary to Proposition 7, an invertible linear map $Q: z \mapsto z^*$ from $\underline{C}^n$ to itself would map Γ onto a lattice Γ^* with basis $\omega_1^* = Q\omega_1, \dots, \omega_{2n}^* = Q\omega_{2n}$. Thinking of Q as a matrix in $GL_n(\underline{C})$, we get the period matrix associated with Γ^* and with the ω_j^*'s as

$$\Omega^* = (\omega_1^* \ \omega_2^* \ \dots \ \omega_{2n}^*) = Q\Omega$$

These and the proposition below motivate the following definitions.

Definition. A **Riemann matrix** is a matrix $\Omega \in M_{n,2n}(\underline{C})$ the columns of which are vectors in $\underline{C}^n$ linearly independent over $\underline{R}$. Two Riemann matrices Ω and Ω^* in $M_{n,2n}(\underline{C})$ are **equivalent** if there exist $Q \in GL_n(\underline{C})$ and $M \in GL_{2n}(\mathbb{Z})$ such that $\Omega^* = Q\Omega M$.

Proposition 11

(i) The relation defined above is an equivalence relation between Riemann matrices.

(ii) Let Ω and Ω^* be two period matrices associated with two lattices Γ and Γ^* in $\underline{C}^n$, furnished respectively with basis $\{\omega_1, \dots, \omega_{2n}\}$ and $\{\omega_1^*, \dots, \omega_{2n}^*\}$. Then Ω and Ω^* are equivalent if and only if the complex tori $T = \underline{C}^n/\Gamma$ and $T^* = C^n/\Gamma^*$ are isomorphic as complex Lie groups.

Proof. Assertion (i) is clear. For (ii), if $\Omega^* = Q\Omega M$ for some $Q \in GL_n(\underline{C})$ and $M \in GL_{2n}(\mathbb{Z})$, then the linear map $Q: \underline{C}^n \to \underline{C}^n$ factors as an isomorphism $C^n/\Gamma \to C^n/\Gamma^*$. Conversely, let $q: T \to T^*$ be an isomorphism; then q lifts to a linear map $Q: \underline{C}^n \to \underline{C}^n$ which maps Γ onto Γ^* (this is a repeat from the proof of Proposition 7). In particular $\{Q\omega_1, \dots, Q\omega_{2n}\}$ and $\{\omega_1^*, \dots, \omega_{2n}^*\}$ are two bases of Γ^*, so that there exists $M = (M_{j,k}) \in GL_{2n}(\mathbb{Z})$ with

$$\omega_1^* = \sum_{j=1}^{2n} (Q\omega_j)M_{j,1} \ \dots \ \omega_{2n}^* = \sum_{j=1}^{2n} (Q\omega_j)M_{j,2n}$$

It follows that $\Omega^* = Q\Omega M$. ■

Definition. Let Γ be a lattice in $\underline{C}^n$ furnished with a basis $\{\omega_1, \dots, \omega_{2n}\}$. The **big period matrix** associated with this situation is the matrix

$$P = \begin{pmatrix} \Omega \\ \overline{\Omega} \end{pmatrix} \in M_{2n}(\underline{C})$$

Proposition 12. Let P be as in the definition above.

(i) P is invertible: $P \in GL_{2n}(\underline{C})$.

(ii) P^{-1} can be written as $(G \ \overline{G})$ for some $G \in M_{2n,n}(\underline{C})$ satisfying $\Omega G = I$ and $\Omega \overline{G} = 0$. (I denotes the $(n \times n)$-unit matrix.)

(iii) If $\Omega^* = Q\Omega M$ for some $Q \in GL_n(\underline{C})$ and $M \in GL_{2n}(\mathbb{Z})$, then $P^* = \begin{pmatrix} Q & 0 \\ 0 & \overline{Q} \end{pmatrix} PM$.

(iv) Let $\gamma:\underline{R}^{2n} \to \underline{C}^n$ be the map described by Ω, as earlier in this section, then γ^{-1} is given by

$$\begin{cases} \underline{C}^n \to \underline{R}^{2n} \\ z \mapsto P^{-1}\begin{pmatrix} z \\ \bar{z} \end{pmatrix} \end{cases}$$

Proof

(i) Consider $\Omega' = \frac{1}{2}(\Omega + \bar{\Omega})$ and $\Omega'' = \frac{1}{2i}(\Omega - \bar{\Omega})$, both in $M_{n,2n}(\underline{R})$. One has

$$P = \begin{pmatrix} \Omega \\ \bar{\Omega} \end{pmatrix} = \begin{pmatrix} I & iI \\ I & -iI \end{pmatrix}\begin{pmatrix} \Omega' \\ \Omega'' \end{pmatrix}$$

and

$$\mathrm{Det}\begin{pmatrix} I & iI \\ I & -iI \end{pmatrix} = \mathrm{Det}\begin{pmatrix} 2I & 0 \\ I & -iI \end{pmatrix} = \mathrm{Det}(2I)\ \mathrm{Det}\ (-iI) = (-2i)^n \neq 0$$

Hence it suffices to check that $\begin{pmatrix} \Omega' \\ \Omega'' \end{pmatrix} \in GL_{2n}(\underline{R})$; but the columns of the matrix $\begin{pmatrix} \Omega' \\ \Omega'' \end{pmatrix}$ are clearly $\underline{R}$-linearly independent, because the vectors ω_k' s are $\underline{R}$-linearly independent, so that

(i) follows. [Notice that Det $\begin{pmatrix} \Omega' \\ \Omega'' \end{pmatrix}$ is the "volume" of the parallelotop spanned by the ω_k' s

(ii) Write $P^{-1} = (G_1 G_2)$ with G_1 and G_2 in $M_{2n,n}(\underline{C})$. Then $P^{-1}P = G_1\Omega + G_2\bar{\Omega} = \mathrm{id} = \begin{pmatrix} I & 0 \\ 0 & I \end{pmatrix}$ so that $\bar{G}_1\bar{\Omega} + \bar{G}_2\Omega = \begin{pmatrix} I & 0 \\ 0 & I \end{pmatrix}$ (because $\bar{I} = I$), which reads $(\bar{G}_2\bar{G}_1)P = \mathrm{id}$. Hence $P^{-1} = (\bar{G}_2\bar{G}_1) = (G_1G_2)$ and $\bar{G}_2 = G_1$, written G from now on. Finally $PP^{-1} = \begin{pmatrix} \Omega \\ \bar{\Omega} \end{pmatrix}(G\ \bar{G})$ $= \begin{pmatrix} \Omega G & \Omega\bar{G} \\ \bar{\Omega}G & \bar{\Omega}\bar{G} \end{pmatrix} = \mathrm{id} = \begin{pmatrix} I & 0 \\ 0 & I \end{pmatrix}$, which ends the proof of (ii).

(iii) Is clear.

(iv) For all $z \in M_{n,1}(\underline{C})$: $P^{-1}\begin{pmatrix} z \\ \bar{z} \end{pmatrix} = Gz + \overline{Gz} \in M_{2n,1}(\underline{R}) \approx \underline{R}^{2n}$.

If δ: $\begin{cases} \underline{C}^n \to \underline{R}^{2n} \\ z \mapsto P^{-1}\begin{pmatrix} z \\ \bar{z} \end{pmatrix} \end{cases}$

then one has, for all $x \in \underline{R}^{2n}$, $\delta(\gamma x) = \delta(\Omega x) = P^{-1}\begin{pmatrix} \Omega x \\ \bar{\Omega} x \end{pmatrix} = P^{-1}\begin{pmatrix} \Omega \\ \bar{\Omega} \end{pmatrix} x = x$, and $\delta = \gamma^{-1}$. ■

Exercise 43. Show that any Riemann matrix in $M_{n,2n}(\underline{C})$ is equivalent to one of the form (I T) for some $T \in M_n(\underline{C})$ with

$$\frac{1}{2i}(T - \bar{T}) \in GL_n(\underline{R})$$

i.e. is equivalent to a so-called **reduced Riemann matrix.** (Compare with the map $\mathscr{M} \to P_1$ in Section II.2.) Check that two distinct reduced Riemann matrices may be equivalent. [**Hint:** consider

$$\begin{pmatrix} 1 & 0 & i & 0 \\ 0 & 1 & 0 & i \end{pmatrix} \text{ and } \begin{pmatrix} 1 & 0 & 0 & i \\ 0 & 1 & i & 0 \end{pmatrix}]$$

2. FROBENIUS RELATIONS

Suppose now that Γ is a lattice in $\underline{C}^n$ with basis $\{\omega_1, ..., \omega_{2n}\}$ and let $f = p/q$ be a meromorphic function on $\underline{C}^n$ with $\Gamma \subset G(f)$, as in the fundamental theorem of Section III.3. Write $v_1, ..., v_{2n}$ for the polynomials $v_{\omega_1}, ..., v_{\omega_{2n}}$. The compatibility conditions first express v_ω for an arbitrary $\omega \in \Gamma$ in terms of the v_j's and, second, impose that the v_j's satisfy

(1) $v_j(z + \omega_k) + v_k(z) - v_k(z + \omega_j) - v_j(z) \in i2\pi\mathbb{Z}$ for all $j,k \in \{1, ..., 2n\}$

We shall see that (1) impose on Γ remarkable conditions, at least if we assume that f is nondegenerate; this was discovered by Frobenius (1884). We shall follow Siegel [17], § 20, and use essentially his notation. Many of the steps below may seem quite artificial; they could be made to look more "natural" via cohomological considerations, but this would take us too far from our project and we shall not comment further.

Step one (f may be degenerate): Definition of the matrices B, A *and* S.

Write the polynomials $v_1, ..., v_{2n}$ as

(2) $$v_j(z) = \sum_{k=1}^{2n} b_{k,j} z_k + \beta_j, \quad j = 1, ..., 2n$$

If $b_j = \begin{pmatrix} b_{1,j} \\ \vdots \\ b_{n,j} \end{pmatrix} \in C^n$ this reads

(3) $v_j(z) = \langle b_j | z \rangle + \beta_j, \quad j = 1, ..., 2n$

and the conditions (1) become

(4) $\langle b_j | \omega_k \rangle - \langle b_k | \omega_j \rangle \in i2\pi\mathbb{Z}, \quad j,k = 1, ..., 2n$

Define the matrix B to be

(5) $$B = (b_1 \ b_2 \ ... \ b_{2n}) = \begin{pmatrix} b_{1,1} & \cdots & b_{1,2n} \\ \vdots & & \vdots \\ b_{n,1} & & b_{n,2n} \end{pmatrix} \in M_{n,2n}(\underline{C})$$

so that (4) becomes'

(6) $${}^t B\Omega - {}^t\Omega B = 2A \quad \text{with} \quad \frac{1}{i\pi} A \in M_{2n}(\mathbb{Z})$$

The matrix A is clearly skew-symmetric, and also hermitian because it has purely imaginary entries. Define also

(7) $S = \frac{1}{2}({}^t B\Omega + {}^t\Omega B)$

so that ${}^t B\Omega = A + S$.

Step two (f may be degenerate): There exist a constant μ *and a continuous map* $\phi : \underline{C}^n \to \underline{C}$ *such that* $\phi(z + \omega_j) - \phi(z) \equiv v_j(z)$ *(mod. imaginary terms)* $(j = 1, ..., 2n)$ *and* $|p(z)\exp\{-\phi(z)\}| \leqslant \mu$ *for all* $z \in \underline{C}^n$.

Let s_j be the jth diagonal entry of S (also of ${}^t B\Omega$), let $r_j = \mathrm{Re}(\beta_j - \frac{1}{2}s_j)$, and let

$$r = \begin{pmatrix} r_1 \\ \vdots \\ r_{2n} \end{pmatrix} \in \underline{R}^{2n}$$

Notation being as in Proposition 12(iv), define the quadratic expression

$$(8) \quad \phi: \begin{cases} \underline{C}^n \to \underline{R}^{2n} \\ z \mapsto \frac{1}{2}\langle x|Sx\rangle + \langle r|x\rangle \text{ with } x = \gamma^{-1}(z) \end{cases}$$

Let $e_1, ..., e_{2n}$ be the canonical basis of $\underline{R}^{2n}$, so that $e_1 = \gamma^{-1}(\omega_1), ..., e_{2n} = \gamma^{-1}(\omega_{2n})$. As both the matrix S and the form $\langle\,|\,\rangle$ are symmetric: $\phi(z + \omega_k) - \phi(z) = \frac{1}{2}\langle x + e_k|S(x + e_k)\rangle +$ $+ \langle r|x + e_k\rangle - \frac{1}{2}\langle x|Sx\rangle - \langle r|x\rangle = \langle Se_k|x\rangle + \frac{1}{2}s_k + r_k$ for $k = 1, ..., 2n$. Now $\mathrm{Re}(\langle Se_k|x\rangle) = \mathrm{Re}(\langle (S-A)e$ and $\langle (S-A)e_k|x\rangle = \langle {}^t\Omega Be_k|x\rangle = \langle Be_k|\Omega x\rangle = \langle b_k|z\rangle$ on one hand, and $\mathrm{Re}(\frac{1}{2}s_k + r_k) = \mathrm{Re}(\beta_k)$ on the other hand, so that the first relation stated in step two holds.

Notice that

$$|p(z + \omega_k)\exp\{-\phi(z + \omega_k)\}| = |p(z)\exp\{v_k(z)\}\exp\{-\phi(z) - \mathrm{Re}(v_k(z)) + i(...)\}|$$

$$= |p(z)\exp\{-\phi(z)\}|$$

It follows that the continuous function

$$(9) \quad \begin{cases} \underline{C}^n \to \underline{R}_+ \\ z \mapsto |p(z)\exp\{-\phi(z)\}| \end{cases}$$

is continuous and Γ-invariant, hence bounded by some finite real number, say μ.

Step three (f may be degenerate): There exists a holomorphic map $\omega : \underline{C}^n \to \underline{C}$ *such that* $|p(z)\exp\{-\omega(z)\}| \leqslant \mu\exp\{+\mathrm{Re}(\langle H\bar{z}|z\rangle)\}$ *for all* $z \in \underline{C}^n$, *where* $H = \frac{1}{2}B\bar{G}$ *is a hermitian matrix.*

One has (first equality as in step two)

$$\mathrm{Re}(\langle x|Sx\rangle) = \mathrm{Re}(\langle x|{}^t B\Omega x\rangle) = \mathrm{Re}(\langle Gz + \overline{G}\bar{z}|{}^t Bz\rangle) = \mathrm{Re}(\langle Gz|{}^t Bz\rangle) + \mathrm{Re}(\langle B\overline{G}\bar{z}|z\rangle)$$

and also $\mathrm{Re}(\langle r|x\rangle) = \mathrm{Re}(\langle r|Gz + \overline{G}\bar{z}\rangle) = 2\mathrm{Re}(\langle r|Gz\rangle)$. Define the holomorphic function

$$(10)\ \omega: \begin{cases} \underline{C}^n \to \underline{C} \\ z \mapsto \frac{1}{2}\langle Gz|{}^tBz\rangle + 2\langle r|Gz\rangle \end{cases}$$

and the matrix

$$(11)\ H = \tfrac{1}{2}B\overline{G} \in M_n(\underline{C})$$

One has $\mathrm{Re}(\phi(z)) = \mathrm{Re}(\omega(z) + \langle H\overline{z}\,|z\rangle)$ so that

$$|p(z)\exp\{-\omega(z)\}| = |p(z)\exp\{-\phi(z)\}|\exp\{+\mathrm{Re}(\langle H\overline{z}\,|z\rangle)\} \leqslant \mu\exp\{+\mathrm{Re}(\langle H\overline{z}\,|z\rangle)\}$$

for all $z \in \underline{C}^n$ by step two.

By Proposition 12(ii), H may also be written as

$$H = +\tfrac{1}{2}{}^t(\Omega G)B\overline{G} - \tfrac{1}{2}{}^tG{}^tB(\Omega\overline{G}) = \tfrac{1}{2}{}^tG({}^t\Omega B - {}^tB\Omega)\overline{G} = -{}^tGA\overline{G}$$

As A is hermitian, ${}^t\overline{H} = -{}^tG{}^t\overline{A}\overline{G} = -{}^tGA\overline{G} = H$ and H is also hermitian.

Observe that

$${}^tP^{-1}AP^{-1} = \begin{pmatrix} {}^tG \\ {}^t\overline{G} \end{pmatrix} A(G\overline{G}) = \begin{pmatrix} {}^tGAG & {}^tGA\overline{G} \\ {}^t\overline{G}AG & {}^t\overline{G}A\overline{G} \end{pmatrix}$$

where $2{}^t\overline{G}A\overline{G} = {}^t\overline{G}{}^tB(\Omega\overline{G}) - ({}^t\overline{G}{}^t\Omega)B\overline{G} = 0$ by (6) and Proposition 12(ii), so that also ${}^tGAG = 0$, and where ${}^t\overline{G}AG = {}^t({}^tG{}^tA\overline{G}) = -{}^t({}^tGA\overline{G}) = -{}^tH$. This implies

$$(12)\ {}^tP^{-1}AP^{-1} = \begin{pmatrix} 0 & -H \\ {}^tH & 0 \end{pmatrix}$$

so that, in particular, A is invertible if and only if H is invertible.

Step four (f may be degenerate): effect of a change of coordinates in $\underline{C}^n$.

Suppose U is a unitary (n × n) matrix: ${}^t\overline{U} = U^{-1}$. If we change the co-ordinates in $\underline{C}^n$ according to U, the matrix Ω is replaced by $\Omega^* = U^{-1}\Omega$; similarly $P^* = \begin{pmatrix} U^{-1} & 0 \\ 0 & \overline{U^{-1}} \end{pmatrix} P$ and $P^{*-1} = (G^*\ \overline{G^*}) = (G\ \overline{G})\begin{pmatrix} U & 0 \\ 0 & \overline{U} \end{pmatrix}$. As $\langle b_j|z\rangle$ does not change, $\langle b_j^*|z^*\rangle = \langle b_j^*|U^{-1}z\rangle = \langle {}^tU^{-1}b_j^*|z\rangle = \langle b_j|z\rangle$ for all $z \in \underline{C}^n$, so that $b_j^* = {}^tUb_j$ ($j = 1, \ldots, 2n$) and $B^* = {}^tUB$. (Note that $2A^* = {}^tB^*\Omega^* - {}^t\Omega^*B^* = 2A$.) Finally, $H^* = \tfrac{1}{2}B^*\overline{G}^* = \tfrac{1}{2}{}^tUB\overline{G}\overline{U} = {}^tUH\overline{U} = {}^tUH({}^tU)^{-1}$.

In particular, modulo an appropriate change of co-ordinates, one may always assume H to be a diagonal matrix; its entries are then necessarily real.

Step five (f *is* **non-degenerate**)*: The matrix* H *is positive definite.*

Recall that a matrix is said to be positive definite if it is hermitian and if its (necessarily real) eigenvalues are all strictly positive.

By step four, one may assume that

$$H = \begin{pmatrix} h_1 & & & 0 \\ & h_2 & & \\ & & \ddots & \\ 0 & & & h_n \end{pmatrix}$$

Suppose, e.g., that $h_n \leqslant 0$. Let

$$z = \begin{pmatrix} 0 \\ \vdots \\ \lambda \end{pmatrix} \in \underline{C}^n$$

Then the function $z \mapsto p(z) \exp\{-\omega(z)\}$ defines an entire function of the **one** complex variable λ. By step three, this function is bounded: $|p(\lambda) \exp\{-\omega(\lambda)\}| \leqslant \mu \exp\{h_n |\lambda|^2\} = \mu \exp\{-|h_n| \, |\lambda|^2\}$ for all $\lambda \in \underline{C}$; Liouville's theorem implies that it is a constant. Similarly $z \mapsto q(z) \exp\{-\omega(z)\}$ does not depend on the last variable, and neither does $z \mapsto f(z) = p(z)/q(z)$. But this is impossible if f is non-degenerate, so that step five is proved.

Proposition 13. Let Γ be a lattice in $\underline{C}^n$ furnished with a basis and let Ω be the associated period matrix. Then the following is a **necessary condition** for the existence of an abelian function f on $\underline{C}^n$ with $G(f) = \Gamma$:

There exists a $(2n \times 2n)$-matrix A with

(1) $\frac{1}{i\pi} A \in M_{2n}(\mathbb{Z})$ and A is skew-symmetric.

(2) A is non-singular.

(3) $\Omega A^{-1}\, {}^t\Omega = 0$.

(4) $\bar{\Omega} A^{-1}\, {}^t\Omega < 0$.

Proof. We take A given by (6) above. From (12) and step five, A and H are invertible. From (12) and from the definition of P:

$$PA^{-1}\, {}^tP = \begin{pmatrix} 0 & {}^tH^{-1} \\ -H^{-1} & 0 \end{pmatrix} = \begin{pmatrix} \Omega \\ \bar{\Omega} \end{pmatrix} A^{-1}({}^t\Omega \; {}^t\bar{\Omega}) = \begin{pmatrix} \Omega A^{-1}\, {}^t\Omega & * \\ \bar{\Omega} A^{-1}\, {}^t\Omega & * \end{pmatrix}$$

so that (3) and (4) hold by step five. ■

Caution: we do *not* claim that $(1/i\pi)A \in GL_{2n}(\mathbb{Z})$, e.g. $\begin{pmatrix} 2 & 0 \\ 0 & 2 \end{pmatrix}$ is in $M_2(\mathbb{Z})$ and in $GL_2(Q)$ but not in $GL_2(\mathbb{Z})$.

Exercise 44. Let $\omega_1 = \begin{pmatrix} 1 \\ 0 \end{pmatrix}$ $\omega_2 = \begin{pmatrix} 0 \\ 1 \end{pmatrix}$ $\omega_3 = \begin{pmatrix} \tau_1 \\ 0 \end{pmatrix}$ and $\omega_4 = \begin{pmatrix} 0 \\ \tau_2 \end{pmatrix}$ be as in the example of Section III.2, so that $\Omega = \begin{pmatrix} 1 & 0 & \tau_1 & 0 \\ 0 & 1 & 0 & \tau_2 \end{pmatrix}$. Let (d_1, d_2) be any pair of strictly positive integers.

Check that

$$A = i\pi \begin{pmatrix} 0 & 0 & d_1 & 0 \\ 0 & 0 & 0 & d_2 \\ -d_1 & 0 & 0 & 0 \\ 0 & -d_2 & 0 & 0 \end{pmatrix}$$

satisfies the four conditions of Proposition 13.

Exercise 45. The Riemann matrix $\Omega \in M_{n,2n}(\underline{C})$ being given, any $(2n \times 2n)$ matrix A satisfying the conditions of Proposition 13 is said to be a **principal matrix** for Ω. Check that there exists a principal matrix for Ω if and only if there exist principal matrices for all Riemann matrices equivalent to Ω.

Exercise 46. Let Γ be a lattice in $\underline{C}^n$ furnished with a basis and let Ω be the associated period matrix. Prove that the following conditions are equivalent:

(1) There exists a positive definite hermitian form H on $\underline{C}^n$ with $\mathrm{Im}(H(\Gamma,\Gamma)) \subset \mathbb{Z}$.

(2) There exists a skew-symmetric $\underline{R}$-valued $\underline{R}$-bilinear form A on $\underline{C}^n$ with $A(\Gamma,\Gamma) \subset \mathbb{Z}$ and such that

$$\begin{cases} \underline{C}^n \times \underline{C}^n \to \underline{R} \\ (z, w) \mapsto A(z, iw) \end{cases}$$

is a scalar product on the real vector space underlying $\underline{C}^n$.

(3) There exists a matrix satisfying the Frobenius relations of Proposition 13.

Hint: For (1) $\Leftrightarrow$ (2), see Exercise 42. We shall indicate below (2) $\Rightarrow$ (3) and leave the converse to the reader. We hope that there will be no confusion between the meaning of A here and in Exercise 42 (as in Ref.[20]) on one hand, and the A of Proposition 13 (which is as in [17]) on the other hand.

Suppose (2) holds and let $M \in M_{2n}(\underline{C})$ be defined by

$$(1)\quad A(\Omega x, \Omega y) = \frac{1}{i\pi}\, {}^t xMy, \qquad x,y \in \underline{R}^{2n}$$

i.e. by $M = (1/i\pi)\,{}^t\Omega A\Omega$; it is clearly a skew-symmetric matrix with $(1/i\pi)M \in M_{2n}(\mathbb{Z})$. Choose $x,y \in \underline{R}^{2n}$, let $z = \Omega x$ and $w = \Omega y$; then (1) implies

$$(2)\quad A(z,w) = \frac{1}{i\pi}({}^tz\,{}^tG + {}^t\bar{z}\,{}^t\bar{G})M(Gw + \bar{G}\bar{w})$$

and

$$(3)\quad \pi A(z,iw) = {}^tz\,{}^tGMGw + {}^t\bar{z}\,{}^t\bar{G}MGw - {}^tz\,{}^tGM\bar{G}\bar{w} - {}^tz\,{}^t\bar{G}M\bar{G}\bar{w}.$$

Using ${}^tz({}^tGMG)w = {}^tw\,{}^t({}^tGMG)z = -{}^tw\,{}^tGMGz$ and similar relations, one obtains

$$(4)\quad \pi A(z,iw) - \pi A(w,iz) = 2\,{}^tz\,{}^tGMGw - 2\,{}^t\bar{z}\,{}^t\bar{G}M\bar{G}\bar{w}$$

As (4) must vanish for any z,w $\in \underline{C}^n$, one has

(5) $\quad {}^t GMG = {}^t\overline{G}M\overline{G} = 0$

and (3) simplifies to

(6) $\quad \pi A(z,iw) = {}^t\overline{z}\,{}^t\overline{G}MGw + {}^t\overline{w}\,{}^t\overline{G}MGz$

As A(z,iw) defines a positive definite form, one must have

(7) $\quad {}^t\overline{G}MG > 0$

Relations (5) and (7) imply that there exists a positive definite matrix K with

(8) $$\begin{pmatrix} {}^tG \\ {}^t\overline{G} \end{pmatrix} M(G\ \overline{G}) = \begin{pmatrix} 0 & -{}^tK \\ K & 0 \end{pmatrix}$$

Taking inverses:

(9) $$PM^{-1}\,{}^tP = \begin{pmatrix} 0 & K^{-1} \\ -{}^tK^{-1} & 0 \end{pmatrix}$$

or

(10) $$\begin{pmatrix} \Omega \\ \overline{\Omega} \end{pmatrix} M^{-1}({}^t\Omega\ {}^t\overline{\Omega}) = \begin{pmatrix} \Omega M^{-1}\,{}^t\Omega & * \\ \overline{\Omega}M^{-1}\,{}^t\Omega & * \end{pmatrix} = \begin{pmatrix} 0 & K^{-1} \\ -{}^tK^{-1} & 0 \end{pmatrix}$$

We have proved that M satisfies the conditions of Proposition 13.

3. A COMPLEX TORUS WITHOUT NON-CONSTANT MEROMORPHIC FUNCTIONS

We follow § 31 in Siegel [17].

Consider the lattice Γ generated by $\omega_1 = \begin{pmatrix} 1 \\ 0 \end{pmatrix}$, $\omega_2 = \begin{pmatrix} 0 \\ 1 \end{pmatrix}$, $\omega_3 = i\begin{pmatrix} \sqrt{2} \\ \sqrt{3} \end{pmatrix}$, and $\omega_4 = i\begin{pmatrix} \sqrt{5} \\ \sqrt{7} \end{pmatrix}$ in $\underline{C}^2$. Let us first show that any Γ-invariant meromorphic function on $\underline{C}^2$ is degenerate.

Suppose not. Then there exists a matrix A satisfying the conditions of Proposition 13. Write $K = i\pi A^{-1}$, which is a skew-symmetric matrix in $GL_4(\underline{Q})$. Then condition (3) of Propositio[n 13] reads:

$$\begin{pmatrix} 1 & 0 & i\sqrt{2} & i\sqrt{5} \\ 0 & 1 & i\sqrt{3} & i\sqrt{7} \end{pmatrix} \begin{pmatrix} K_{1,1} & \cdots & K_{1,4} \\ \vdots & & \\ K_{4,1} & & K_{4,4} \end{pmatrix} \begin{pmatrix} 1 & 0 \\ 0 & 1 \\ i\sqrt{2} & i\sqrt{3} \\ i\sqrt{5} & i\sqrt{7} \end{pmatrix} = \begin{pmatrix} r_{1,1} & r_{1,2} \\ r_{2,1} & r_{2,2} \end{pmatrix} = 0$$

with

$$r_{1,2} = K_{1,2} + i\sqrt{3}\,K_{1,3} + i\sqrt{7}\,K_{1,4} - i\sqrt{2}\,K_{2,3} - i\sqrt{5}\,K_{2,4} + (\sqrt{15} - \sqrt{14})K_{3,4} = 0$$

As the $K_{j,k}$'s are rational numbers, it follows from the last expression that the skew-symmetric matrix K is zero. This is impossible because $K = i\pi A^{-1}$ is invertible, and we have proved our first claim.

Let us then show that any Γ-invariant meromorphic function f on $\underline{C}^2$ is a constant. It follows from Corollary 2 to Proposition 10 that G(f) contains a complex line $V \subset C^2$. Let $Q = \begin{pmatrix} \alpha & \beta \\ \gamma & \delta \end{pmatrix} \in GL_2(\underline{C})$ be an invertible linear map which sends V onto $\left\{ \begin{pmatrix} 0 \\ z_2 \end{pmatrix} \in \underline{C}^2 \middle| z_2 \in \underline{C} \right\}$, let $\Gamma^* = Q(\Gamma)$ and let

$$\Omega^* = Q\Omega = \begin{pmatrix} \alpha & \beta & i(\alpha\sqrt{2} + \beta\sqrt{3}) & i(\alpha\sqrt{5} + \beta\sqrt{7}) \\ \gamma & \delta & i(\gamma\sqrt{2} + \delta\sqrt{3}) & i(\gamma\sqrt{5} + \delta\sqrt{7}) \end{pmatrix}$$

Then the function ${}^{Q}f$, defined by ${}^{Q}f(z) = f(Q^{-1}z)$, is invariant by Γ^* and does not depend on z_2; hence this function defines a meromorphic function of one complex variable, say g, which has α, β, $i(\alpha\sqrt{2} + \beta\sqrt{3})$ and $i(\alpha\sqrt{5} + \beta\sqrt{7})$ as periods. Either these four periods generate a lattice in $\underline{C}$, or g and then also f is a constant. We shall show that the first alternative cannot happen.

Suppose the four periods above generate a lattice in $\underline{C}$, and let b_1, b_2 be a basis of this lattice. Then there are integers $n_{1,1}, \ldots, n_{2,4}$ such that

$$\alpha = n_{1,1} b_1 + n_{2,1} b_2 \qquad i(\alpha\sqrt{2} + \beta\sqrt{3}) = n_{1,3} b_1 + n_{2,3} b_2$$

$$\beta = n_{1,2} b_1 + n_{2,2} b_2 \qquad i(\alpha\sqrt{5} + \beta\sqrt{7}) = n_{1,4} b_1 + n_{2,4} b_2$$

By elimination of α and β:

$$b_1(i\sqrt{2}\,n_{1,1} + i\sqrt{3}\,n_{1,2} - n_{1,3}) + b_2(i\sqrt{2}\,n_{2,1} + i\sqrt{3}\,n_{2,2} - n_{2,3}) = 0$$

$$b_1(i\sqrt{5}\,n_{1,1} + i\sqrt{7}\,n_{1,2} - n_{1,4}) + b_2(i\sqrt{5}\,n_{2,1} + i\sqrt{7}\,n_{2,2} - n_{2,4}) = 0$$

This is a homogeneous system of linear equations which possess a non-trivial solution: b_1, b_2. Hence

$$\mathrm{Det}\begin{pmatrix} i\sqrt{2}\,n_{1,1} + i\sqrt{3}\,n_{1,2} - n_{1,3} & i\sqrt{2}\,n_{2,1} + i\sqrt{3}\,n_{2,2} - n_{2,3} \\ i\sqrt{5}\,n_{1,1} + i\sqrt{7}\,n_{1,2} - n_{1,4} & i\sqrt{5}\,n_{2,1} + i\sqrt{7}\,n_{2,2} - n_{2,4} \end{pmatrix} = 0$$

This is a linear expression in $\sqrt{14}$, $i\sqrt{5}$, $i\sqrt{2}$, $i\sqrt{7}$, $i\sqrt{3}$, 1, $\sqrt{15}$, with coefficients in $\mathbb{Z}$. It follows that all coefficients must vanish, which reads (in the order given: $\sqrt{14}$ first, and so on):

$$n_{1,1}n_{2,2} - n_{1,2}n_{2,1} = 0$$

$$n_{1,1}n_{2,3} - n_{1,3}n_{2,1} = 0$$

$$n_{1,1}n_{2,4} - n_{1,4}n_{2,1} = 0$$

$$n_{1,2}n_{2,3} - n_{1,3}n_{2,2} = 0$$

$$n_{1,2}n_{2,4} - n_{1,4}n_{2,2} = 0$$

$$n_{1,3}n_{2,4} - n_{2,3}n_{1,4} = 0$$

...

Written differently, the rank of the matrix

$$\begin{pmatrix} n_{1,1} & n_{2,1} \\ n_{1,2} & n_{2,2} \\ n_{1,3} & n_{2,3} \\ n_{1,4} & n_{2,4} \end{pmatrix}$$

is at most one, and α, β, $i(\alpha\sqrt{2} + \beta\sqrt{3})$, $i(\alpha\sqrt{5} + \beta\sqrt{7})$ do not generate a lattice in $\underline{C}$.

We have proved the following result.

Proposition 14. There exist complex tori of complex dimension 2 without non-constant meromorphic functions.

Corollary 1. There exist complex tori which are not algebraic.

Proof. Meromorphic functions separate points in $P_n(\underline{C})$, hence in any submanifold of $P_n(\underline{C})$, and they certainly do not in the example above. Compare with Exercise 12 and Proposition 9. ■

Corollary 2. There exist Kähler manifolds which are not algebraic.

Proof. See above, and Exercise 18. Compare with Robertson's paper, Section 11. ■

Propositions 9 and 14 imply evidently that the Corollary to Proposition 4 does not hold in the complex domain.

4. FINAL REMARKS

Let Γ again be a lattice in $\underline{C}^n$ with basis $\{\omega_1, \ldots, \omega_{2n}\}$, and let Ω be the associated period matrix. It turns out that the conditions of Proposition 13 are not only necessary but also sufficient to ensure the existence of an abelian function with Γ as group of periods.

Theorem. Let Ω be as above. Suppose that there exist a $(2n \times 2n)$-matrix satisfying conditions (1) to (4) of Proposition 13. Then there exists an abelian function f on $\underline{C}^n$ with $G(f) = \Gamma$.

Proof. See Siegel [17], §21–26, or Conforto [16], most of Ch.III, or Swinnerton-Dyer [11], §6.

It is also known that the conditions of Proposition 13 are necessary and sufficient for a complex torus to be algebraic. We refer to the literature for a proof and for further comments, but we emphasize that escaping this proof is escaping one of the main points of the subject.

Two topics are not included in this paper which were covered in the course of lectures delivered: (a) A report by Hefez on some open problems and on a recent result (this will appear in his thesis); (b) Siegel's generalized half-planes (see Siegel [21] for a classical exposition, and Siegel [17] for connection with algebraic complex tori).

ACKNOWLEDGEMENTS

I am grateful to J. Guenot for some discussions, and to the *Fonds national suisse de la recherche scientifique* for partially supporting the writing-up of these notes.

SOME PROBLEMS IN THE THEORY OF SINGLE-VALUED ANALYTICAL FUNCTIONS AND HARMONIC ANALYSIS IN THE COMPLEX DOMAIN

M.M. DJRBASHIAN
Institute of Mathematics,
Academy of Sciences of the Armenian SSR,
Yerevan, USSR

Abstract

THEORY OF SINGLE-VALUED ANALYTICAL FUNCTIONS AND HARMONIC ANALYSIS IN THE COMPLEX DOMAIN.

Part I: 1. The integral formulae of Schwartz and Poisson. 2. The Dirichlet problem for the disc. Part II: The class of harmonic functions representable by the Poisson-Stieltjes integral. Part III: 1. The operator $L^{(\omega)}$ and its properties. 2. Formulae of the Cauchy, Schwartz and Poisson type. Part IV: 1. General classes of functions harmonic in the disc and their representations. 2. A generalization of the theorem of Herglotz. Part V: Theory of factorization and boundary properties of functions meromorphic in the disc. Part VI: Harmonic analysis in the complex domain and its applications in the theory of analytical and infinitely differentiable functions. Part VII: Some open problems.

INTRODUCTION

Parts I–IV deal with representation and boundary values of certain classes of analytical and harmonic functions in the disc $|z| < 1$ and contain results obtained by the author as well as those known for a considerable time. Part V presents a review of the author's principal results in the theory of factorization of the functions meromorphic in the disc $|z| < 1$ and their boundary values. As a rule, the proofs are omitted for the sake of brevity. Part VI consists of a review of the author's results in the theory of harmonic analysis in the strictly complex domain and, in this connection, to the representation of some other classes of single-valued analytical functions. Here also, limitations of space prevent the presentation of proofs. In Part VII some open problems are formulated.

Part I

1. THE INTEGRAL FORMULAE OF SCHWARTZ AND POISSON

Let the function f(z) be analytical in the disc $D(R) = \{z; |z| < R\}$ and continuous in its closure $\overline{D}(R)$.

From the Cauchy integral formula we have

$$\frac{1}{2\pi i}\int_{|\zeta|=R}\frac{f(\zeta)}{\zeta - z}d\zeta = \begin{cases} f(z), & z \in D(R) \\ 0, & z \in \mathbb{C} - \overline{D}(R) \end{cases} \tag{1.1}$$

Therefore, at each point $z^* = R^2/\bar{z}$, symmetrical with the point $z \in D(R)$,

$$\frac{1}{2\pi i}\int_{|\zeta|=R} \frac{f(\zeta)}{\zeta - \frac{R^2}{\bar{z}}}\, d\zeta = 0, \quad z \in D(R) \tag{1.2}$$

In view of $\zeta = Re^{i\varphi}$, $d\zeta = i\,Re^{i\varphi}d\varphi$, passing over to conjugate values, we get

$$\int_{|\zeta|=R} \frac{\overline{f(\zeta)\, d\zeta}}{\overline{\zeta - \frac{R^2}{\bar{z}}}} = i\int_{|\zeta|=R} \overline{\frac{f(\zeta)\, z\, d\varphi}{\zeta - z}} = -i\,\overline{\int_{|\zeta|=R} f(\zeta)\, d\varphi} + \int_{|\zeta|=R} \overline{\frac{f(\zeta)\, d\zeta}{\zeta - z}}$$

$$= -i\,\overline{\int_{|\zeta|=R} \frac{f(\zeta)\, d\zeta}{i\zeta}} + \int_{|\zeta|=R} \overline{\frac{f(\zeta)\, d\zeta}{\zeta - z}} = -2\pi i\, \overline{f(0)} + \int_{|\zeta|=R} \frac{\overline{f(\zeta)}\, d\zeta}{\zeta - z}$$

$$= 0, \quad z \in D(R)$$

Hence for any point $z \in D(R)$

$$\frac{1}{2\pi i}\int_{|\zeta|=R} \frac{\overline{f(\zeta)}\, d\zeta}{\zeta - z} = \overline{f(0)} \tag{1.3}$$

Summing Eqs (1.1) and (1.3), we get

$$f(z) + \overline{f(0)} = \frac{1}{\pi i}\int_{|\zeta|=R} \frac{u(\zeta)\, d\zeta}{\zeta - z} \tag{1.4}$$

where $u(\zeta) = \operatorname{Re} f(\zeta)$.

When $z = 0$, this yields

$$\operatorname{Re} f(0) = \frac{1}{2\pi i}\int_{|\zeta|=R} \frac{u(\zeta)}{\zeta}\, d\zeta$$

Hence, from (1.4), the following identical formulae are obtained:

$$f(z) = i\,\operatorname{Im} f(0) + \frac{1}{\pi i}\int_{|\zeta|=R} \left\{\frac{1}{\zeta - z} - \frac{1}{2\zeta}\right\} u(\zeta)\, d\zeta$$

$$= i\,\operatorname{Im} f(0) + \frac{1}{2\pi i}\int_{|\zeta|=R} \frac{\zeta + z}{\zeta - z}\, u(\zeta)\frac{d\zeta}{\zeta}$$

and

$$f(z) = i\,\operatorname{Im} f(0) + \frac{1}{2\pi}\int_0^{2\pi} \frac{Re^{i\vartheta} + z}{Re^{i\vartheta} - z}\left\{\operatorname{Re} f(Re^{i\vartheta})\right\} d\vartheta, \qquad z \in D(R) \tag{1.5}$$

This formula, expressing the values of the function f(z) analytical in the disc $|z| < R$ through the random values of its real part on the circumference $|\zeta| = R$, is called the **Schwartz integral formula.**

Writing $z = re^{i\varphi}$, from the Schwartz formula (1.5) we get

$$u(x,y) \equiv u(re^{i\varphi}) = \operatorname{Re} f(re^{i\varphi})$$

$$= \frac{1}{2\pi}\int_0^{2\pi}\left\{\operatorname{Re}\frac{Re^{i\vartheta}+re^{i\varphi}}{Re^{i\vartheta}-re^{i\varphi}}\right\}\left\{\operatorname{Re} f(Re^{i\vartheta})\right\}d\vartheta$$

or, since

$$\operatorname{Re}\frac{Re^{i\vartheta}+re^{i\varphi}}{Re^{i\vartheta}-re^{i\varphi}} = \frac{R^2-r^2}{R^2-2Rr\cos(\varphi-\vartheta)+r^2} \tag{1.6}$$

we have

$$u(re^{i\varphi}) = \frac{1}{2\pi}\int_0^{2\pi}\frac{R^2-r^2}{R^2-2Rr\cos(\varphi-\vartheta)+r^2}\,u(Re^{i\vartheta})\,d\vartheta \tag{1.7}$$

Thus we arrived at the Poisson formula (1.7), expressing the values of a function u(z) continuous on the disc $|z| \leqslant R$ and harmonic in its interior, through its values on the boundary of the disc.

The function

$$P\left(\varphi-\vartheta;\frac{r}{R}\right) \equiv \frac{R^2-r^2}{R^2-2Rr\cos(\varphi-\vartheta)+r^2} \tag{1.8}$$

itself harmonic in the disc $|re^{i\varphi}| < R$ for any $\vartheta \in [0, 2\pi]$ by virtue of (1.6), is called the **Poisson kernel.**

2. THE DIRICHLET PROBLEM FOR THE DISC

The classical problem of Dirichlet reads as follows.

To find a function $u(x, y) = u(z)$ harmonic in the disc $|z| < R$ and continuous on its closure $|z| \leqslant R$ such that $u(\zeta) = h(\zeta)$, $|\zeta| = R$ where $h(\zeta)$ is a previously given function continuous on the circumference $|\zeta| = R$.

If it is also known that the function $f(z) = u(z) + iv(z)$ is continuous on the closed disc $|z| \leqslant R$, then, by virtue of the Poisson formula (1.7), the required solution of the Dirichlet problem can be represented in the form:

$$u(z) = \frac{1}{2\pi}\int_0^{2\pi} P\left(\varphi-\vartheta;\frac{r}{R}\right) h(Re^{i\vartheta})\,d\vartheta, \quad |z| < R \tag{1.9}$$

We shall prove now that the Poisson formula (1.9) solves the Dirichlet problem in the given formulation.

With no loss of generality we can assume that R = 1 and write the relation (1.9) in the form

$$u(z) = \frac{1}{2\pi} \int_0^{2\pi} \frac{1-|z|^2}{|e^{i\vartheta}-z|^2} h(e^{i\vartheta})\, d\vartheta, \quad |z|<1 \tag{1.9'}$$

noting that according to (1.6) (when R = 1)

$$P(\varphi-\vartheta; r) = \operatorname{Re} \frac{e^{i\vartheta}+z}{e^{i\vartheta}-z} = \frac{1-|z|^2}{|e^{i\vartheta}-z|^2}$$

Since $P(\varphi - \vartheta; r)$ is harmonic for any $\vartheta \in [0, 2\pi]$, then

$$\Delta P(\varphi-\vartheta; r) \equiv \frac{\partial^2 P}{\partial x^2} + \frac{\partial^2 P}{\partial y^2} = 0, \quad \vartheta \in [0, 2\pi]$$

and, therefore, we also have

$$\Delta u = \frac{1}{2\pi} \int_0^{2\pi} \Delta \left\{ \frac{1-|z|^2}{|e^{i\vartheta}-z|^2} \right\} h(e^{i\vartheta})\, d\vartheta = 0$$

i.e. u(z) is harmonic in the disc $|z| < 1$. It is also easy to see that

$$\lim_{\substack{z \to e^{i\vartheta_0} \\ |z|<1}} u(z) = h(e^{i\vartheta_0}) \tag{1.10}$$

Indeed, for the harmonic function $u(z) \equiv 1$, according to the Poisson formula (1.7) (when R = 1), we have

$$1 = \frac{1}{2\pi} \int_0^{2\pi} \frac{1-|z|^2}{|e^{i\vartheta}-z|^2}\, d\vartheta, \quad |z|<1 \tag{1.11}$$

Hence

$$u(z) - h(e^{i\vartheta_0}) = \frac{1}{2\pi} \int_0^{2\pi} \frac{1-|z|^2}{|e^{i\vartheta}-z|^2} \left[h(e^{i\vartheta}) - h(e^{i\vartheta_0}) \right] d\vartheta \tag{1.12}$$

Since the function $h(\zeta)$ is continuous on the circumference $|\zeta| = 1$, it is 2π-periodical and uniformly continuous for all $\vartheta \in (-\infty, +\infty)$. Therefore for any $\epsilon > 0$ there is a $\delta = \delta(\epsilon) > 0$ such that

$$|h(e^{i\vartheta}) - h(e^{i\vartheta_0})| < \frac{\varepsilon}{2} \tag{1.13}$$

for any pair ϑ, ϑ_0 for which $|\vartheta - \vartheta_0| < \delta$.

We write (1.12) in the form

$$u(z) - h(e^{i\vartheta_0}) = I_1 + I_2$$

where

$$I_1 = \frac{1}{2\pi}\int_{\vartheta_0-\delta}^{\vartheta_0+\delta} \frac{1-|z|^2}{|e^{i\vartheta}-z|^2}\left[h(e^{i\vartheta}) - h(e^{i\vartheta_0})\right] d\vartheta$$

$$I_2 = \frac{1}{2\pi}\left(\int_0^{\vartheta_0-\delta} + \int_{\vartheta_0+\delta}^{2\pi}\right) \frac{1-|z|^2}{|e^{i\vartheta}-z|^2}\left[h(e^{i\vartheta}) - h(e^{i\vartheta_0})\right] d\vartheta$$

and here, by virtue of (1.11) and (1.13), we have $|I_1| < \epsilon/2$. Further, denoting

$$M = \max_{0\leq\vartheta\leq 2\pi} |h(e^{i\vartheta})|$$

we shall have

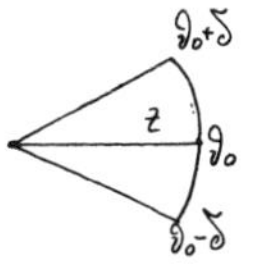

$$|I_2| \leq 2M \frac{1-|z|^2}{\rho(z)}$$

where

$$\rho(z) = \min |e^{i\vartheta} - z| \, , \quad \vartheta \,\overline{\in}\, (\vartheta_0-\delta, \vartheta_0+\delta)$$

and hence

$$\lim_{z\to e^{i\vartheta_0}} \rho(z) = |e^{i\delta} - 1| > 0$$

Consequently, we can take z so close to $e^{i\vartheta_0}$ as to have also $|I_2| < \epsilon/2$.

This completes the proof of the equality (1.10).

3. THE INTEGRAL OF POISSON-STIELTJES AND ITS BOUNDARY VALUES

Let us remember that the real function $\psi(\vartheta)$ defined on the interval $[0, 2\pi]$ is called a function of finite variation, if

$$\sup_{n\geq 1}\left\{\sum_{k=0}^{n-1} |\psi(\vartheta_{k+1}^{(n)}) - \psi(\vartheta_k^{(n)})|\right\} \equiv \bigvee_0^{2\pi}(\psi) < +\infty$$

where the upper bound is taken over the set of all possible divisions of the interval:

$$0 = \vartheta_0^{(n)} < \vartheta_1^{(n)} < \cdots < \vartheta_{n-1}^{(n)} < \vartheta_n^{(n)} = 2\pi \quad (n \geq 1)$$

Assume that a real function $\psi(\vartheta)$ of finite variation is given, defined on the points $e^{i\vartheta}$ $(0 \leqslant \vartheta \leqslant 2\pi)$ of the unit circumference. An expression of the type

$$u(re^{i\varphi}) = \frac{1}{2\pi}\int_0^{2\pi} \frac{1-r^2}{1-2r\cos(\vartheta-\varphi)+r^2}\, d\psi(\vartheta) \tag{1.14}$$

is usually called a Poisson-Stieltjes integral.

As in the case of ordinary Poisson integrals, it is easily seen that the function u(z) is harmoni in the disc $|z| < 1$.

Theorem 1.1. *At any point* $e^{i\vartheta_0}$ *where the derivative* $\psi'(\vartheta_0)$ *exists and is finite we shall have*

$$\lim_{r\to 1-0} u(re^{i\vartheta_0}) = \psi'(\vartheta_0) \tag{1.15}$$

Proof. We put $\varphi = \vartheta_0$ in (1.14) and integrate it by parts:

$$u(re^{i\vartheta_0}) = \left[\frac{1}{2\pi}\,\frac{(1-r^2)\,\psi(\vartheta)}{1-2r\cos(\vartheta-\vartheta_0)+r_0}\right]_0^{2\pi} - \frac{1}{2\pi}\int_0^{2\pi}\psi(\vartheta)\frac{d}{d\vartheta}\,\frac{1-r^2}{1-2r\cos(\vartheta-\vartheta_0)+r^2}\,d\vartheta$$

or, assuming that the function $\psi(\vartheta)$ is continued to a 2π-periodical function, defined on the entire axis $-\infty < \vartheta < +\infty$, we shall have

$$u(re^{i\vartheta_0}) = \frac{1}{2\pi}\,\frac{1-r^2}{1-2r\cos\vartheta_0+r^2}\left[\psi(2\pi-0)-\psi(+0)\right]$$
$$+\frac{1}{2\pi}\int_{-\pi}^{\pi}\psi(\vartheta_0+\vartheta)\frac{2r(1-r^2)\sin\vartheta}{(1-2r\cos\vartheta+r^2)^2}\,d\vartheta \tag{1.16}$$

If $\vartheta_0 = 0$, then $\psi(2\pi - 0) - \psi(+0) = 0$, since by the assumption $\psi(\vartheta)$ has finite derivative $\psi'(\vartheta_0)$, and if $\vartheta_0 \neq 0$ then

$$\frac{1-r^2}{1-2r\cos\vartheta+r^2}$$

tends to zero when $r \to 1-0$. Hence, when $r \to 1-0$, from (1.16) we have in both cases

$$u(re^{i\vartheta_0}) = \frac{1}{2\pi}\int_{-\pi}^{\pi}\psi(\vartheta_0+\vartheta)\frac{2r(1-r^2)\sin\vartheta}{(1-2r\cos\vartheta+r^2)^2}\,d\vartheta + o(1) \tag{1.16'}$$

Further, since $\psi'(\vartheta_0)$ exists, it is obvious that, putting

$$\psi(\vartheta_0+\vartheta) = \psi(\vartheta_0)+\psi'(\vartheta_0)\vartheta+\varepsilon(\vartheta;\vartheta_0)\vartheta \tag{1.17}$$

for any $\epsilon > 0$ we can specify $\eta = \eta(\epsilon) > 0$ $(\eta < \pi)$ such that

$$|\varepsilon(\vartheta;\vartheta_0)| < \varepsilon\,, \quad \text{when } |\vartheta| < \eta \tag{1.18}$$

Divide the integral (1.16′) into three terms I_1, I_2 and I_3, performing the integration over the intervals $[-\pi, -\eta]$, $[-\eta, \eta]$ and $[\eta, \pi]$ respectively, i.e.

$$u(re^{i\vartheta_0}) = o(1) + I_1 + I_2 + I_3\,, \quad r \to 1-0 \tag{1.19}$$

where

$$I_1 = \frac{1}{2\pi}\int_{-\pi}^{-\eta} \psi(\vartheta_0+\vartheta)\frac{2r(1-r^2)\sin\vartheta}{(1-2r\cos\vartheta+r^2)^2}\,d\vartheta$$

while I_2 and I_3 are defined by similar integrals over $[-\eta, \eta]$ and $[\eta, \pi]$ respectively.

Let $\sup\limits_{[0,2\pi]} |\psi(\vartheta)| = M$. Then it is obvious that

$$|I_1|,\ |I_3| \leqslant \frac{(1-r^2)M}{(1-2r\cos\eta+r^2)^2}$$

i.e. for $r \to 1-0$, $I_1 = I_3 = 0(1)$. Hence Eq.(1.19) takes the following form:

$$u(re^{i\vartheta_0}) = I_2 + o(1),\quad r \to 1-0 \tag{1.19'}$$

Now applying (1.17), rewrite I_2 in the form

$$\begin{aligned} I_2 &= \frac{1}{2\pi}\int_0^{\eta}[\psi(\vartheta_0+\vartheta) - \psi(\vartheta_0-\vartheta)]\frac{2r(1-r^2)\sin\vartheta}{(1-2r\cos\vartheta+r^2)^2}\,d\vartheta \\ &= -\frac{1}{2\pi}2\psi'(\vartheta_0)\int_0^{\eta}\vartheta\frac{d}{d\vartheta}\,\frac{1-r^2}{1-2r\cos\vartheta+r^2}\,d\vartheta \\ &\quad -\frac{1}{2\pi}\int_0^{\eta}[\varepsilon(\vartheta;\vartheta_0)-\varepsilon(-\vartheta;\vartheta_0)]\,\vartheta\frac{d}{d\vartheta}\,\frac{1-r^2}{1-2r\cos\vartheta+r^2}\,d\vartheta \\ &\equiv \psi'(\vartheta_0)\,I_2^{(1)} + I_2^{(2)} \end{aligned} \tag{1.20}$$

By virtue of (1.18) and of the fact that, as easily seen,

$$\vartheta\frac{d}{d\vartheta}\,\frac{1-r^2}{1-2r\cos\vartheta+r^2} \leqslant 0,\quad 0<\vartheta<\eta$$

we shall have $|I_2^{(2)}| < \epsilon I_2^{(1)}$ and hence, by virtue of (1.20),

$$I_2 = [\psi'(\vartheta_0)+\alpha]\,I_2^{(1)} \tag{1.20'}$$

where $|\alpha| < \epsilon$.

Returning to Eq.(1.19′), by virtue of (1.20′), we shall have

$$u(re^{i\vartheta_0}) = [\psi'(\vartheta_0)+\alpha]\,I_2^{(1)} + o(1),\quad r \to 1-0$$

The passage to the limit in the latter relation when r → 1 − 0 yields the statement (1.15) of the theorem if we show that

$$\lim_{r\to 1-0} I_2^{(1)} = 1$$

For this purpose let us note that

$$I_2^{(1)} \equiv -\frac{1}{\pi}\int_0^{\eta} \vartheta \frac{d}{d\vartheta}\frac{1-r^2}{1-2r\cos\vartheta+r^2}\,d\vartheta$$

$$= -\frac{1}{\pi}\eta\frac{1-r^2}{1-2r\cos\eta+r^2} + \frac{1}{\pi}\int_0^{\eta}\frac{1-r^2}{1-2r\cos\eta+r^2}\,d\vartheta$$

or

$$I_2^{(1)} = \frac{1}{\pi}\int_0^{\pi}\frac{1-r^2}{1-2r\cos\vartheta+r^2}\,d\vartheta + o(1),\ r\to 1-0 \tag{1.21}$$

since

$$\lim_{r\to 1-0}\int_{\eta}^{\pi}\frac{1-r^2}{1-2r\cos\eta+r^2}\,d\vartheta = 0$$

Noting finally that according to (1.11),

$$1 = \frac{1}{2\pi}\int_{-\pi}^{\pi}\frac{1-r^2}{1-2r\cos\vartheta+r^2}\,d\vartheta = \frac{1}{\pi}\int_0^{\pi}\frac{1-r^2}{1-2r\cos\vartheta+r^2}\,d\vartheta$$

we can represent (1.21) in the form $I_2^{(1)} = 1 + 0(1)$, $r \to 1-0$, which completes the proof of the theorem.

Since the function $\psi(\vartheta)$ of finite variation on $[0, 2\pi]$ has finite derivative almost everywhere and $\psi'(\vartheta) \in L(0, 2\pi)$, then from Theorem 1.1 follows:

Theorem 1.2. *The harmonic function* u(z) *representable by Poisson-Stieltjes integral, has radial boundary values*

$$\lim_{r\to 1-0} u(re^{i\vartheta}) = u(e^{i\vartheta}) = \psi'(\vartheta) \in L(0,2\pi) \tag{1.22}$$

almost everywhere on the circumference $z = e^{i\vartheta}$.

In particular, the Poisson-Lebesgue integral

$$u(z) = \frac{1}{2\pi}\int_0^{2\pi}\frac{1-r^2}{1-2r\cos\vartheta+r^2}f(\vartheta)\,d\vartheta \tag{1.23}$$

with an arbitrary $f(\vartheta) \in L(0, 2\pi)$, can be regarded as a Poisson-Stieltjes integral with the function

$$\psi(\vartheta) = \int_0^{\vartheta} f(\vartheta)\, d\vartheta$$

which is absolutely continuous on $[0, 2\pi]$.

Therefore the preceding theorems yield the theorem of P. Fatou (1906).

Theorem 1.3. *At every point* $z = e^{i\vartheta}$, *where* $\psi'(\vartheta) = f(\vartheta)$, *i.e. almost everywhere on* $|z| = 1$

$$\lim_{r \to 1-0} u(re^{i\vartheta}) \equiv u(e^{i\vartheta}) = f(\vartheta) \tag{1.24}$$

Part II

THE CLASS OF HARMONIC FUNCTIONS REPRESENTABLE BY THE POISSON-STIELTJES INTEGRAL

Let us first prove the Herglotz theorem (1911):

Theorem 2.1. *Let the function* $u(z)$ *be harmonic in the disc* $|z| < 1$ *and*

$$u(z) \geq 0, \quad |z| < 1 \tag{2.1}$$

Then $u(z)$ *is representable by the Poisson-Stieltjes integral:*

$$u(re^{i\varphi}) = \frac{1}{2\pi} \int_0^{2\pi} \frac{1 - r^2}{1 - 2r\cos(\vartheta - \varphi) + r^2}\, d\psi(\vartheta) \tag{2.2}$$

where $\psi(\vartheta)$ *is a bounded non-decreasing function on* $[0, 2\pi]$.

Proof. Let

$$\varphi_\rho(\vartheta) = \int_0^{\vartheta} u(\rho e^{i\alpha})\, d\alpha, \quad 0 < \rho < 1, \quad 0 \leq \vartheta \leq 2\pi \tag{2.3}$$

In view of (2.1), $\{\varphi_\rho(\vartheta)\}$ $(0 < \rho < 1)$ is a family of absolutely continuous functions non-decreasing on $[0, 2\pi]$, for which

$$0 \leq \varphi_\rho(\vartheta) \leq \int_0^{2\pi} u(\rho e^{i\alpha})\, d\alpha = 2\pi\, u(0)$$

and obviously

$$\bigvee_0^{2\pi} (\varphi_\rho) = 2\pi\, u(0)$$

According to the first theorem of Helly, this family is compact, i.e. there is a sequence $\rho_n \uparrow 1$ such that

$$\lim_{n\to\infty} \varphi_{\rho_n}(\vartheta) = \psi(\vartheta)$$

for all $\vartheta \in [0, 2\pi]$, except for a countable set of discontinuity of the function $\psi(\vartheta)$. Thus, the limit function $\psi(\vartheta)$ will be bounded and non-decreasing on $[0, 2\pi]$.

According to the second theorem of Helly,

$$\begin{aligned} J_n(re^{i\varphi}) &\equiv \frac{1}{2\pi}\int_0^{2\pi} \frac{1-r^2}{1-2r\cos(\vartheta-\varphi)+r^2}\, d\varphi_{\rho_n}(\vartheta) \\ &= \frac{1}{2\pi}\int_0^{2\pi} \frac{1-r^2}{1-2r\cos(\vartheta-\varphi)+r^2}\, d\psi(\vartheta) + o(1) \end{aligned} \tag{2.4}$$

when $n \to +\infty$.

But in view of (2.3) and the Poisson formula,

$$\begin{aligned} J_n(re^{i\varphi}) &= \frac{1}{2\pi}\int_0^{2\pi} \frac{1-r^2}{1-2r\cos(\vartheta-\varphi)+r^2}\, u(\rho_n e^{i\vartheta})\,d\vartheta \\ &= u(\rho_n r e^{i\varphi}) \end{aligned} \tag{2.4'}$$

Finally, passing to the limit in (2.4) and (2.4′) when $n \to +\infty$, we get the representation (2.2) of the theorem.

Denote by U the class of functions u(z) harmonic in the disc $|z| < 1$ and subject to condition:

$$\sup_{0<r<1}\left\{\int_0^{2\pi} |u(re^{i\varphi})|\,d\varphi\right\} \equiv J[u] < +\infty \tag{2.5}$$

Now prove the following general

Theorem 2.2. *The class* U *coincides with the set of functions* u(z) *representable by the Poisson-Stieltjes integral and also with the set of functions* u(z) *representable in the form*

$$u(z) = u_1(z) - u_2(z), \quad |z| < 1 \tag{2.6}$$

where $u_j(z) \geqslant 0$ $(j = 1, 2)$ *are harmonic functions in* $|z| < 1$.

Proof. If $u(z) \in U$, letting again

$$\varphi_\rho(\vartheta) = \int_0^{\vartheta} u(\rho e^{i\alpha})\,d\alpha, \quad 0 < \rho < 1$$

we obtain a family $\{\varphi_\rho(\vartheta)\}$ $(0 < \rho < 1)$ for which

$$|\varphi_\rho(\vartheta)| \leqslant 2\pi J[u] \quad \text{and} \quad \bigvee_0^{2\pi}(\varphi_\rho) \leqslant 2\pi J[u]$$

Therefore, repeating literally the proof of Theorem 2.1, we shall obtain the representation

$$u(re^{i\vartheta}) = \frac{1}{2\pi}\int_0^{2\pi} \frac{1-r^2}{1-2r\cos(\vartheta-\varphi)+r^2}\, d\psi(\vartheta)$$

now only with $\overset{2\pi}{\underset{0}{V}}(\psi) < +\infty$. Further, representing

$$\psi(\vartheta) = \psi_1(\vartheta) - \psi_2(\vartheta)$$

where $\psi_j(\vartheta)$ (j = 1, 2) are now non-decreasing and bounded functions on $[0, 2\pi]$, for u(z) we obtain the representation (2.6).

Finally, if u(z) permits a representation of the form (2.6), then

$$\int_0^{2\pi} |u(re^{i\varphi})|\, d\varphi \leq \int_0^{2\pi} u_1(re^{i\varphi})\, d\varphi + \int_0^{2\pi} u_2(re^{i\varphi})\, d\varphi$$

$$= 2\pi\left[u_1(0) + u_2(0)\right], \quad (0 < r < 1)$$

i.e. $u(z) \in U$.

This concludes the proof because, according to the Theorem 2.1, the functions of the type (2.6) are representable by the Poisson-Stieltjes integral.

Note in conclusion that, while proving the preceding theorems, we saw that

$$\psi(\vartheta) = \lim_{\rho_n \to 1} \int_0^{\vartheta} u(\rho_n e^{i\alpha})\, d\alpha$$

at any point $\vartheta \in [0, 2\pi]$, where the function $\psi(\vartheta)$ is continuous.

It can be proved that, if $u(z) \in U$ is also representable by means of another function $\psi_1(\vartheta)$, then

$$\psi_1(\vartheta) = \psi(\vartheta) + \text{const}$$

for all points, where the functions are both continuous. In other words, *the representation of a function* $u(z) \in U$ *by Poisson-Stieltjes integral is unique.*

Theorem 2.3 (the theorem of P. Fatou). Below, we shall lean upon a statement derived from the theorem of Vitaly:

Let the function f(z) *be analytical and bounded in a sector*

$$\Delta(\alpha) = \{z;\ |\arg z| < \alpha,\ 0 < |z| < R\}$$

and the limit

$$\lim_{r \to +0} f(r) = A$$

exists. Then for any $0 < \alpha_1 < \alpha$ *we also have*

$$\lim_{\substack{z \to 0 \\ z \in \Delta(\alpha_1)}} f(z) = A \tag{2.7}$$

From now on if the limit A in (2.7) exists, let us agree to call it an **angular boundary value of** f(z) **at** z = 0. The same term will be used if the vertex of the sector is at an arbitrary point of the boundary of the disc $|z| < 1$.

We now prove the theorem of Fatou (1906).

Theorem 2.4. *Let the function* f(z) *be analytical in the disc* $|z| < 1$ *and in addition*

$$\sup_{|z|<1} |f(z)| \leq M < +\infty$$

Then, for almost all points of the circumference $|\zeta| = 1$, *the angular boundary values*

$$f^*(\zeta) = \lim_{z \to \zeta} f(z) \tag{2.8}$$

exist, when $z \to \zeta$ *along an arbitrary way, belonging to a sector of the type*

$$\Delta_\zeta = \{z; |\arg(\zeta - z)| \leq \frac{\pi}{2} - \eta \;\; (0 < \eta < \frac{\pi}{2}), \; |\zeta - z| \leq \frac{1}{2}\} \tag{2.9}$$

Proof. If f(z) = u(z) + iv(z), then

$$\sup_{|z|<1} \left\{ \begin{matrix} |u(z)| \\ |v(z)| \end{matrix} \right\} \leq \sup_{|z|<1} |f(z)| \leq M < +\infty$$

Hence the harmonic functions u(z) and v(z) belong to the class U and are representable by Poisson-Stieltjes integrals.

Therefore, by virtue of Theorem 1.2, we can assert that almost everywhere on $|\zeta| = 1$ the radial limits

$$\lim_{r \to 1-0} f(r, \zeta) = f^*(\zeta) \tag{2.8'}$$

exist and $f^*(e^{i\vartheta}) \in L(0, 2\pi)$. But, since any radius 0ζ can be regarded as a bisectrix of a sector Δ_ζ of the type (2.9), by virtue of the corollary of Vitaly's theorem mentioned, from (2.8′) follows also (2.8), i.e. the statement of the theorem. This theorem of Fatou allows us to supplement Theorem 1.1 about the existence of boundary values of the Poisson-Stieltjes integrals.

Let $u(z) \in U$ and $u(z) = u_1(z) - u_2(z)$, $u_j(z) \geqslant 0$ (j = 1, 2), $v_j(z)$ be the harmonic functions, conjugate to $u_j(z)$. Then the functions

$$f_j(z) = \frac{1}{1 + u_j(z) + i v_j(z)} \quad (j = 1, 2)$$

are analytical and obviously bounded in the disc $|z| < 1$, and hence, according to the Theorem 2.4, it has definite angular boundary values almost everywhere on the circumference. Hence the functions $u_j(z)$ (j = 1, 2) must possess the same property, whence, with respect to the assertion

$$\lim_{r \to 1-0} u_j(re^{i\vartheta}) = \psi_j'(\vartheta) \quad (j = 1, 2)$$

of Theorem 1.2, it follows that $u_j(z)$ has angular boundary values $\psi'(\vartheta)$ almost everywhere on the circumference. Consequently the angular boundary values of the Poisson-Stieltjes integral $u(z) = u_1(z) - u_2(z)$ exist almost everywhere on the circumference $\zeta = e^{i\vartheta}$, are finite and equal to $\psi'(\vartheta) = \psi_1'(\vartheta) - \psi_2'(\vartheta)$.

In particular, the angular boundary values of the Poisson-Lebesgue integral (1.23) also exist almost everywhere on the circumference $\zeta = e^{i\vartheta}$ and are equal to $f(\vartheta)$.

Part III

1. THE OPERATOR $L^{(\omega)}$ AND ITS PROPERTIES

Denote by Ω the class of functions satisfying the following conditions:

(1) $\omega(x) > 0$ and continuous on $[0, 1)$,

(2) $\omega(0) = 1$, $\int_0^1 \omega(x)\,dx < +\infty$,

(3) $\int_0^1 |1 - \omega(x)| x^{-1}\,dx < +\infty$.

Further, let us agree to denote by P_ω the class of functions $p(\tau)$ for which

$$p(0) = 1 \text{ and } p(\tau) = \tau \int_\tau^1 \frac{\omega(x)}{x^2}\,dx, \quad \tau \in (0, 1] \tag{3.1}$$

for some $\omega(x) \in \Omega$.

In view of (1) and (2) we shall have $p(\tau) \in C[0, 1]$, and it is easily seen that

$$p(+0) = p(0) = 1, \quad p(1) = 0 \tag{3.2}$$

For $p(\tau) \in P_\omega$ we introduce the operator

$$L^{(\omega)}[\varphi] \equiv -\frac{d}{dx}\left\{x \int_0^1 \varphi(x\tau)\,dp(\tau)\right\} \quad (0, 1) \tag{3.3}$$

assuming that in proper classes of acceptable functions $\varphi(x)$ on $(0, 1)$, the right-hand part in (3.3) exists at least almost everywhere.

Note that in the simplest case, when $\omega(x) \equiv 1$ for any function $\varphi(x) \in L(0, 1)$ we shall have

$$L^{(\omega)}[\varphi(x)] = \varphi(x)$$

almost everywhere on (0, 1). In fact, in this case

$$p(\tau) = \tau \int_{\tau}^{1} \frac{dx}{x^2} = 1 - \tau$$

and therefore for almost all $x \in (0, 1)$

$$L^{(\omega)}[\varphi(x)] = \frac{d}{dx}\left\{x \int_0^1 \varphi(x\tau)\,d\tau\right\} = \frac{d}{dx}\left\{\int_0^x \varphi(\tau)\,d\tau\right\} = \varphi(x)$$

In certain classes of acceptable functions $\varphi(x)$ and under additional restrictions on the function $\omega(x) \in \Omega$, the operator $L^{(\omega)}$ also permits other representations simpler than (3.3).

Lemma 3.1. (a) *Let* $\omega(x) \in \Omega$, $\omega(x) \in C[0, 1]$ *and* $\omega'(x) \in L(0, 1)$. *Then for the class of functio* $\varphi(x)$, *piecewise continuous on* $[0, 1]$, *the formula*

$$L^{(\omega)}[\varphi(x)] = \omega(1)\varphi(x) - \int_0^1 \varphi(x\tau)\omega'(\tau)\,d\tau\,, \quad x \in (0,1) \tag{3.4}$$

holds.

(b) *For the class of functions* $\varphi(x)$, *possessing a bounded and piecewise continuous derivativ* $\varphi'(x)$ *on* $[0, 1]$,

$$L^{(\omega)}[\varphi(x)] = \varphi(0) + x \int_0^1 \varphi'(x\tau)\omega(\tau)\,d\tau, \quad x \in (0,1) \tag{3.5}$$

Proof

(a) In the case considered, we get from (3.1) by means of integration by parts

$$p(\tau) = -\omega(1)\tau + \int_{\tau}^{1} \frac{\omega'(x)}{x}\,dx + \omega(\tau)$$

and therefore

$$p'(\tau) = -\omega(1) + \int_{\tau}^{1} \frac{\omega'(x)}{x}\,dx$$

Hence, from definition (3.3) of the operator $L^{(\omega)}[\varphi]$, we find

$$L^{(\omega)}[\varphi(x)] = -\frac{d}{dx}\left\{x \int_0^1 \varphi(x\tau)\left(-\omega(1) + \int_{\tau}^{1} \frac{\omega'(t)}{t}\,dt\right)d\tau\right\}$$

$$= -\frac{d}{dx}\left\{\int_0^x \varphi(\tau)\left(-\omega(1) + \int_{\tau/x}^{1} \frac{\omega'(t)}{t}\,dt\right)d\tau\right\}, \quad x \in (0,1)$$

i.e.

$$L^{(\omega)}[\varphi(x)] = -\omega(1)\varphi(x) - \frac{1}{x}\int_0^x \varphi(\tau)\,\omega'\left(\frac{\tau}{x}\right)d\tau, \quad x \in (0,1)$$

from which Eq.(3.4) follows.

(b) By virtue of (3.2), integrating by parts we shall obtain

$$-x\int_0^1 \varphi(x\tau)\,dp(\tau) = \varphi(0)x + x^2\int_0^1 \varphi'(x\tau)\,p(\tau)\,d\tau$$

$$= \varphi(0)x + x\int_0^x \varphi'(\tau)\,p\left(\frac{\tau}{x}\right)d\tau, \quad x \in (0,1]$$

This yields

$$L^{(\omega)}[\varphi(x)] = \varphi(0) + \frac{d}{dx}\left\{x\int_0^x \varphi'(\tau)\cdot p\left(\frac{\tau}{x}\right)d\tau\right\}$$

$$= \varphi(0) + \int_0^x \varphi'(\tau)\left\{p\left(\frac{\tau}{x}\right) - \frac{\tau}{x}\,p'\left(\frac{\tau}{x}\right)\right\}d\tau, \quad x \in (0,1)$$

But from definition (3.1) of the function $p(\tau)$ it follows that, in general, $p(\tau) - \tau p'(\tau) = \omega(\tau)$. Therefore

$$L^{(\omega)}[\varphi(x)] = \varphi(0) + \int_0^x \varphi'(\tau)\,\omega\left(\frac{\tau}{x}\right)d\tau, \quad x \in (0,1)$$

which yields the correlation (3.5) of the lemma.

As is known, the Riemann-Liouville operator $D^{-\alpha}\varphi(x)$ $(-1 < \alpha < \infty)$ is defined as follows. If $0 < \alpha < +\infty$,

$$D^{-\alpha}\varphi(x) \equiv \frac{1}{\Gamma(\alpha)}\int_0^x (x-t)^{\alpha-1}\varphi(t)\,dt, \quad x \in (0,\ell)$$

and when $\varphi(x) \in L(0,\ell)$, the right-hand part exists almost everywhere and again belongs to $L(0,\ell)$. And for $-1 < \alpha < 0$

$$D^{-\alpha}\varphi(x) \equiv \frac{d}{dx}D^{-(1+\alpha)}\varphi(x), \quad x \in (0,\ell)$$

under assumption that the right-hand part exists almost everywhere. Finally, it is assumed that

$$D^{0}\varphi(x) \equiv \varphi(x), \quad x \in (0,\ell)$$

The operator $L^{(\omega)}[\varphi(x)]$ is an essential generalization of the Riemann-Liouville operator, as the following lemma shows.

Lemma 3.2. *When*

$$\omega(x) = (1-x)^{\alpha} \quad (-1<\alpha<\infty) \tag{3.6}$$

the formula

$$L^{(\omega)}[\varphi(x)] \equiv \Gamma(1+\alpha)x^{-\alpha}D^{-\alpha}\varphi(x) \tag{3.7}$$

holds almost everywhere on (0, 1). *And for the values* $0 \leqslant \alpha < +\infty$ *it is valid in the class of functions* $\varphi(x) \in L(0,1)$, *and at least in the class* $C_1[0,1]$ *of continuously differentiable functions on* [0, 1] *if* $-1<\alpha<0$.

These assertions follow directly from the assertions (a) and (b) of Lemma 3.1 if we note that

$$\omega'(x) = -\alpha(1-x)^{\alpha-1} \in L(0,1)$$

when $0<\alpha<+\infty$.

2. FORMULAE OF THE CAUCHY, SCHWARTZ AND POISSON TYPE

Let us introduce the function

$$\Delta(r) = \begin{cases} 1, & \text{when } r=0 \\ r\int_0^1 \omega(\tau)\tau^{r-1}d\tau, & 0<r<+\infty \end{cases} \tag{3.8}$$

obviously continuous on $[0,+\infty)$. Note that

$$L^{(\omega)}[x^r] = \Delta(r)x^r, \quad \begin{pmatrix} 0 \leq r < +\infty \\ 0 < x \leq 1 \end{pmatrix} \tag{3.9}$$

Indeed, according to (3.5)

$$L^{(\omega)}[x^0] = L^{(\omega)}[1] = 1 = \Delta(0)x^0$$

and for $0<r<+\infty$

$$L^{(\omega)}[x^r] = x\int_0^1 r(x\tau)^{r-1}\omega(\tau)d\tau = x^r\Delta(r)$$

Putting $\Delta_k = \Delta(k)$ $(k \geqslant 0)$, from definition (3.8) of the function $\Delta(r)$ we get

$$\Delta_0 = 1, \quad \Delta_k = k\int_0^1 \omega(\tau)\tau^{k-1}d\tau \quad (k \geq 1) \tag{3.10}$$

Since $\omega(x)>0$, $x \in [0,1)$, for any δ $(0<\delta<1)$ we have

$$0 < q(\delta) = \int_\delta^1 \omega(x)dx \leq \int_0^1 \omega(x)dx = Q < +\infty$$

and therefore it is obvious that

$$1 < q(\delta)\delta^{K-1} \leq \Delta_K \leq QK \quad (K \geq 1)$$

This yields

$$\lim_{K\to+\infty} \sqrt[K]{\Delta_K} = 1 \tag{3.11}$$

Now, consider the power series

$$C(z;\omega) \equiv \sum_{K=0}^{\infty} \frac{z^K}{\Delta_K} \quad (|z|<1) \tag{3.12}$$

and

$$S(z;\omega) = 2C(z;\omega) - 2C(0;1)$$
$$= 1 + 2\sum_{K=1}^{\infty} \frac{z^K}{\Delta_K} \quad (|z|<1) \tag{3.13}$$

By virtue of (3.11) the functions C(z; ω) and S(z; ω) are analytic in the disc $|z| < 1$ and have singularity at the point $z = 1$ and,in addition,

$$\lim_{z\to 1-0} C(z;\omega) = \lim_{z\to 1-0} S(z;\omega) = +\infty$$

Note the special but important case when

$$\omega(x) = (1-x)^\alpha \quad (-1 < \alpha < +\infty)$$

It is easy to see that in this case

$$\Delta_0 = 1, \quad \Delta_K = \frac{\Gamma(1+\alpha)\Gamma(1+K)}{\Gamma(1+\alpha+K)}$$

and therefore

$$C(z;\omega) \equiv \Gamma^{-1}(1+\alpha)\, C_\alpha(z) = \frac{1}{(1-z)^{1+\alpha}}$$

$$S(z;\omega) = \Gamma^{-1}(1+\alpha)\, S_\alpha(z) = \frac{2}{(1-z)^{1+\alpha}} - 1$$

In particular, when $\alpha = 0$, hence $\omega(x) \equiv 1$, we get

$$C(z;1) = C_0(z) = \frac{1}{1-z}$$

$$S(z;1) = S_0(z) = \frac{1+z}{1-z}$$

Direct applications of the operator $L^{(\omega)}$ result in analogues to the integral formulae of Cauchy, Schwartz and Poisson.

Theorem 3.1. *Let the function*

$$f(z)=\sum_{k=0}^{\infty} a_k z^k \tag{3.14}$$

be regular in the disc $|z|<1$. *Then*

(a) *The function*

$$L^{(\omega)}[f(re^{i\varphi})]=f_\omega(re^{i\varphi})=\sum_{k=0}^{\infty} \Delta_k a_k (re^{i\varphi})^k \tag{3.15}$$

is also regular in the disc $|z|<1$.

(b) *For any* $r\ (0<r<1)$ *the integral formulae*

$$f(z)=\frac{1}{2\pi}\int_0^{2\pi} C\left(e^{-i\vartheta}\frac{z}{r};\omega\right) f_\omega(re^{i\vartheta})\,d\vartheta \quad (|z|<r) \tag{3.16}$$

$$f(z)=i\,\mathrm{Im}\,f(0)+\frac{1}{2\pi}\int_0^{2\pi} S\left(e^{-i\vartheta}\frac{z}{r};\omega\right)\mathrm{Re}\,f_\omega(re^{i\vartheta})\,d\vartheta \quad (|z|<r) \tag{3.17}$$

hold.

Proof

(a) Since the function f(z) is regular in the disc $|z|<1$, according to (3.11) and the Cauchy-Hadamard theorem,

$$\overline{\lim_{k\to+\infty}}\sqrt[k]{|a_k|}=\overline{\lim_{k\to+\infty}}\sqrt[k]{\Delta_k|a_k|}\leq 1$$

Hence the function $f_\omega(z)$ is also regular in the disc $|z|<1$. But, since according to (3.9)

$$L^{(\omega)}[z^k]=\Delta_k z^k \quad (k=0,1,2,\ldots)$$

then, using Eq.(3.5) we get

$$L^{(\omega)}[f(re^{i\varphi})]=a_0+r\int_0^1\left\{\sum_{k=1}^{\infty} k a_k (\tau e^{i\varphi})^{k-1} e^{i\varphi}\right\}\omega(\tau)\,d\tau$$

$$=\Delta_0 a_0+\sum_{k=1}^{\infty}\left(k\int_0^1 \tau^{k-1}\omega(\tau)\,d\tau\right) a_k\,(re^{i\varphi})^k$$

$$=\sum_{k=0}^{\infty}\Delta_k a_k (re^{i\varphi})^k$$

(b) For any ρ $(0<\rho<1)$ the expansion

$$C\left(e^{-i\vartheta}\frac{z}{\rho};\omega\right)=\sum_{n=0}^{\infty}\Delta_n^{-1}\left(\frac{z}{\rho}\right)^n e^{-in\vartheta}\quad(|z|<\rho<1) \tag{3.18}$$

converges uniformly with respect to the parameter $\vartheta \in [0, 2\pi]$. Therefore, using the expansions (3.15) and (3.18) we obtain

$$\frac{1}{2\pi}\int_0^{2\pi} C\left(e^{-i\vartheta}\frac{z}{\rho};\omega\right) f_\omega(\rho e^{i\vartheta})\,d\vartheta$$

$$=\sum_{k=0}^{\infty}\sum_{n=0}^{\infty}\delta(k;n)\Delta_k a_k \Delta_n^{-1}\rho^{k-n}z^n$$

where

$$\delta(k;n)=\frac{1}{2\pi}\int_0^{2\pi} e^{ik\vartheta}e^{-in\vartheta}\,d\vartheta=\begin{cases}0, & k\neq n\\ 1, & k=n\end{cases}$$

This yields Eq.(3.16) of the theorem.

To establish the second formula (3.17) we note that, as is easily seen

$$\frac{1}{2\pi}\int_0^{2\pi} C\left(e^{-i\vartheta}\frac{z}{\rho};\omega\right)\overline{f_\omega(\rho e^{i\vartheta})}\,d\vartheta$$

$$=\overline{f(0)}\Delta_0+\sum_{k=1}^{\infty}\sum_{n=0}^{\infty}\delta(-k;n)\Delta_k\bar{a}_k\Delta_n^{-1}\rho^{k-n}z^n$$

$$=\overline{f(0)}\quad(|z|<\rho)$$

since $\delta(-k;n)=0\quad(k=1,2,\ldots;\ n=0,1,2,\ldots)$. Thus

$$\overline{f(0)}=\frac{1}{2\pi}\int_0^{2\pi} C\left(e^{-i\vartheta}\frac{z}{\rho};\omega\right)\overline{f_\omega(\rho e^{i\vartheta})}\,d\vartheta\quad(|z|<\rho)$$

Adding this to Eq.(3.16), we get

$$f(z)=-\overline{f(0)}+\frac{1}{\pi}\int_0^{2\pi} C\left(e^{-i\vartheta}\frac{z}{\rho};\omega\right)\mathrm{Re}\,f_\omega(\rho e^{i\vartheta})\,d\vartheta\quad(|z|<\rho)$$

or, since

$$C(z;\omega)=\frac{1}{2}S(z;\omega)+\frac{1}{2}$$

and

$$\frac{1}{2\pi}\int_0^{2\pi} \operatorname{Re} f_\omega(\rho e^{i\vartheta})\,d\vartheta = \operatorname{Re}\sum_{k=0}^{\infty} \delta(k;0)\Delta_k a_k$$

$$= \operatorname{Re}\Delta_0 a_0 = \operatorname{Re} f(0)$$

we arrive at Eq.(3.17) of our theorem. In the simplest case, when $\omega(x) \equiv 1$, Eqs (3.16) and (3.17) become the known formulae of Cauchy and Schwartz, described in Part I.

Finally, introduce the function

$$P_\omega(\vartheta;r) = \operatorname{Re} S(re^{i\vartheta};\omega) = 1 + 2\sum_{k=1}^{\infty} r^k \cos k\vartheta$$

Note that in the special case when $\omega(x) \equiv 1$, hence $\Delta_k = 1$ $(k = 0, 1, \ldots)$, this formula is identical to the Poisson kernel, since

$$P_1(\vartheta;r) = 1 + 2\sum_{k=1}^{\infty} r^k \cos k\vartheta$$

$$= P(\vartheta;r) = \frac{1-r^2}{1-2r\cos\vartheta + r^2} > 0 \quad \begin{pmatrix} 0 < r < 1 \\ 0 \le \vartheta \le 2\pi \end{pmatrix}$$

From Theorem 3.1 follows immediately

Theorem 3.2. *Let* u(z) *be a harmonic function in the disc* $|z| < 1$. *Then the function*

$$u_\omega(re^{i\vartheta}) = L^{(\omega)}[u(re^{i\vartheta})]$$

is also harmonic in the disc $|z| < 1$. *Also, for any* ρ $(0 < \rho < 1)$ *the integral formula*

$$u(re^{i\varphi}) = \frac{1}{2\pi}\int_0^{2\pi} P_\omega\left(\varphi-\vartheta;\frac{r}{\rho}\right) u_\omega(\rho e^{i\vartheta})\,d\vartheta \quad \begin{pmatrix} 0 \le r < \rho \\ 0 \le \varphi \le 2\pi \end{pmatrix}$$

holds.

Part IV

1. GENERAL CLASSES OF FUNCTIONS HARMONIC IN THE DISC AND THEIR REPRESENTATIONS

Denote by U_ω the set of functions u(z) harmonic in the disc $|z| < 1$, subject to condition:

$$\sup_{0<r<1}\left\{\int_0^{2\pi} |u_\omega(re^{i\varphi})|\,d\varphi\right\} < +\infty \tag{4.1}$$

where, as usual, $\omega(x) \in \Omega$ and

$$u_\omega(re^{i\varphi}) = L^{(\omega)}[u(re^{i\varphi})]$$

Since for $\omega(x) \equiv 1$

$$u_\omega(re^{i\varphi}) \equiv u(re^{i\varphi})$$

in this case the class U_ω coincides with the class U of functions u(z) harmonic in the disc $|z| < 1$, satisfying the condition

$$\sup_{0<r<1} \left\{ \int_0^{2\pi} |u(re^{i\varphi})| d\varphi \right\} < +\infty$$

As established in Part II, the class U coincides with the set of functions, permitting representation as the Poisson-Stieltjes-type integral. The following theorem gives the representation for the class U_ω with an arbitrary generator $\omega(x) \in \Omega$.

Theorem 4.1

(a) *The class* U_ω *coincides with the set of functions* u(z), *representable in the form*

$$u(re^{i\varphi}) = \frac{1}{2\pi} \int_0^{2\pi} P_\omega(\varphi - \vartheta; r)\, d\psi(\vartheta), \quad (0 \leq r < 1,\ 0 \leq \varphi \leq 2\pi) \tag{4.2}$$

where

$$\psi(\vartheta) = \lim_{n \to +\infty} \int_0^{\vartheta} u_\omega(\rho_n e^{i\alpha})\, d\alpha \tag{4.2'}$$

and $\rho_n \uparrow 1$ *is a sequence.*

(b) *The class* $U_\omega^* \subset U_\omega$ *of functions* u(z) *harmonic in the disc* $|z| < 1$, *such that*

$$u_\omega(z) \geq 0 \quad (|z| < 1)$$

coincides with the set of functions of the type (4.2) with the function $\psi(\vartheta)$ *non-decreasing on* $[0, 2\pi]$.

Proof

(a) According to Theorem 3.2, we have for any ρ $(0 < \rho < 1)$

$$u(re^{i\varphi}) = \frac{1}{2\pi} \int_0^{2\pi} P_\omega\left(\varphi - \vartheta; \frac{r}{\rho}\right) d\psi_\rho(\vartheta) \quad \begin{pmatrix} 0 \leq r < \rho \\ 0 \leq \varphi \leq 2\pi \end{pmatrix} \tag{4.3}$$

where

$$\psi_\rho(\vartheta) = \int_0^{\vartheta} u_\omega(\rho e^{i\alpha})\, d\alpha$$

On the other hand, since $u(z) \in U_\omega$, it follows from (4.1) that

$$|\psi_\rho(\vartheta)| \leqslant K, \quad \overset{2\pi}{\underset{0}{V}}(\psi_\rho) \leqslant K, \quad (0<\rho<1)$$

where

$$K = \sup_{0<\rho<1}\left\{\int_0^{2\pi} |u(\rho e^{i\alpha})|\,d\alpha\right\} < +\infty$$

Thus, for the family of functions $\Psi = \{\psi_\rho(\vartheta)\}\,(0<\rho<1)$ the conditions of the first theorem of Helly are satisfied. Hence there exists a real function $\psi(\vartheta)$ of finite variation on $[0, 2\pi]$ and a sequence $\rho_n \uparrow 1$ such that

$$\psi(\vartheta) = \lim_{n\to+\infty} \psi_{\rho_n}(e^{i\vartheta}), \quad \vartheta \in [0, 2\pi]$$

By virtue of Helly's second theorem this yields

$$\lim_{n\to+\infty} \int_0^{2\pi} P_\omega\left(\varphi-\vartheta; \frac{r}{\rho_n}\right) d\psi_{\rho_n}(\vartheta) = \int_0^{2\pi} P_\omega(\varphi-\vartheta; r)\, d\psi(\vartheta)$$

Finally, fixing the value $z = re^{i\varphi}$ ($|z| < 1$), we shall write Eq.(4.3) for $\rho = \rho_n > r$ ($n \geqslant n_0$). Then, passing to the limit in (4.3), when $n \to +\infty$, we get the representation (4.2) of our theorem.

There remains to be proved the inverse assertion that any integral of the type (4.2) represen some function $u(z) \in U_\omega$. Indeed, the harmonicity of this integral in the disc $|z| < 1$ follows from that of the kernel

$$P_\omega(\varphi-\vartheta; r) = 1 + 2\sum_{k=1}^{\infty} \Delta_k^{-1} r^k \cos k(\varphi-\vartheta)$$

for all the values of the parameter $\vartheta \in [0, 2\pi]$. Further, since

$$L^{(\omega)}[r^k] = \Delta_k r^k \quad (k = 0, 1, 2, \ldots)$$

then

$$L^{(\omega)}[P_\omega(\varphi-\vartheta; r)] = 1 + 2\sum_{k=1}^{\infty} r^k \cos k(\varphi-\vartheta)$$

$$= P(\varphi-\vartheta; r)$$

Therefore, if u(z) is representable in the form (4.2),

$$u_\omega(re^{i\varphi}) \equiv L^{(\omega)}[u(re^{i\varphi})]$$

$$= \frac{1}{2\pi}\int_0^{2\pi} P(\varphi-\vartheta; r)\, d\psi(\vartheta) \qquad (4.4$$

This means that $u_\omega(z) \in U$; in other words, $u(z) \in U_\omega$.

(b) Since $u_\omega(z) \geqslant 0$, it is obvious that for each function $u(z) \in U^*_\omega$ the function $\psi(\vartheta)$ does not decrease. Conversely, if $\psi(\vartheta) \geqslant 0$, then from (4.2) and (4.4) it follows that $u_\omega(z) \geqslant 0$, i.e. $u(z) \in U^*_\omega$,

The following theorem shows that the classes U_ω for all possible $\omega(x) \in \Omega$ exhaust the whole set of functions harmonic in the disc $|z| < 1$.

Theorem 4.2. *For any function* $u(z)$, *harmonic in the disc* $|z| < 1$, *there exists a function* $\omega_0(x) \in \Omega$ *such that* $u(z) \in U_{\omega_0}$.

Proof. Note that

$$L^{(\omega)}[u(re^{i\varphi})] = \omega(1)u(re^{i\varphi}) - \int_0^1 u(r\tau e^{i\varphi})\,\omega'(\tau)\,d\tau \tag{4.5}$$

if $\omega'(\tau) \in L(0, 1)$. Put

$$U(\tau) = \int_0^{2\pi} |u(\tau e^{i\varphi})|\,d\varphi \quad (0 < \tau < 1)$$

and assume that

$$\sup_{0<\tau<1} U(\tau) = +\infty$$

Then there exists at least one function $h(\tau)$ continuous on $[0, 1]$ such that

$$\int_0^1 U(\tau)\,h(\tau)\,d\tau < +\infty$$

For example, one can take

$$h(\tau) = \frac{1}{(1 + U(\tau))(1-\tau)\log^2 \frac{2-\tau}{1-\tau}}$$

Introduce now the function

$$\omega_0(x) = c_0 \int_x^1 h(\tau)\,d\tau, \quad \omega_0(1) = 0$$

where

$$c_0 = \left(\int_0^1 h(\tau)\,d\tau\right)^{-1}$$

It is easy to see that $\omega_0(x) \in \Omega$; hence, using Eq.(4.5), we get

$$\int_0^{2\pi} |L^{(\omega_0)}[u(re^{i\varphi})]| d\varphi \leq \int_0^1 |\omega_0'(\tau)| \left(\int_0^{2\pi} |u(r\tau e^{i\varphi})| d\varphi \right) d\tau$$

$$= \int_0^1 U(\tau) h(\tau) d\tau < +\infty \quad (0 < r < 1)$$

i.e. $u(z) \in U_{\omega_0}$ and the proof is complete.

This brings us at last to the theorem about the comparison between the classes U_ω and the class U.

Theorem 4.3

(a) *If the function* $\omega(x) \in \Omega$ *does not decrease on* $[0, 1)$, *then* $U_\omega \subset U$.

(b) *If* $\omega(x) \in \Omega$ *does not increase on* $[0, 1)$, *then* $U \subset U_\omega$.

(c) *If the* $\omega(x) \uparrow +\infty$ *or* $\omega(x) \downarrow +0$ *when* $x \uparrow 1$, *then the corresponding inclusions are strict.*

We shall not dwell on the proof of this theorem. Note that it is based on the following lemm which is easy to prove.

Lemma

(a) *If* $\omega(x) \downarrow 0$, $x \uparrow 1$, *then*

$$\lim_{r \to 1-0} (1-r) C(r; \omega) = +\infty$$

(b) *If* $\omega(x) \uparrow +\infty$, $x \uparrow 1$, *then*

$$\lim_{r \to 1-0} (1-r) C(r; \omega) = 0$$

where

$$C(r; \omega) = 1 + \sum_{k=1}^{\infty} \frac{r^k}{k \int_0^1 \omega(x) x^{k-1} dx}$$

2. A GENERALIZATION OF THE THEOREM OF HERGLOTZ

Denote by C_ω the class of functions f(z), analytical in the disc $|z| < 1$, such that

$$\operatorname{Re} f_\omega(z) \geq 0, \quad |z| < 1 \tag{4.6}$$

where $\omega(x) \in \Omega$ and $f_\omega(re^{i\varphi}) = L^{(\omega)}[f(re^{i\varphi})]$.

Further, denote by R_ω the class of functions, analytical in the disc $|z| < 1$, such that

$$\sup_{0 \leq r < 1} \left\{ \int_0^{2\pi} |\operatorname{Re} f_\omega(re^{i\varphi})| d\varphi \right\} < +\infty \tag{4.7}$$

Since the integrals

$$\int_0^{2\pi} |\operatorname{Re} f_\omega(re^{i\varphi})| d\varphi = \int_0^{2\pi} \operatorname{Re} f_\omega(re^{i\varphi}) d\varphi = 2\pi \operatorname{Re} f_\omega(0), \quad 0<r<1$$

are bounded for $f(z) \in C_\omega$, it is obvious that $C_\omega \subset R_\omega$.

In the case $\omega(x) \equiv 1$ the representations for corresponding classes have been obtained by Herglotz and Riesz.

For an arbitrary $\omega(x) \in \Omega$ there holds

Theorem 4.4

(a) *The class* C_ω *coincides with the set of functions* f(z) *representable in the form*

$$f(z) = iC + \frac{1}{2\pi}\int_0^{2\pi} S(e^{-i\vartheta}z; \omega) d\psi(\vartheta) \quad (|z|<1) \tag{4.8}$$

where $\operatorname{Im} C = 0$, $\psi(\vartheta)$ *is an arbitrary non-decreasing function, bounded on* $[0, 2\pi]$.

(b) *The class* R_ω *coincides with the set of functions representable in the form* (4.6), *where* $\psi(\vartheta)$ *is an arbitrary function of finite variation on* $[0, 2\pi]$.

Proof. If the function f(z) allows the representation (4.8), then

$$\operatorname{Re} f(re^{i\varphi}) = u(re^{i\varphi}) = \frac{1}{2\pi}\int_0^{2\pi} P_\omega(\varphi-\vartheta; r) d\psi(\vartheta)$$

and according to Theorem 4.1, $u(z) \in U_\omega$. This means that the function

$$u_\omega(re^{i\varphi}) = L^{(\omega)}[\operatorname{Re} f(re^{i\varphi})] = \operatorname{Re} L^{(\omega)}[f(re^{i\varphi})] = \operatorname{Re} f_\omega(re^{i\varphi})$$

satisfies the condition (4.6), i.e. $f(z) \in C_\omega$ if $\psi(\vartheta)$ does not decrease and is bounded on $[0, 2\pi]$ and the condition (4.7), i.e. $f(z) \in R_\omega$ if $\psi(\vartheta)$ is a function of bounded variation on $[0, 2\pi]$.

Conversely, let $f(z) \in R_\omega$; then, denoting

$$\psi_\rho(\vartheta) = \int_0^{\vartheta} \operatorname{Re} f_\omega(\rho e^{i\alpha}) d\alpha, \quad 0<\rho<1$$

according to Eq.(3.17) of Theorem 3.1, we have

$$f(z) = i \operatorname{Im} f(0) + \frac{1}{2\pi}\int_0^{2\pi} S\left(e^{-i\vartheta}\frac{z}{\rho}; \omega\right) d\psi_\rho(\vartheta) \quad (|z|<\rho)$$

Now, in the same way as in the proof of Theorem 4.1, we shall get the required representation (4.8) and with the function

$$\psi(\vartheta) = \lim_{n\to+\infty} \int_0^{\vartheta} \operatorname{Re} f_\omega(\rho_n e^{i\alpha}) d\alpha, \quad \vartheta \in [0, 2\pi]$$

for which $\overset{2\pi}{\underset{0}{V}}(\psi) < +\infty$. In particular, if $f(z) \in C_\omega$, i.e. $\mathrm{Re} f_\omega(z) \geqslant 0$, then $\psi(\vartheta)$ will be non-decreasing. The theorem is proved.

Part V

THEORY OF FACTORIZATION AND BOUNDARY PROPERTIES OF FUNCTIONS MEROMORPHIC IN THE DISC

Part V consists of a brief survey of basic results obtained in the period 1964–1972 in the theory of factorization of meromorphic functions. A detailed account of these results would require too much space. For closer acquaintance with the methods of these investigations we refer to the Bibliography for Part V, at the end of this paper.

(A) The new classical formula of Jensen-Nevanlinna,and the most important notion of the characteristic function which is deduced from it, constitute the basis of the modern theory of meromorphic functions. We shall remind you this formula and the definition of the characteristic function.

Let F(z) be a function meromorphic in the disc $|z| < 1$, $\{a_\mu\}$ and $\{b_\nu\}$ be the sequences of zeros and poles of F(z), respectively, different from $z = 0$, numbered in the order of non-decreasing modulus and in accordance with their multiplicity. Then, for any ρ $(0 < \rho < 1)$ the Jensen-Nevanlinna formula

$$\ln F(z) = ic + \lambda \ln \frac{z}{\rho} + \sum_{0<|a_\mu|\leq\rho} \ln \frac{\rho(a_\mu - z)}{\rho^2 - \bar{a}_\mu z} \frac{|a_\mu|}{a_\mu} - \sum_{0<|b_\nu|\leq\rho} \ln \frac{\rho(b_\nu - z)}{\rho^2 - \bar{b}_\nu z} \frac{|b_\nu|}{b_\nu}$$

$$+ \frac{1}{2\pi} \int_0^{2\pi} \frac{\rho e^{i\vartheta} + z}{\rho e^{i\vartheta} - z} \ln |F(\rho e^{i\vartheta})| d\vartheta \quad (|z| < \rho) \tag{5.1}$$

holds, where $|\lambda|$ is the multiplicity of the origin (a zero of the function F(z) if $\lambda \geqslant 1$ and a pole if $\lambda \leqslant -1$); C is a real constant, defined exactly up to the items of the type $2\pi m$.

The problem of complete description of the class of functions F(z) meromorphic in the disc $|z| < 1$ for which it is possible to perform the termwise passage to the limit in the Jensen-Nevanlinna formula,and thus obtain a representation for the function $\ln F(z)$ in the entire open disc $|z| < 1$, was posed and solved originally by R. Nevanlinna in the mid-twenties.

As in the investigations in the value distribution theory of meromorphic functions, the solution of this problem is also based on the determination of the characteristic function:

$$T(\rho; F) = \frac{1}{2\pi} \int_0^{2\pi} \ln^+ |F(\rho e^{i\vartheta})| d\vartheta + \int_0^{+\infty} \frac{n(t;\infty) - n(0;\infty)}{t} dt + n(0;\infty) \ln \rho \tag{5.2}$$

where $n(t; \infty)$ is the number of poles $\{b_\nu\}$ in the disc $|z| \leqslant t$ $(0 \leqslant t < 1)$;

$$\ln^+ x = \max\{\ln x \,;\, 0\} \quad (0 < x < +\infty)$$

Defining the class N as the set of functions meromorphic in $|z| < 1$ for which

$$\sup_{0<r<1} T(r; F) < +\infty ,$$

Nevanlinna showed that the termwise passage to the limit in Eq.(5.1) mentioned is possible only for the functions of the class N. Thus the following theorem about the parametric representation and factorization of the class N was proved.

The class N *coincides with the set of functions* F(z) *representable in the form*

$$F(z) = e^{i\gamma} z^{\lambda} \frac{B(z; a_\mu)}{B(z; b_\nu)} \exp\left\{ \frac{1}{2\pi} \int_0^{2\pi} \frac{e^{i\vartheta} + z}{e^{i\vartheta} - z} d\psi(\vartheta) \right\} \tag{5.3}$$

where

$$B(z; a_\mu) = \prod_{(\mu)} \frac{a_\mu - z}{1 - \bar{a}_\mu z} \frac{|a_\mu|}{a_\mu} , \quad B(z; b_\nu) = \prod_{(\nu)} \frac{b_\nu - z}{1 - \bar{b}_\nu z} \frac{|b_\nu|}{b_\nu}$$

are convergent Blaschke products, $\psi(\vartheta)$ *is a real function of bounded total variation on* $[0, 2\pi]$, $\lambda \gtrless 0$ *is an arbitrary integer, and* γ *is an arbitrary real number.*

Essentially, this fundamental result by Nevanlinna became the cornerstone and the base of the construction of an elegant theory of classes A, H_δ, etc., and their boundary properties. This result has also had substantial applications in other areas of function theory, particularly in approximation theory of functions of the complex variable, in the theory of operators.

(B) In connection with this basic theorem of Nevanlinna and its method of proof, the following problem arises naturally. Do there exist other, more general formulae of the Jensen-Nevanlinna type, permitting the establishment of parametric representation and factorization for wider as well as for more restricted classes than N of functions meromorphic in the disc?

In the early papers of the present author (1945), the first attempt in the direction of solving a problem of this kind was apparently undertaken, although the result obtained there was far from being completed. A generalization of the Jensen-Nevanlinna formula was obtained there, being, as a matter of fact, connected with the integrals of fractional order. However, we obtained only a **canonical** but **not a parametric** representation for the classes N_α^* $(0 < \alpha < +\infty)$ of functions meromorphic in the disc $|z| < 1$, whose characteristics satisfy a condition of the type

$$\int_0^1 (1-r)^{\alpha-1} T(r; F) dr < +\infty$$

So, up to 1964, there had in fact been neither natural inner characteristics nor factorization theorems of the Nevanlinna type, covering his theorem or making it more exact for wider as well as for more restricted classes than N of meromorphic functions.

Considerably later (1964), this author returned to the problem of factorization of meromorphic functions in a general form, constructing a general theory of parametric representation of the classes N_α $(-1 < \alpha < +\infty)$ of functions meromorphic in the disc $|z| < 1$, as complete as the now classical theory of the class N. These results were presented in detail in the monograph by the author (1966). The peculiarity of the method of this investigation was the systematic application of the apparatus of the Riemann-Liouville fractional integro-differential operators $D^{-\alpha}$ $(-1<\alpha<+\infty)$.

By means of these operators an important class of new formulae of Jensen-Nevanlinna type was discovered, depending on an arbitrary parameter α $(-1<\alpha<+\infty)$, coinciding with the classical formula (5.1) for the special value $\alpha=0$. These principally new formulae lead naturally to the considerably more general notion of the α-characteristic function $T_\alpha(r;\ F)$ also coinciding with the usual characteristic $T(r; F)$ for $\alpha=0$.

These classes N_α $(-1<\alpha<+\infty)$, possessing the important property of inclusion $N_{\alpha_1}\subset N_{\alpha_2}$ $(-1<\alpha_1<\alpha_2<+\infty)$, and coinciding with the class N of Nevanlinna only in the case $\alpha=0$, were defined by means of the condition

$$\sup_{0<r<1} T_\alpha(r; F) < +\infty$$

The basic theorem of parametric representation of the classes N_α reads:

The class N_α $(-1<\alpha<+\infty)$ *coincides with the set of functions, representable in the form*

$$F(z)=C z^\lambda \frac{B_\alpha(z;a_\mu)}{B_\alpha(z;b_\nu)} \exp\left\{\frac{1}{2\pi}\int_0^{2\pi}\left[\frac{2}{(1-e^{-i\vartheta}z)^{1+\alpha}}-1\right]d\psi(\vartheta)\right\} \tag{5.4}$$

where C *is a constant,* $\psi(\vartheta)$ *a real function of finite total variation on* $[0, 2\pi]$, *and* $B_\alpha(z; a_\mu)$ *and* $B_\alpha(z; b_\nu)$ *are certain convergent products in the disc* $|z|<1$ *with zeros* $\{a_\mu\}$ *and* $\{b_\nu\}$ *subject to conditions:*

$$\sum_{(\mu)}(1-|a_\mu|)^{1+\alpha}<+\infty,\quad \sum_{(\nu)}(1-|b_\nu|)^{1+\alpha}<+\infty \tag{5.5}$$

In the case $\alpha=0$ this theorem is merely reduced to the theorem of Nevanlinna formulated above.

The theory of factorization of the classes N_α, although being considerably general, could not be regarded as final. The point is that in the case when $0<\alpha<+\infty$ $(N_\alpha\supset N)$, they were still not able to include the functions meromorphic in the disc $|z|<1$, having characteristics $T(r; F)$ growing at an arbitrarily rapid rate when $r\to 1-0$. And in the case when $-1<\alpha<0$ $(N_\alpha\subset N)$, the constructed theory also needed to be made more precise, since for any α $(-1<\alpha<+\infty)$ the classes N_α covered only the meromorphic functions whose distribution of zeros and poles was characterized only by means of a condition of the type (5.5). So, e.g., the functions from N_α without zeros and poles were being represented only by the formula:

$$\exp\left\{\frac{1}{2\pi}\int_0^{2\pi}\frac{d\psi(\vartheta)}{(1-ze^{-i\vartheta})^{1+\alpha}}\right\}$$

Thus, the problem naturally arose of constructing the theory of factorization of functions meromorphic in the disc $|z|<1$ in the sense that it could simultaneously include the functions with characteristics $T(r)$ arbitrarily increasing when $r\to 1-0$, as well as the functions with arbitrary rare or dense distribution of zeros and poles.

ization of functions meromorphic in the
a special case, was constructed in the recent

ts. Denote by Ω the set of functions $\omega(r)$,

r), $r \in [0, 1)$, the operator $L^{(\omega)}\{\varphi(r)\}$ will

$\varphi(z)$

In particular, in the class of functions $\varphi(r)$ with bounded and piecewise continuous derivative $\varphi'(r)$, the operator $L^{(\omega)}\{\varphi(r)\}$ is representable in the form

$$L^{(\omega)}\{\varphi(z)\} = \varphi(0) + z\int_0^1 \varphi'(z\tau)\,\omega(\tau)\,d\tau, \quad z \in (0,1]$$

The operator $L^{(\omega)}\{\varphi(r)\}$ is an essential generalization of the Riemann-Liouville operator $D^{-\alpha}$ $(-1 < \alpha < +\infty)$, since, as in the special case, the identity

$$L^{(\omega)}\{\varphi(z)\} \equiv \Gamma(1+\alpha)\, z^{-\alpha} D^{-\alpha}\varphi(z)$$

holds, when $\omega(x) = (1-x)^{\alpha}$ $(-1<\alpha<+\infty)$. The functions

$$C(z;\omega) = \sum_{k=0}^{\infty} \frac{z^k}{\Delta(k)}, \quad S(z;\omega) = 1 + 2\sum_{k=1}^{\infty} \frac{z^k}{\Delta(k)}$$

where

$$\Delta(0)=1\,,\ \Delta(\lambda)=\lambda\int_0^1 \omega(\tau)\tau^{\lambda-1}d\tau\,,\ \lambda\in(0,+\infty)$$

are associated with the operator $L^{(\omega)}$.

The evaluation of the simplest functions r^λ and $\log r$ by the operator $L^{(\omega)}$ results in the important correlations:

$$L^{(\omega)}\{r^\lambda\}=\Delta(\lambda)r^\lambda,\quad L^{(\omega)}\{\log r\}=-K_\omega+\log r$$

With the operator $L^{(\omega)}$ are also associated the following three functions, playing an important role in the entire theory:

$$V_\omega(re^{i\varphi};\zeta)=L^{(\omega)}\left\{\log\left|1-\frac{re^{i\varphi}}{\zeta}\right|\right\},\qquad W_\omega(re^{i\varphi};\zeta)=\frac{1}{2\pi}\int_0^{2\pi}S(e^{i\vartheta}z;\omega)V_\omega(e^{i\vartheta};\zeta)d\vartheta$$

$$A_\omega(z;\zeta)=\left(1-\frac{z}{\zeta}\right)e^{-W_\omega(z;\zeta)},\quad (0<|\zeta|\leq 1)$$

Theorem 5.1. *For an arbitrary function* F(z) *meromorphic in the disc* $|z|<1$ *having the zeros* $\{a_\mu\}$ *and poles* $\{b_\nu\}$ *and for arbitrary* $\omega(r)\in\Omega$ *and* ρ $(0<\rho<1)$, *the formula*

$$\ln F(z)=i\operatorname{Arg}c_\lambda+\lambda K_\omega+\lambda\ln\frac{z}{\rho}+\sum_{0<|a_\mu|\leq\rho}\ln A_\omega\left(\frac{z}{\rho};\frac{a_\mu}{\rho}\right)-\sum_{0<|b_\nu|\leq\rho}\ln A_\omega\left(\frac{z}{\rho};\frac{b_\nu}{\rho}\right)$$

$$+\frac{1}{2\pi}\int_0^{2\pi}S\left(e^{-i\vartheta}\frac{z}{\rho};\omega\right)L^{(\omega)}\{\ln|F(\rho e^{i\vartheta})|\}d\vartheta\quad(|z|<\rho) \tag{5.6}$$

holds, where λ *and* C_λ *are defined from the development* $F(z)=c_\lambda z^\lambda+c_{\lambda+1}z^{\lambda+1}+\dots$ $(c_\lambda\neq 0)$ *in a neighbourhood of the point* $z=0$.

Note that in the special case $\omega(r)\equiv 1$, when Eq.(5.4) coincides with that of Jensen-Nevanlinna here, as in the theory of Nevanlinna, the general formula (5.4) leads naturally to the definition of the functions:

$$m_\omega(r;F)\equiv\frac{1}{2\pi}\int_0^{2\pi}L_+^{(\omega)}\{\ln|F(re^{i\varphi})|\}d\varphi \tag{5.7}$$

and

$$N_\omega(r;F)\equiv\int_0^1\frac{n(t;\infty)-n(0;\infty)}{t}\,\omega\left(\frac{t}{r}\right)dt+n(0;\infty)\{\log r-K_\omega\} \tag{5.8}$$

where $L_+^{(\omega)}=\max\{L^{(\omega)},0\}$ and $n(t;\infty)$ is the number of poles of the function F(z) in the disc $|z|\leq t$, counted in accordance with their multiplicity.

By means of these functions, passing to fundamental functions m(r; F) and N(r; F) of the value distribution theory of Nevanlinna, when $\omega(r) \equiv 1$, the function

$$T_\omega(r;F) \equiv m_\omega(r;F) + N_\omega(r;F) \tag{5.9}$$

is at last defined. We shall call this function the ω characteristic. It coincides with the function T(r; F) of Nevanlinna when $\omega(r) \equiv 1$ and for it an equilibrium correlation of the type

$$T_\omega(r;F) = \text{const} + T_\omega\left(r;\frac{1}{F}\right) \tag{5.10}$$

also holds.

Finally, with every function $\omega(r) \in \Omega$ a class $N\{\omega\}$ is associated as the set of functions F(z) meromorphic in the disc $|z| < 1$ subject to the condition

$$\sup_{0<r<1} T_\omega(r;F) < +\infty$$

Note that the class $N\{\omega\}$ coincides with the class N of Nevanlinna in the special case when $\omega(r) \equiv 1$. And in the case $\omega(r) = (1-r)^\alpha (-1 < \alpha < +\infty)$ the class $N\{\omega\}$ coincides with the class N_α mentioned above. The comparison between the classes $N\{\omega\}$, when $\omega(r) \not\equiv 1$, and the class N is given in the following theorem.

Theorem 5.2. *Let $\omega(r) \in \Omega$. Then*

(a) *If $\omega(r)$ is non-decreasing on $[0, 1)$, then the inclusion $N\{\omega\} \subset N$ holds. Moreover it is strict if $\omega(r) \uparrow +\infty$ as $r \uparrow 1-0$.*

(b) *If $\omega(r)$ is non-increasing on $[0, 1)$, then the inclusion $N \subset N\{\omega\}$ holds, and it is strict if $\omega(r) \downarrow 0$ as $r \uparrow 1-0$.*

(D) Passing to the most fundamental theorems of the theory of factorization of the classes $N\{\omega\}$ brings us first to one of the central theorems of that theory.

Theorem 5.3. *Let $\omega(r) \in \Omega$ and $\{z_k\}_1^\infty (0 < |z_k| \leqslant |z_{k+1}| < 1)$ be an arbitrary sequence of complex numbers.*

(a) *In order that the infinite product*

$$B_\omega(z) \equiv B_\omega(z;z_k) = \prod_{k=1}^{\infty} A_\omega(z;z_k) \tag{5.11}$$

converges and defines a function $B_\omega(z) \not\equiv 0$ analytical in the disc $|z| < 1$, vanishing on the sequence $\{z_k\}_1^\infty$, it is necessary and sufficient that the condition

$$\sum_{k=1}^{\infty} \int_{|z_k|}^{1} \omega(x)\,dx < +\infty \tag{5.12}$$

holds.

(b) *For any convergent product $B_\omega(z)$ we have*

$$B_\omega(z) \in N\{\omega\}, \quad \frac{1}{B_\omega(z)} \in N\{\omega\} \tag{5.13}$$

The function $B_\omega(z)$ is the natural analogue of the classical Blaschke product

$$B(z)=\prod_{k=1}^{\infty}\frac{z_k-z}{1-\bar{z}_k z}\,\frac{|z_k|}{z_k}$$

when instead of the known condition

$$\sum_{k=1}^{\infty}(1-|z_k|)<+\infty$$

the sequence $\{z_k\}_1^\infty$ satisfies the condition (5.12). In addition,

$$[B_\omega(z)]_{\omega\equiv 1}\equiv B(z)$$

Formulate now the basic theorem about factorization and the parametric representation of the classes $N\{\omega\}$.

Theorem 5.4. *The class* $N\{\omega\}$ *coincides with the set of functions representable in the disc* $|z|<1$ *in the form*

$$F(z)=e^{i\gamma+\lambda K_\omega}z^\lambda\frac{B_\omega(z;a_\mu)}{B_\omega(z;b_\nu)}\exp\left\{\frac{1}{2\pi}\int_0^{2\pi}S(e^{-i\vartheta}z;\omega)\,d\psi(\vartheta)\right\} \tag{5.14}$$

where $B_\omega(z;a_\mu)$ *and* $B_\omega(z;b_\nu)$ *are arbitrary convergent products of the form (5.11),* $\psi(\vartheta)$ *is a real function on* $[0,2\pi]$ *with* $\bigvee_0^{2\pi}(\psi)<+\infty$, $\lambda\gtrless 0$ *is an arbitrary integer,* γ *is an arbitrary real number and, finally,*

$$K_\omega=\int_0^1\frac{1-\omega(x)}{x}\,dx$$

In the special case when $\omega(r)=(1-r)^\alpha(-1<\alpha<+\infty)$ from the representation (5.14) of this theorem, the representation classes N_α are obtained.

In the case when the function $\omega(r)\in\Omega$ is non-decreasing on $[0,1)$ and thus the inclusion $N\{\omega\}\subset N$ is true, the following theorem holds.

Theorem 5.5

(a) *if the sequence* $\{z_k\}_1^\infty$ *satisfies the condition*

$$\sum_{k=1}^{\infty}\int_{|z_k|}^{1}\omega(x)\,dx<+\infty$$

then the representation

$$B_\omega(z;z_k)=B(z;z_k)\exp\left\{\frac{1}{2\pi}\int_0^{2\pi}S(e^{-i\vartheta}z;\omega)\,d\mu(\vartheta)\right\} \tag{5.15}$$

holds, where $\mu(\vartheta)$ *is a certain non-increasing and bounded function on* $[0,2\pi]$.

(b) *The class* $N\{\omega\} \subset N$ *coincides with the set of functions representable in the disc* $|z| < 1$ *in the form*

$$F(z) = e^{i\gamma + \lambda k_\omega} z^\lambda \frac{B(z; a_\mu)}{B(z; b_\nu)} \exp\left\{\frac{1}{2\pi}\int_0^{2\pi} S(e^{-i\vartheta}z; \omega)\, d\psi(\vartheta)\right\} \tag{5.16}$$

where the parameters γ, $\lambda \gtrless 0$ *and* k_ω *are the same as in Theorem 5.4,* $B(z; a_\mu)$ *and* $B(z; b_\nu)$ *are the Blaschke functions with zeros satisfying the conditions*

$$\sum_{(\mu)} \int_{|a_\mu|}^{1} \omega(x)\, dx < +\infty, \qquad \sum_{(\nu)} \int_{|b_\nu|}^{1} \omega(x)\, dx < +\infty \tag{5.17}$$

(c) *Any function* $F(z) \in N\{\omega\} \subset N$ *is representable in the form*

$$F(z) = \frac{f_1(z)}{f_2(z)} \quad (|z| < 1) \tag{5.18}$$

where $f_k(z) \in N\{\omega\}$, $|f_k(z)| \leq 1$ $(|z| < 1)$ *are analytical in the disc* $|z| < 1$.

We now come to the theorem giving the complete solution of the problem of factorization of the entire family of functions meromorphic in the disc $|z| < 1$.

Theorem 5.6. *For any function* $F(z)$ *meromorphic in the disc* $|z| < 1$ *there exists a function* $\omega_F(r) \in \Omega$ *such that* $F(z) \in N\{\omega_F\}$.

Note, in conclusion, that the passage from the theory of the classes N_α $(-1 < \alpha < +\infty)$ to the construction of the theory of the classes $N\{\omega\}$, containing the completion of the factorization theory, required considerable efforts in the case of the disc. Here, at first, the theory of operators $L^{(\omega)}$ associated with the given function $\omega(r) \in \Omega$ had to be constructed. This theory permitted the family of Eq.(5.6) of Theorem 1 to be discovered, which became the basis of the entire cycle of our investigations.

Just as an example, we point out the following results, which played a principal role in the theory of classes $N\{\omega\} \subset N$ and their boundary properties.

Theorem 5.7. *Let* $\omega(r) \in \Omega$ *be non-decreasing on* $[0, 1)$. *Then*

(a) *The problem of the Hausdorff moments*

$$\mu_n = \frac{1}{\Delta(n)} = \int_0^1 x^n\, d\alpha(x) \qquad (n = 0, 1, 2, \ldots)$$

has a solution $\alpha(x)$ *in the class of non-decreasing functions bounded on* $[0, 1]$.

(b) *In the disc* $|z| < 1$

$$\operatorname{Re} S(z; \omega) \geq 0, \quad \operatorname{Re} C(z; \omega) \geq 0$$

(E) As is well known, the class N possesses important boundary properties. For any function $F(z) \in N$ the limit

$$F(e^{i\vartheta}) = \lim_{r \to 1-0} F(re^{i\vartheta}) \tag{5.19}$$

exists for almost all $\vartheta \in [0, 2\pi]$ and if $F(z) \not\equiv 0$ then

$$\int_0^{2\pi} \left| \ln |F(e^{i\vartheta})| \right| d\vartheta < +\infty \tag{5.20}$$

But according to Theorem 5.2 we have

$$N\{\omega\} \supset N \text{ if } \omega(r) \downarrow 0,\ r \uparrow 1 \quad \text{and} \quad N\{\omega\} \subset N \text{ if } \omega(r) \uparrow +\infty,\ r \uparrow 1$$

The boundary properties of the classes $N\{\omega\}$ have been investigated for the two following cases:

(1) The case $N\{\omega\} \supset N$. Here the following theorems are established.

Theorem 5.8. *Let* $F(z) \in N\{\omega\}$ *where* $\omega(r)$ *is non-increasing and satisfying the condition Lip 1 on every interval* $[0, \Delta]$.[1] *Then*

(a) *The limit*

$$\lim_{r \to 1-0} \operatorname{Re} L^{(\omega)}\{\ln F(re^{i\vartheta})\} = \psi'(\vartheta) \in L(0, 2\pi) \tag{5.21}$$

exists almost everywhere on $[0, 2\pi]$.

(b) *We also have*

$$\lim_{r \to 1-0} \operatorname{Re} L^{(\omega)}\{\ln B_\omega(re^{i\varphi})\} = 0 \tag{5.22}$$

almost everywhere on $[0, 2\pi]$.

The following theorem of uniqueness is an enlargement of the now classical theorem of Szegö for the classes $N\{\omega\} \supset N$.

Theorem 5.9. *Let* $f(z) \in N\{\omega\}$ *be analytical in the disc* $|z| < 1$. *Then*

(a) *If* $\omega(r) \in \tilde{\Omega}$ *is non-increasing on* $[0, 1)$, *then its boundary values*

$$L^{(\omega)}\{\log |f(e^{i\vartheta})|\} = \lim_{r \to 1-0} L^{(\omega)}\{\log |f(re^{i\vartheta})|\}$$

are such that

$$\int_0^{2\pi} L^{(\omega)}\{\log |f(e^{i\vartheta})|\} d\vartheta > -\infty \tag{5.23}$$

(b) *There is no function* $f(z) \not\equiv 0$ *analytical in the disc* $|z| < 1$ *for which*

$$\int_0^{2\pi} L^{(\omega)}\{\log |f(e^{i\vartheta})| d\vartheta\} = -\infty \tag{5.24}$$

[1] Such functions will be further attributed to the class $\tilde{\Omega}$.

or

$$\sum_{(\mu)} \int_{|a_\mu|}^{1} \omega(x)\,dx = +\infty, \quad \{f(a_\mu)=0\} \tag{5.25}$$

(2) The case N ⊃ N{ω}. In connection with the known boundary properties of classes N ⊃ N{ω} noted above, the following questions arise naturally.

Does the exceptional set of linear measure zero $E \subset [0, 2\pi]$ get thinner for the classes $N\{\omega\} \subset N$ where the limit $F(e^{i\vartheta})$ of a function $F(z) \in N\{\omega\}$ perhaps does not exist? Can anything else be stated about the boundary values $F(e^{i\vartheta})$ of a function $F(z) \in N\{\omega\} \subset N$ besides the boundedness of the integral (5.20)? The author and V.S. Zakharian succeeded in obtaining the positive solution of both these problems in Refs [4] and [5].

To formulate the results obtained here, it is necessary to introduce a definition.

Let $\omega(r) \in \Omega$ be non-decreasing on (0, 1) and

$$C(z;\omega) = \sum_{k=0}^{\infty} \frac{z^k}{k\int_0^1 \omega(x)\,x^{k-1}\,dx}$$

be a function associated with it.

We shall assume that the B-measurable set $E \subset [0, 2\pi]$ has positive ω-capacity if there exists a measure $\mu \prec E$ such that the integral

$$U_\omega(z;\mu) = \int_0^{2\pi} |C(ze^{-i\vartheta};\omega)|\,d\mu(\vartheta) \tag{5.26}$$

satisfies the condition

$$\sup_{|z|\le 1} U_\omega(z;\mu) < +\infty$$

If there is no such measure, i.e. if for any measure $\mu \prec E$, $\mu(E) = 1$

$$\sup_{|z|\le 1} U_\omega(z;\mu) = +\infty$$

we shall assume that the ω-capacity of the set E is equal to zero. By this, we shall write $C_\omega(E) > 0$ or $C_\omega(E) = 0$. Note that in the special case when $\omega(r) = (1-r)^\alpha(-1 < \alpha < 0)$, $C_\omega(E)$ is nothing but the $(1+\alpha)$-capacity of the set E in the Frostman sense.

Theorem 5.10. *For any function* $F(z) \in N\{\omega\} \subset N$ *the bounded limit*

$$F(e^{i\vartheta}) = \lim_{r\to 1-0} F(re^{i\vartheta})$$

exists for all $\vartheta \in [0, 2\pi]$ *except perhaps an exceptional set* $E \subset [0, 2\pi]$ *such that* $C_\omega(E) = 0$.

Theorem 5.11. *Let* $F(z) \in N\{\omega\} \subset N$ *and* $E \subset [0, 2\pi]$ *be any set for which* $C_\omega(E) > 0$. *Further, let* $\mu \prec E$, $\mu(E) = 1$ *be any measure with the property*

$$U_\omega(\mu) \equiv \sup_{|z|\le 1} U_\omega(z;\mu) < +\infty$$

Then the boundary values $F(e^{i\vartheta})$ *of the function* $F(z)$ *satisfy the condition*

$$\int_E |\log|F(e^{i\vartheta})||\,d\mu(\vartheta) = \int_0^{2\pi} |\log|F(e^{i\vartheta})||\,d\mu(\vartheta) < +\infty$$

The solution of the above problem about the boundary properties of the functions of classes $N\{\omega\} \subset N$ is contained in Theorems 5.10 and 5.11.

As is well known, if $f(z)$ is an analytical function from the class N, then $f(z) \equiv 0$ $(|z|<1)$ if $f(e^{i\vartheta}) = 0$, $\vartheta \in E$, $\text{mes}\, E > 0$.

Theorem 5.12. *Let* $f(z) \in N\{\omega\}$ *be analytical in the disc* $|z| < 1$ *and* $f(e^{i\vartheta}) = 0$, $\vartheta \in E$ *where* $\text{mes}\, E = 0$ *but* $C_\omega(E) > 0$. *Then* $f(z) \equiv 0$ $(|z| < 1)$.

The proofs of Theorems 5.10 and 5.11 lean essentially upon a series of fine lemmas and, in particular, upon the following theorems about the functions of Blaschke and $S(z; \omega)$.

Theorem 5.13. *Let* $\omega(r) \in \Omega$ *be non-decreasing on* $[0, 1)$. *Then the Blaschke function*

$$B(z) = \prod_{k=1}^{\infty} \frac{z_k - z}{1 - \bar{z}_k z} \frac{|z_k|}{z_k}$$

belongs to the class $N\{\omega\}$ *if and only if*

$$\sum_{k=1}^{\infty} \int_{|z_k|}^{1} \omega(x)\,dx < +\infty$$

Theorem 5.14. *If* $\omega(r) \in \Omega$ *does not decrease on* $[0, 1)$ *then the integral formula*

$$S(z;\omega) = \int_0^1 \frac{1+tz}{1-tz}\,d\alpha(t)$$

holds, where $\alpha(t)$ *is a non-decreasing function bounded on* $[0, 1]$.

Part VI

HARMONIC ANALYSIS IN THE COMPLEX DOMAIN AND ITS APPLICATIONS IN THE THEORY OF ANALYTICAL AND INFINITELY DIFFERENTIABLE FUNCTIONS

Part VI is devoted to a survey of a large cycle of investigations by the author and his students concerning harmonic analysis in the complex domain and the most important applicatior of the theory developed here to the solution of certain fundamental problems of the classical theory of functions. The major part of these investigations has been published in different mathe matical journals and also in a monograph by the author [7–10].

1. THEORY OF INTEGRAL TRANSFORMS WITH MITTAG-LEFFLER AND VOLTERRA KERNELS (Plancherel-type theorems in the complex domain)

(A) An important landmark in the development of functional analysis in general and of the theory of harmonic analysis in particular was the fundamental theorem of Fourier transforms in the classes L_2 established in 1910 by M. Plancherel.

It is well known that for the complex form of Fourier transforms this theorem reads:

(a) *For any function* $f(x) \in L_2(-\infty, +\infty)$ *the following limit in the mean exists:*

$$F(u) = \underset{\sigma \to +\infty}{\text{l.i.m.}} \frac{1}{\sqrt{2\pi}} \int_{-\sigma}^{\sigma} f(t) e^{-iut} dt \in L_2(-\infty, +\infty)$$

defining the Fourier transform $\mathcal{F}[f] = F$.,

(b) *The reverse limit relation*

$$f(x) = \underset{\sigma \to +\infty}{\text{l.i.m.}} \frac{1}{\sqrt{2\pi}} \int_{-\sigma}^{\sigma} F(u) e^{ixu} du \equiv \mathcal{F}^{-1}[F]$$

and the equality

$$\int_{-\infty}^{+\infty} |f(x)|^2 dx = \int_{-\infty}^{+\infty} |F(u)|^2 du$$

hold.

Thus, this theorem has for the first time constructed a Fourier operator in L_2 and established a complete equivalence between the function and its Fourier transform, giving a unitary mapping of the whole space L_2 upon itself.

At a considerably later date G. Watson (1933) constructed the general theory of the Fourier-type transforms in $L_2(0, +\infty)$ performing unitary or quasi-unitary mapping of L_2 upon itself with the aid of formulae of the form

$$g(x) = \frac{d}{dx} \int_0^{+\infty} \frac{K(x,y)}{y} f(y) dy, \quad f(x) = \frac{d}{dx} \int_0^{+\infty} \frac{h(x,y)}{y} g(y) dy$$

Notice should also be taken of the significant generalizations of Plancherel's theorems discovered within the theory of singular boundary problems of the Sturm-Liouville type on the half-axis $(0, +\infty)$.

(B) The theory of integral transform developed in our investigations is very substantially supported by the remarkable asymptotic properties of the two families of functions; the functions of the Mittag-Leffler type:

$$E_\rho(z; \mu) = \sum_{k=0}^{\infty} \frac{z^k}{\Gamma(\mu + k\rho^{-1})} \quad (\mu > 0, \rho > 0)$$

and their continual analogues – the functions of Volterra:

$$\nu_\rho(z;\mu) = \int_0^{+\infty} \frac{z^t}{\Gamma(\mu + \frac{t}{\rho})}\, dt \quad (\mu > -1, \rho > 0)$$

Note that the function $E_\rho(z;\mu)$ (with $\mu = 1$) was first introduced into analysis by Mittag-Leffler at the beginning of this century (1903–1905) in connection with his discovery of a new rather strong method of summing the diverging series. In this connection he was also the first to discover asymptotic formulae for this function when $|z| \to \infty$ in the domain

$$\Delta_\rho = \{z;\, |\arg z| < \frac{\pi}{2\rho}\} \quad (\rho > \frac{1}{2})$$

and in its complement $\Delta_\rho^* = C\Delta_\rho$. As to the function $\nu_\rho(z;\mu)$ (for $\rho = 1$, $\mu = 0$), it was introduce into the analysis by V. Volterra (1916) in connection with the solution of a special integral equation occurring in the theory of heredity.

The considerable time that has elapsed since then has not enriched the mathematical literatu with any more or less serious investigations on the properties or, even more important, on applications of these functions to contemporary analysis.

The asymptotic properties of the functions $E_\rho(z;\mu)$ and $\nu_\rho(z;\mu)$ are given by the following theorem.

Theorem 6.1

(a) *For each* $\mu \in (-\infty, +\infty)$ *and* $\rho > 1/2$ *the function* $E_\rho(z;\mu)$ *is entire of the order* ρ *and type* $\sigma = 1$, and, if α *is any number satisfying the condition*

$$\frac{\pi}{2\rho} < \alpha < \min\{\pi, \frac{\pi}{\rho}\}$$

then for $|z| \to \infty$

(i) $E_\rho(z;\mu) = \rho\, z^{\rho(1-\mu)} + O(\frac{1}{z})$ for $|\arg z| \leq \alpha$ (6.1)

(ii) $E_\rho(z;\mu) = O(\frac{1}{z})$ for $\alpha \leq |\arg z| \leq \pi$ (6.2)

(b) *For each* $\mu \in (-\infty, +\infty)$ *and* $\rho > 0$ *the function* $\nu_\rho(z;\mu)$ *is analytical on the whole Riemann surface*

$$G_\infty = \{z;\, |\operatorname{Arg} z| < \infty,\ 0 < |z| < \infty\}$$

(we call such functions quasi-entire), has the order ρ *and type* $\sigma = 1$ *and, for any* $\alpha \in (\pi/2\rho, \pi/\rho)$ *when* $|z| \to \infty$

(i) $\nu_\rho(z;\mu) = \rho\, z^{\rho(1-\mu)} e^{z^\rho} + O(\frac{1}{\log|z|})$ for $|\operatorname{Arg} z| \leq \alpha_0$ (6.3)

(ii) $\nu_\rho(z;\mu) = O(\frac{1}{\log|z|})$ for $\alpha_0 \leq |\operatorname{Arg} z| < \infty$ (6.4)

The asymptotic properties of $E_\rho(z;\mu)$ for $0<\rho\leqslant 1/2$ have a different formulation. For us it was particularly important to reveal these properties for $\rho = 1/2$.

Theorem 6.2. *If* $0<\mu<3$, *then*

(i) *For* $0\leqslant \arg z\leqslant\pi$ *or* $-\pi\leqslant \arg z\leqslant 0$, *respectively, when* $|z|\to\infty$ *we have*

$$E_{\frac{1}{2}}(z;\mu)=\frac{1}{2}z^{\frac{1}{2}(1-\mu)}\left\{e^{z^{\frac{1}{2}}}+e^{\mp i\pi(1-\mu)}e^{-z^{\frac{1}{2}}}\right\}+O\left(\frac{1}{|z|}\right) \tag{6.5}$$

(ii) *For* $0<x<+\infty$ *when* $x\to+\infty$ *we have*

$$E_{\frac{1}{2}}(-x;\mu)=x^{\frac{1}{2}(1-\mu)}\cos\left(\sqrt{x}+\frac{\pi}{2}(1-\mu)\right)+O\left(\frac{1}{x}\right) \tag{6.5'}$$

(C) It will be assumed below that by a given arbitrary $\rho\geqslant 1/2$ the parameter μ is subjected to the condition

$$\frac{1}{2}<\mu<\frac{1}{2}+\frac{1}{\rho} \tag{6.6}$$

Henceforth we shall write $g(y)\in L_{2,\mu}(0,+\infty)$, if $g(y)y^{\mu-1}\in L_2(0,+\infty)$. Finally, if $g(y)\in L_{2,\mu}(0,+\infty)$ and the family of functions $\{g_\sigma(y)\}$, $g_\sigma(y)\in L_{2,\mu}(0,+\infty)$ depending on the parameter $\sigma\ (0<\sigma<+\infty)$ is such that

$$\lim_{\sigma\to+\infty}\int_0^{+\infty}|g(y)-g_\sigma(y)|^2 y^{2(\mu-1)}dy=0$$

then we shall write

$$g(y)=\underset{\sigma\to+\infty}{\overset{(\mu)}{\text{l.i.m.}}}\ g_\sigma(y)$$

The first basic theorem on the transforms with Mittag-Leffler kernels reads:

Theorem 6.3. *Let* $g(y)\in L_{2,\mu}(0,+\infty)$

$$f(x)=\frac{1}{\sqrt{2\pi\rho}}\frac{d}{dx}\int_0^{+\infty}\frac{e^{-ixy}-1}{-iy}g(y)y^{\mu-1}dy,\quad x\in(-\infty,+\infty) \tag{6.7}$$

and thus

$$\int_0^{+\infty}|g(y)|y^{2(\mu-1)}dy=\rho\int_{-\infty}^{+\infty}|f(x)|^2dx \tag{6.8}$$

(a) *Putting*

$$g(y;\varphi) \equiv \frac{y^{1-\mu}}{\sqrt{2\pi}\,\rho} \frac{d}{dy}\left\{ y^{\mu-1} \int_{-\infty}^{+\infty} E_\rho\left((ix)^{\frac{1}{\rho}} y^{\frac{1}{\rho}} e^{i\varphi}; \mu+1\right)(ix)^{\mu-1} f(x)\,dx \right\} \tag{6.9}$$

we shall have almost everywhere on the half-axis $y \in (0, +\infty)$

$$g(y;\varphi) = \begin{cases} g(y), \ \rho \geq \frac{1}{2}, \ \varphi = 0 \\ 0, \ \rho \geq 1, \ \frac{\pi}{\rho} \leq |\varphi| \leq \pi \end{cases} \tag{6.10}$$

(b) *Also putting*

$$g(y;\varphi;\sigma) = \frac{1}{\sqrt{2\pi}\,\rho} \int_{-\sigma}^{\sigma} E_\rho\left((ix)^{\frac{1}{\rho}} y^{\frac{1}{\rho}} e^{i\varphi}; \mu\right)(ix)^{\mu-1} f(x)\,dx \tag{6.11}$$

with respect to the conditions (6.10), we shall also have

$$g(y;\varphi) = \underset{\sigma \to +\infty}{\overset{(\mu)}{l.i.m.}}\, g(y;\varphi;\sigma) \tag{6.10'}$$

Thus, this theorem contains two substantially new statements of harmonic analysis:

(1) For the usual Fourier transform of functions from $L_{2,\mu}(0, +\infty)$ there is not one but a whole family of inversion formulae with $E_\rho(z;\mu)$-type kernels, where $\rho \geqslant 1/2$ is any number. Only the special selection of parameters ρ and μ (when $\rho = 1/2$, $\mu = 1, 2$ or $\rho = \mu = 1$) leads to the inversion formulae well known in the Fourier-Plancherel theory.

(2) When the parameter $\rho \geqslant 1$, the general inversion formula (6.10) has an important property: representing the function $g(y)$ on the ray $y = 0$, it simultaneously represents the identical zero in the angular domain $\pi/\rho \leqslant |\varphi| \leqslant \pi$.

This permits the construction of the apparatus of integral transforms and their inversions for an arbitrary finite system of rays proceeding from one point of the complex plane.

Before formulating the theorem we shall introduce some notation.

Denote by $L\{\varphi_1, \ldots, \varphi_p\}$ the set of rays

$$\ell_k : \arg z = \varphi_k \quad (k = 1, 2, \ldots, p)$$

$$0 \leq \varphi_1 < \varphi_2 < \cdots < \varphi_p < \varphi_{p+1} = 2\pi + \varphi_1$$

emerging from the origin. The system of these rays divides the z-plane into p angular domains with a common vertex at $z = 0$. Denote

$$\omega = \max_{1 \leq k \leq p} \left\{ \frac{\pi}{\varphi_{k+1} - \varphi_k} \right\}$$

noting that π/ω is the value of the smallest of these angles.

Theorem 6.4. *Let* $\rho \geqslant \omega$ *and* g(z) *be an arbitrary function defined on the set of rays* $L\{\varphi_1, \ldots, \varphi_p\}$ *and such that*

$$\int_{L\{\varphi_1,\ldots,\varphi_p\}} |g(z)|^2 |z|^{2(\mu-1)} |dz| < +\infty \tag{6.12}$$

(a) *Putting*

$$f_K(x) = \frac{1}{\sqrt{2\pi\rho}} \frac{d}{dx} \int_0^{+\infty} \frac{e^{-ix\tau}-1}{i\tau} g(\tau e^{i\varphi_K}) \tau^{\mu-1} d\tau \in L_2(0,+\infty) \quad (K=1,2,\ldots,p) \tag{6.13}$$

we shall have for almost any $\tau e^{i\varphi} \in L\{\varphi_1,\ldots,\varphi_p\}$

$$g(\tau e^{i\varphi}) = \frac{\tau^{1-\mu}}{\sqrt{2\pi\rho}} \sum_{K=1}^{p} \frac{d}{d\tau} \left\{ \tau^{\mu} \int_{-\infty}^{+\infty} E_\rho\left((ix)^{\frac{1}{\rho}} \tau^{\frac{1}{\rho}} e^{\frac{i}{\rho}(\varphi-\varphi_K)}; \mu+1\right) x^{\mu-1} f(x) dx \right\} \tag{6.14}$$

(b) *The Parseval type equality*

$$\int_{L\{\varphi_1,\ldots,\varphi_p\}} |g(z)|^2 |z|^{2(\mu-1)} |dz| = \rho \sum_{K=1}^{p} \int_{-\infty}^{+\infty} |f_K(x)|^2 dx \tag{6.15}$$

holds.

It is easy to see that the basic theorem of Plancherel is a special case of this theorem when a system of two rays $L\{0, \pi\}$ is considered and $\mu = 1$, $\rho = 1$.

Finally we shall note that the inverse of Theorem 6.3 holds.

Theorem 6.5

(a) *For any function* $f(x) \in L_2(0, +\infty)$ *the transforms*

$$g^{(\pm)}(y) = \frac{y^{1-\mu}}{\sqrt{2\pi\rho}} \frac{d}{dy} \left\{ y^{\mu} \int_0^{+\infty} E_\rho\left(e^{\pm i\frac{\pi}{2\rho}} y^{\frac{1}{\rho}} x^{\frac{1}{\rho}}; \mu+1\right) x^{\mu-1} f(x) dx \right\} \tag{6.16}$$

determine functions $g^{(\pm)}(y) \in L_{2,\mu}(0,+\infty)$ *and the inversion formula*

$$f(x) = \frac{1}{\sqrt{2\pi\rho}} \left\{ e^{-i\frac{\pi}{2}(1-\mu)} \frac{d}{dx} \int_0^{+\infty} \frac{e^{-ixy}-1}{iy} g^{(+)}(y) y^{\mu-1} dy \right.$$

$$\left. + e^{i\frac{\pi}{2}(1-\mu)} \frac{d}{dx} \int_0^{+\infty} \frac{e^{ixy}-1}{iy} g^{(-)}(y) y^{\mu-1} dy \right\}, \quad x \in (0,+\infty) \tag{6.17}$$

holds.

(b) *The inequalities*

$$\int_0^{+\infty} |g^{(\pm)}(y)|^2 y^{2(\mu-1)} dy \leqslant \frac{M\mu^2}{2\pi\rho} \int_0^{+\infty} |f(x)|^2 dx \tag{6.18}$$

$$\int_0^{+\infty} |f(x)|^2 dx \leq \rho^{-1} \left\{ \int_0^{+\infty} |g^{(+)}(y)|^2 y^{2(\mu-1)} dy + \int_0^{+\infty} |g^{(-)}(y)|^2 y^{2(\mu-1)} dy \right\} \tag{6.18'}$$

hold.

(D) The results of Theorems 6.3, 6.4 and 6.5 remain valid if the function $E_\rho(z; \mu)$ is replaced by $\nu_\rho(z; \mu)$, putting everywhere $\mu = 1/2$ and $\rho > 0$.

For this reason we do not present here the reformulations of these theorems, but only give the analogue of Theorem 6.4 which is a rather extensive generality. Let $\mathscr{L}\{v_1, \ldots, v_p\}$ denote the set of parallel straight lines

$$w = u + iv_k, \quad -\infty < u < +\infty \quad (k = 1, 2, \ldots, p)$$

$$-\infty < v_1 < v_2 < \cdots < v_p < +\infty$$

Denote

$$\omega = \max_{1 \leq k \leq p-1} \left\{ \frac{\pi}{v_{k+1} - v_k} \right\}$$

See that the value π/ω is equal to the width of the smallest of the strips generated on the z-plane by the system of lines $\mathscr{L}\{v_1, \ldots, v_p\}$.

Theorem 6.6. *Let the function* $G(w)$ *be defined on the system of straight lines* $\mathscr{L}\{v_1, \ldots, v_p\}$ *and satisfy the condition*

$$\int_{\mathscr{L}\{v_1, \ldots, v_p\}} |G(w)|^2 d(\mathrm{Re}\, w) < +\infty \tag{6.19}$$

Then for any $\rho \geqslant \omega$ *the following statements hold:*

(a) *For almost any* $w \in \mathscr{L}\{v_1, \ldots, v_p\}$

$$G(w) = e^{-\frac{\rho w}{2}} \sum_{k=1}^{p} \frac{1}{\sqrt{2\pi\rho}} \frac{d}{dw} \left\{ e^{\frac{\rho w}{2}} \int_{-\infty}^{+\infty} \nu_\rho\left(e^{i\left(\frac{\pi}{2\rho} - v_k\right) + w} x^{\frac{1}{\rho}}; \frac{3}{2}\right) x^{-\frac{1}{2}} f_k(x)\, dx \right\} \tag{6.20}$$

where

$$f_k(x) = \frac{e^{-i\frac{\pi}{4}}}{\sqrt{2\pi\rho}} \frac{d}{dx} \int_{-\infty}^{+\infty} \left(e^{-ixe^{\rho u}} - 1\right) e^{-\frac{u\rho}{2}} G(u + iv_k)\, du \quad (k = 1, 2, \ldots, p) \tag{6.21}$$

(b) *The equality*

$$\int_{\mathscr{L}\{v_1, \ldots, v_p\}} |G(w)|^2 d(\mathrm{Re}\, w) = \rho \sum_{k=1}^{p} \int_{-\infty}^{+\infty} |f_k(x)|^2 dx \tag{6.22}$$

holds.

In conclusion, we should like to note that Theorems 6.3, 6.4 and 6.6 may be considered as theorems of approximation by entire functions on the half-axis $(0, +\infty)$ or on the system of rays $L\{\varphi_1, \ldots, \varphi_p\}$ or on the system of parallel straight lines $\mathscr{L}\{v_1, \ldots, v_p\}$.

2. CLASSES OF ENTIRE, QUASI-ENTIRE AND ANALYTICAL FUNCTIONS AND THEIR INTEGRAL REPRESENTATIONS (Wiener-Paley-type theorems)

(A) In their famous monograph "Fourier transforms in the complex domain" (1934), Wiener and Paley established two fundamental theorems: on representations of exponential-type entire functions from $L_2(-\infty, +\infty)$ and functions from H_2 analytical in the half-plane. We shall formulate these famous theorems, since they served as a starting point for our investigations in this area. The first of these theorems reads:

The class of entire functions from $L_2(-\infty, +\infty)$ *of exponential type* $\leqslant \sigma$ *coincides with the set of functions admitting representation of the form*

$$f(z) = \int_{-\sigma}^{\sigma} e^{itz} \varphi(t)\, dt$$

where $\varphi(t)$ *is an arbitrary function from* $L_2(-\sigma, \sigma)$.

The second theorem reads:

The class H_2 *of functions* $F(z)$ *analytical in the half-plane* $\operatorname{Re} z > 0$ *and satisfying the condition*

$$\sup_{0<x<+\infty} \left\{ \int_{-\infty}^{+\infty} |F(x+iy)|^2\, dy \right\} < +\infty$$

coincides with the set of functions admitting representation of the form

$$F(z) = \int_0^{+\infty} e^{-tz} \varphi(t)\, dt\,, \quad \operatorname{Re} z > 0$$

where $\varphi(t) \in L_2(0, +\infty)$.

Our results on representations of analytical functions are far-reaching substantial generalizations of these two theorems.

(B) Let us first give the formulation of the most general theorem on entire functions. To this end we shall first introduce some preliminary notation.

We shall assume that the natural number $\varkappa = \varkappa(\rho) \geqslant 0$ satisfies the condition

$$\varkappa \geq [2\rho] - 1 \tag{6.23}$$

for any $\rho \geqslant 1/2$. Then, for a given $\rho \geqslant 1/2$ we shall assume that the set of numbers

$\{\vartheta_0, \vartheta_1, \ldots, \vartheta_{\varkappa+1}\}$ satisfies the conditions

$$
\begin{aligned}
&-\pi < \vartheta_0 < \vartheta_1 < \cdots < \vartheta_\varkappa \leq \pi < \vartheta_{\varkappa+1} = \vartheta_0 + 2\pi \\
&\max_{0 \leq k \leq \varkappa} \{\vartheta_{k+1} - \vartheta_k\} = \frac{\pi}{\rho}
\end{aligned}
\tag{6.24}
$$

Starting with the set $\{\vartheta_0, \vartheta_1, \ldots, \vartheta_{\varkappa+1}\}$ we form a sequence of pairs

$$
(\vartheta_k, \vartheta_{k+1})_1^{\varkappa} = \{(\vartheta_0, \vartheta_1), (\vartheta_1, \vartheta_2), \ldots, (\vartheta_\varkappa, \vartheta_{\varkappa+1})\} \tag{6.25}
$$

and then, preserving the mutual order of their succession, we shall isolate all pairs for which the equality

$$
\vartheta_{\tau_{k+1}} - \vartheta_{\tau_k} = \frac{\pi}{\rho} \quad (k = 0, 1, \ldots, p \leq \varkappa) \tag{6.26}
$$

holds. Here, if $p < \varkappa$, let us denote the remaining pairs of (6.25) by $(\vartheta_{s_k}, \vartheta_{s_{k+1}})_1^q$ $(q = \varkappa -$ Denote further

$$
\theta_k = \frac{1}{2}\{\vartheta_{\tau_k} + \vartheta_{\tau_{k+1}}\} \quad (k = 0, 1, \ldots, p) \tag{6.27}
$$

and assuming that $-1 < \omega < 1$, $\sigma_k \geqslant 0$ $(k = 0, 1, \ldots, p)$, associate with the set of numbers $\{\vartheta_0, \ldots, \vartheta_{k+1}\}$ the class

$$
W_\sigma^{(\rho)}(\omega; \{\vartheta_k\}; \{\sigma_k\})
$$

of entire functions of order ρ and normal type $\leqslant \sigma$ satisfying the conditions

$$
\begin{aligned}
&\int_0^{+\infty} |f(te^{-i\vartheta_k})|^2 t^\omega dt < +\infty \quad (k = 0, 1, \ldots, \varkappa) \\
&\overline{\lim_{r \to +\infty}} \frac{\log |f(re^{-i\theta_k})|}{r^\rho} = h(-\theta_k; f) \leq \sigma_k \leq \sigma \quad (k = 0, 1, \ldots, p)
\end{aligned}
\tag{6.28}
$$

We have established the following general theorem, based substantially on our theory of integral transforms with Mittag-Leffler-type kernels.

Theorem 6.7. *The class* $W_\sigma^{(\rho)}(\omega; \{\vartheta_k\}; \{\sigma_k\})$ *coincides with the set of functions* f(z) *admitting the representation*

$$
f(z) = \sum_{k=0}^{p} \int_0^{\sigma_k} E_\rho\{e^{i\theta_k} z \tau^{\frac{1}{\rho}}; \mu\} \varphi_k(\tau) \tau^{\mu - 1} d\tau \tag{6.29}
$$

where

$$
\mu = \frac{\omega + \rho + 1}{2\rho}, \quad \varphi_k(\tau) \in L_2(0, \sigma_k) \quad (k = 0, 1, \ldots, p)
$$

Here, if

$$\Phi_k(\tau) = \frac{1}{\sqrt{2\pi}} \frac{d}{d\tau} \int_0^{+\infty} f(e^{-i\vartheta_k} v^{\frac{1}{\rho}}) \frac{e^{-i\tau v} - 1}{-iv} v^{\mu-1} dv \tag{6.30}$$

then almost everywhere we have

$$\frac{i}{\sqrt{2\pi}\,\rho}\left\{e^{-i\frac{\pi}{2}\mu}\Phi_{\tau_{k+1}}(-\tau) - e^{i\frac{\pi}{2}\mu}\Phi_{\tau_k}(\tau)\right\}$$

$$= \begin{cases} \varphi_k(\tau), & \tau \in (0, \sigma_k) \\ 0, & \tau \in (\sigma_k, +\infty) \end{cases} \qquad (k = 0, 1, \ldots, p). \tag{6.31}$$

By special selection of the system of rays $\{\arg z = \vartheta_k\}_0^{\varkappa}$ along which the conditions of the form (6.28) are imposed, some corollaries of a more special character, including the Wiener-Paley theorem itself, may be obtained. For example, the following statement is true.

Theorem 6.8

(a) *The class of entire functions* f(z) *of order* $\rho \geqslant 1/2$ *and type* $\leqslant \sigma$ *subject to the condition*

$$\sup_{\frac{\pi}{2\rho} < \vartheta \leqslant \pi} \left\{\int_0^{+\infty} |f(te^{-i\vartheta})|^2 t^{\omega} dt\right\} < +\infty \qquad (-1 < \omega < 1)$$

coincides with the set of functions of the form

$$f(z) = \int_0^{\sigma} E_\rho\{z\tau^{\frac{1}{\rho}}; \mu\} \varphi(\tau) \tau^{\mu-1} d\tau$$

where $\mu = (\omega + 1 + \rho)/2\rho$ *and* $\varphi(t) \in L_2(0, \sigma)$.

(b) *The class of entire functions* f(z) *of order one and type* $\leqslant \sigma$ *for which*

$$\int_{-\infty}^{+\infty} |f(x)|^2 |x|^{\omega} dx < +\infty$$

coincides with the set of functions of the form

$$f(z) = \int_{-\sigma}^{\sigma} E_1\{i\tau z; \mu\} \varphi(\tau) |\tau|^{\mu-1} d\tau$$

where $\mu = 1 + \omega/2$ *and* $\varphi(\tau) \in L_2(-\sigma, \sigma)$.

Note that the theorem of Wiener-Paley is contained in statement (b) of this theorem as a special case when $\omega = 0$ because then $\mu = 1$ and $E_1(z; 1) = e^z$.

(C) A function analytical on the whole Riemann surface G_∞ with

(1) $\tilde{M}_f(r) = \sup_\varphi |f(re^{i\varphi})| < +\infty \quad (0 < r < +\infty)$

(2) $\lim_{r \to +0} \tilde{M}_f(r) < +\infty$

is called quasi-entire. The order ρ and type σ of quasi-entire functions are determined in a way analogous to that in the theory of entire functions. The result similar to Theorem 6.7 is also established for the quasi-entire functions with the function $E_\rho(z;\mu)$ replaced by $\nu_\rho(z; 1/2)$ and $\omega = -1$, $\rho > 0$ may be arbitrary.

In order to state the result,some new notations are needed. For the given value of ρ $(0 < \rho < +\infty)$, assume that the set of number $\{\vartheta_k\}_{-p}^{q}$ $(p \geq 0, q \geq 0)$ satisfies the following conditions:

$$-\alpha_0 \leq \vartheta_{-p} < \vartheta_{-p+1} < \cdots < \vartheta_{-1} < \vartheta_0 < \vartheta_1 < \cdots < \vartheta_q \leq \alpha_0$$

where $\alpha_0 > 0$ and

$$\max_{-(p+1) \leq k \leq q-1} \{\vartheta_{k+1} - \vartheta_k\} = \frac{\pi}{\rho}$$

Further, forming the successive pairs

$$\{(\vartheta_k, \vartheta_{k+1})\}_{-p}^{q-1}$$

we shall isolate, retaining their mutual order of succession, all pairs

$$(\vartheta_{\tau_k}, \vartheta_{\tau_{k+1}})_1^m$$

for which

$$\vartheta_{\tau_{k+1}} - \vartheta_{\tau_k} = \frac{\pi}{\rho} \quad (k = 1, 2, \ldots, m \leq p + q)$$

Finally denote by

$$\tilde{W}_\sigma^{(\rho)}(\{\vartheta_k\}; \{\sigma_k\})$$

the set of quasi-entire functions f(z) of the order ρ $(0 < \rho < +\infty)$ and type $\leqslant \sigma$ subject to the conditions:

(1) $\int_0^{+\infty} |f(re^{i\vartheta_k})|^2 r^{-1} dr < +\infty \quad (-p \leq k \leq q)$

(2) $\sup_{|\varphi| \geq \alpha_0} \left\{ \int_0^{+\infty} |f(re^{i\varphi})|^2 r^{-1} dr \right\} < +\infty$

(3) $h(-\theta_K; f) = \overline{\lim_{r\to+\infty}} \frac{\log|f(re^{-i\theta_K})|}{r^\rho} \leq \sigma_K \leq \sigma \qquad (K = 1, 2, \ldots, m)$

where

$$\theta_K = \frac{1}{2}\{\vartheta_{\tau_K} + \vartheta_{\tau_{K+1}}\}$$

Theorem 6.9. *The class*

$$\widetilde{W}_\sigma^{(\rho)}(\{\vartheta_K\}; \{\sigma_K\})$$

coincides with the set of functions f(z) *representable in the form*

$$f(z) = \sum_{K=0}^{m} \int_0^{\sigma_K} \mathcal{V}_\rho\{e^{i\vartheta_K} z \tau^{\frac{1}{\rho}}; \tfrac{1}{2}\}\tau^{-\frac{1}{2}} \varphi_K(\tau)d\tau \tag{6.32}$$

where $\varphi_K(\tau) \in L_2(0, \sigma_K) \quad (1 \leq K \leq m)$.

(D) As to the results of the type of the Wiener-Paley second basic theorem, here we have

Theorem 6.10

(a) *The class* $\mathcal{H}_2[\alpha; \omega]$ $(\frac{1}{2} < \alpha < +\infty; -1 < \omega < 1)$ *of functions* F(z) *analytical in the angle*

$$\Delta_\alpha = \{z; |\arg z| < \frac{\pi}{2\alpha}, 0 < |z| < \infty\}$$

for which

$$\sup_{|\varphi| < \frac{\pi}{2\alpha}} \left\{\int_0^{+\infty} |F(re^{i\varphi})|^2 r^\omega dr\right\} < +\infty$$

coincides with the set of functions of the form

$$F(z) = \int_0^{+\infty} E_\rho(e^{i\frac{\pi}{2\gamma}} z t^{1/\rho}; \mu) V_{(-)}(t) t^{\mu-1} dt$$
$$+ \int_0^{+\infty} E_\rho(e^{-i\frac{\pi}{2\gamma}} z t^{1/\rho}; \mu) V_{(+)}(t) t^{\mu-1} dt, \qquad z \in \Delta_\alpha \tag{6.33}$$

where

$$V_{(\pm)}(t) \in L_2(0, +\infty), \ \rho \geq \frac{\alpha}{2\alpha - 1}$$
$$\gamma = \frac{\rho\alpha}{\alpha + \rho}, \quad \mu = \frac{1 + \omega + \rho}{2\rho} \tag{6.34}$$

(b) *The class* $\mathcal{H}_2[\alpha]$ $(0<\alpha<+\infty)$ *of functions analytical in the domain* $\Delta_\alpha \subset G_\infty$ *and subject to the condition*

$$\sup_{|\varphi|<\frac{\pi}{2\alpha}}\left\{\int_0^{+\infty}|F(re^{i\varphi})|^2 dr\right\}<+\infty$$

coincides with the set of functions representable in the form

$$F(z)=z^{-\frac{1}{2}}\int_0^{+\infty}\nu_\rho\left(e^{i\frac{\pi}{2\gamma}}zt^{1/\gamma};\frac{1}{2}\right)t^{-\frac{1}{2}}V_{(-)}(t)\,dt$$
$$+\,z^{-\frac{1}{2}}\int_0^{+\infty}\nu_\rho\left(e^{-i\frac{\pi}{2\gamma}}zt^{1/\gamma};\frac{1}{2}\right)t^{-\frac{1}{2}}V_{(+)}(t)\,dt,\quad z\in\Delta_\alpha \tag{6.35}$$

where $\rho>0$ *is any number,*

$$\gamma=\frac{\alpha\rho}{\alpha+\rho},\quad V_{(\pm)}(t)\in L_2(0,+\infty)$$

An important feature of the integral formulae (6.33) and (6.35) is that they also represent the identical zero in the domain complementary to Δ_α, i.e. for (6.33) this is the case when $\rho\geqslant 2\alpha/(2\alpha-1)$ in the domain

$$|\pi-\arg z|<\frac{\pi}{2\varkappa},\quad \varkappa=\frac{\alpha\rho}{(2\alpha-1)\rho-2\alpha}$$

and for (6.35) this is the case in the domain of values

$$\frac{\pi}{2\alpha}+\frac{\pi}{\rho}<|\operatorname{Arg} z|<+\infty,\quad 0<|z|<+\infty$$

This remarkable fact permits the construction of the apparatus and the development of the theory of Fourier-Plancherel-type integrals for the sets consisting of a finite number of rays and angular domains lying on the z plane or on the Riemann surface G_∞. Precise formulations of the theorems in question will not be dwelt upon.

3. UNIQUENESS OF CERTAIN GENERAL CLASSES OF INFINITELY DIFFERENTIABLE FUNCTIONS (Denjoi-Carleman-type theorems)

(A) As early as 1912, J. Hadamard posed the problem of determining conditions for a sequence of positive numbers $\{M_n\}_1^\infty$ ensuring the uniqueness of the class $C\{M_n\}$ of functions infinitely differentiable on some interval $\mathcal{T}\doteq(a,b)$ and for values $\{\varphi^{(n)}(x_0)\}_0^\infty$, $x_0\in\mathcal{T}$ satisfying the conditions

$$|\varphi^{(n)}(x)|\leqslant AB^n M_n\quad (n=1,2,\ldots)$$

In 1921, A. Denjoi established for the first time the existence of such classes substantially wide as compared with the usual classes C{n!} of analytical functions. In particular, he proved that the class $C\{M_n\}$ is quasi-analytical if e.g.

$$M_n = (n \log n \cdots \log_p n)^n \quad , \quad n \geq N_p$$

T. Carleman (1923–1926) gave a comprehensive solution of Hadamard's problem by laying down the necessary and sufficient condition for the class $C\{M_n\}$ to possess the property of uniqueness; in other words, for its quasi-analyticity.

In the formulation by A. Ostrowsky, Carleman's result (usually called the theorem of Denjoi-Carleman) reads:

The condition

$$\int_1^{+\infty} \frac{\log T(r)}{r^2} dr = +\infty \, , \quad T(r) = \sup_{n \geq 1} \frac{r^n}{M_n} \tag{6.36}$$

is necessary and sufficient for the class $C\{M_n\}$ *to be quasi-analytical.*

In the five decades that followed, many original investigations dealing with the theory of quasi-analytical functions have come into being. Here a special note should be given to the study by S. Mandelbrojt on the generalized quasi-analyticity in the sense of Denjoi-Carleman where, however, the condition (6.36) is always assumed to be fulfilled.

(B) According to the theorem of Denjoi-Carleman, by the condition

$$\int_1^{+\infty} \frac{\log T(r)}{r^2} dr < +\infty \tag{6.37}$$

the class $C\{M_n\}$ of functions infinitely differentiable on the half-axis $[0, +\infty)$ or on a segment $[0, \ell]$ will certainly not be quasi-analytical. That is to say, it is well known that by the conditions (6.37), say, in the case of the half-axis $[0, +\infty)$, there exist non-trivial functions $\varphi(x) \in C\{M_n\}$ satisfying, moreover, the conditions

$$|\varphi^{(n)}(x)| \leq AB^n M_n e^{-\nu x} \quad (\nu > 0, n \geq 1, x \in [0, +\infty)) \tag{6.38}$$

$$\varphi^{(n)}(0) = 0 \quad (n = 0, 1, 2, \ldots) \tag{6.38'}$$

In this connection it would be natural to put the following question:

If the class $C\{M_n\}$ *is not quasi-analytical on* $[0, +\infty)$ *or on* $[0, \ell]$, *then which are functionals* $\{L_n(\varphi)\}_0^\infty$ *that can determine the functions of this class in a unique way instead of the values* $\{\varphi^{(n)}(0)\}_0^\infty$?

It turned out that it is possible to introduce a new general notion of α-quasi-analyticity also covering the notion of classical quasi-analyticity and to obtain a complete solution of this problem in the spirit of the classical theorem of Denjoi-Carleman.

(C) As a preliminary, we introduce some notation and definitions.

Let us consider a set $C_\alpha^{(\infty)}$ $(0 \leqslant \alpha < 1)$ of functions $\varphi(x)$ infinitely differentiable on $[0, +\infty)$ and satisfying the conditions:

$$\sup_{0 \leq x < +\infty} |(1+x^{\alpha m})\varphi^{(n)}(x)| < +\infty \quad (n, m = 0, 1, 2, \ldots)$$

Assuming that $\varphi(x) \in C_\alpha^{(\infty)}$ $(0 \leqslant \alpha < 1)$ and putting $1/\rho = 1 - \alpha$ $(\rho \geqslant 1)$, let us consider the operator of successive differentiation of the function $\varphi(x)$ in the sense of Weyl of orders n/ρ $(n = 0, 1, \ldots)$

$$D_\infty^{\frac{0}{\rho}} \varphi(x) \equiv \varphi(x), \quad D_\infty^{\frac{n}{\rho}} \varphi(x) = D_\infty^{\frac{1}{\rho}} D_\infty^{\frac{n-1}{\rho}} \varphi(x) \quad (n \geq 1)$$

where

$$D_\infty^{\frac{1}{\rho}} \varphi(x) = \frac{d}{dx} D_\infty^{-\alpha} \varphi(x), \quad D_\infty^{-\alpha} \varphi(x) = \frac{1}{\Gamma(\alpha)} \int_x^{+\infty} (t-x)^{\alpha-1} \varphi(t)\,dt$$

Note that in a special case when $\alpha = 0$ $(\rho = 1)$ we shall have

$$D_\infty^{n} \varphi(x) \equiv \varphi^{(n)}(x) \quad (n = 0, 1, 2, \ldots)$$

Finally, for an arbitrary sequence of positive numbers $\{M_n\}_1^\infty$ we introduce two classes of infinitely differentiable functions:

The class $C_\alpha^*\{[0, +\infty);\ M_n\}$ is the set of functions $\varphi(x)$ from $C_\alpha^{(\infty)}$ for which

$$\sup_{0 \leq x < +\infty} |D_\infty^{\frac{n}{\rho}} \varphi(x)| \leq A B^n M_n \quad (n = 1, 2, \ldots) \tag{6.39}$$

and the class $C_\alpha\{[0, +\infty);\ M_n\}$ is the set of functions $\varphi(x)$ from $C_\alpha^{(\infty)}$ for which

$$\sup_{0 \leq x < +\infty} (1+\alpha x^2)|\varphi^{(n)}(x)| \leq A B^n M_n \quad (n = 1, 2, \ldots) \tag{6.40}$$

For both these classes a question is put similar to Hadamard's problem and reducible to this problem when the parameter $\alpha = 0$.

What must the sequence of numbers $\{M_n\}_1^\infty$ be in order that for each function $\varphi(x)$ from the corresponding class the equalities

$$D_\infty^{\frac{n}{\rho}} \varphi(0) = \frac{1}{\Gamma(\alpha n)} \int_0^{+\infty} x^{\alpha n - 1} \varphi^{(n)}(x)\,dx = 0 \quad (n = 0, 1, 2, \ldots) \tag{6.41}$$

yield the identity

$$\varphi(x) \equiv 0, \quad 0 \leq x < +\infty\ ?$$

Classes of this kind are called "α-quasi-analytical", and it is easily seen that the 0-quasi-analytical classes $C_0^*\{[0, +\infty); M_n\}$ or $C_0\{[0, +\infty); M_n\}$ are identical to the classically quasi-analytical class $C\{M_n\}$.

The following basic theorem has been established.

Theorem 6.11

(a) *The class* $C_\alpha^*\{[0, +\infty); M_n\}$ *is* α-quasi-analytical if and only if

$$\int_1^{+\infty} \frac{\log T(\tau)}{\tau^{1+\frac{1}{1+\alpha}}}\, d\tau = +\infty \tag{6.42}$$

(b) *The class* $C_\alpha\{[0, +\infty); M_n\}$ *is α-quasi-analytical if and only if*

$$\int_1^{+\infty} \frac{\log T(\tau)}{\tau^{1+\frac{1-\alpha}{1+\alpha}}}\, d\tau = +\infty \tag{6.43}$$

In both these statements

$$T(\tau) = \sup_{n \geq 1} \frac{\tau^n}{M_n}$$

is the function of Carleman-Ostrowsky. Each of the statements (a) and (b) is reduced to the classical theorem of Denjoi-Carleman when $\alpha = 0$.

Elementary estimates show that if

$$M_n = \left(n^{\frac{1+\alpha}{1-\alpha}} \log n \cdots \log_p n\right)^n, \quad n \geq N_p$$

where $p \geqslant 1$ is any integer, then the condition (6.43) is fulfilled.

This example shows that in the α-quasi-analytical class $C_\alpha\{[0, +\infty); M_n\}$ $(0 < \alpha < 1)$ the successive derivatives of functions may have a substantially faster growth (as $(1+\alpha)/(1-\alpha) > 1$ when $0 < \alpha < 1$) than is possible for 0-quasi-analytical classes: this can be observed from the original results of Denjoi.

The notion of α-quasi-analyticity is also introduced for the classes of functions infinitely differentiable on a finite segment $[0, \ell]$. Here the classes $C_\alpha^*\{[0, \ell]; M_n\}$ and $C_\alpha\{[0, \ell]; M_n\}$ are determined as subclasses of functions $\varphi(x)$ from the corresponding classes on $[0, +\infty)$ satisfying the additional condition

$$\varphi(x) \equiv 0, \quad \ell \leq x < +\infty.$$

The corresponding theorem on α-quasi-analyticity of these classes has exactly the same formulation as statements (a) and (b) of Theorem 6.11.

(D) Let us consider briefly the method of proving these theorems.

As is also the case with the original proof of the Denjoi-Carleman theorem, the problem of α-quasi-analyticity is solved by reducing it to the problem of Watson. In our case, such reduction is possible only by making use of the apparatus of integral transforms and representations with Mittag-Leffler kernels $E_\rho(z; \mu)$. Here a substantial role is played by the following basic theorem.

Theorem 6.12

(a) *Let the function* f(z) *be analytical in the interior and continuous on the closed angular domain*

$$\Delta_\rho^* = \{z;\ \frac{\pi}{2\rho} < |\arg z| \leq \pi,\ 0 < |z| < +\infty\}$$

and in the neighbourhood of $z = \infty$

$$\max_{\frac{\pi}{2\rho} \leq |\varphi| \leq \pi} \{|f(re^{i\varphi})|\} = O(r^{-\omega}) \quad (\omega > 1) \tag{6.44}$$

Then we have an integral representation of the form

$$f(z) = \int_0^{+\infty} E_\rho(zt^{1/\rho}; \frac{1}{\rho})\, t^{\frac{1}{\rho}-1} \varphi(t)\, dt,\ z \in \Delta_\rho^* \tag{6.45}$$

where

$$\varphi(t) = \frac{1}{2\pi i} \int_{L_\rho} e^{-t\zeta^\rho} f(\zeta)\, d\zeta,\ 0 < t < +\infty \tag{6.46}$$

and L_ρ *is the boundary of the domain* Δ_ρ^* *advanced in the positive direction.*

(b) *If* f(z) *is an entire function of the order* $\rho > 1/2$ *and type* ℓ $(0 < \ell < +\infty)$ *satisfying the condition (6.44), then in the representation (6.45) we have*

$$\varphi(t) \equiv 0,\ \ell < t < +\infty$$

These statements permit us to establish the necessity of conditions (6.42) and (6.43) of Theorem 6.11. Here we rely substantially upon our discovery of an important property of the function

$$\mathcal{E}_\rho(x; \lambda) = E_\rho(\lambda x^{\frac{1}{\rho}}; \frac{1}{\rho})\, x^{\frac{1}{\rho}-1}$$

to be a solution of the Cauchy-type problem for a differential operator of fractional order on the half-axis $[0, +\infty)$:

$$\begin{aligned} &D_0^{\frac{1}{\rho}} \mathcal{E}_\rho(x;\lambda) - \lambda \mathcal{E}_\rho(x;\lambda) = 0 \\ &D_0^{-\alpha} \mathcal{E}_\rho(x;\lambda)\Big|_{x=0} = 1 \end{aligned} \tag{6.47}$$

where

$$D_0^{-\alpha} f(x) \equiv \frac{1}{\Gamma(\alpha)} \int_0^x (x-t)^{\alpha-1} f(t)\, dt$$

and

$$D_0^{\frac{1}{\rho}} f(x) \equiv \frac{d}{dx} D_0^{-\alpha} f(x)$$

is the derivative in the sense of Riemann-Liouville of the order $1/\rho$. It is also important that the function

$$e_\rho(x;\lambda) = e^{-\lambda^\rho x} \quad \left(|\arg \lambda| \leq \frac{\pi}{2\rho},\ \rho \geq 1\right)$$

is also a solution of the Cauchy problem, but this time of another kind:

$$\begin{aligned} &D_\infty^{\frac{1}{\rho}} e_\rho(x;\lambda) + \lambda e_\rho(x;\lambda) = 0 \\ &e_\rho(0;\lambda) = 0 \end{aligned} \qquad (6.48)$$

As to the sufficiency of the conditions in Theorem 6.11, here an essential role is played by the fact that if

$$\varphi(x) = C_\alpha^*\{[0,+\infty); M_n\},\quad D_\infty^{\frac{n}{\rho}} \varphi(0) = 0 \quad (n \geq 0)$$

then the function

$$f(z) = \int_0^{+\infty} E_\rho\left(z t^{\frac{1}{\rho}}; \frac{1}{\rho}\right) t^{\frac{1}{\rho}-1} \varphi(t)\, dt,\quad z \in \Delta_\rho^*$$

admits the representation

$$f(z) = \frac{(-1)^n}{z^n} \int_0^{+\infty} E_\rho\left(z t^{\frac{1}{\rho}}; \frac{1}{\rho}\right) t^{\frac{1}{\rho}-1} D_\infty^{\frac{n}{\rho}} \varphi(t)\, dt,\quad z \in \Delta_\rho$$

for any $n \geqslant 1$.

This representation permits us to establish that the fulfilment of the conditions of Theorem 6.11 results in $f(z) \equiv 0$. Then using the theorem on inverse transforms with the Mittag-Leffler kernel already cited, we come to the conclusion that $\varphi(t) \equiv 0$.

It should be noted in conclusion that during the last two years the reporter and his students have been continuing their investigations in this direction. First, we succeeded in obtaining an analogous theorem of uniqueness in the case when the function $\varphi(x)$ is analytical in the domain of arbitrary angle

$$\Delta_\gamma = \left\{ z;\ |\arg z| < \frac{\pi}{2\gamma},\ 0 < |z| < +\infty \right\}$$

of the span π/γ $(0 < \gamma < +\infty)$ on the Riemann surface of the logarithm. Second, we have extended Theorem 6.11 to the case when the parameter α lies within the limits of $-1 < \alpha < 0$. This case yields results of a completely new quality this time for the classes of Denjoi-Carleman. Formulations of the theorems in question will not be dwelt upon owing to the lack of space.

Part VII

SOME OPEN PROBLEMS

In conclusion let us formulate some open problems. Except for the last one, they are directl connected with the subject of this paper.

Problem I. Theorem 4.3 only compares the classes U_ω, $\omega \in \Omega$, with the class U. A similar situati is found in Theorem 5.2, which compares the classes $N\{\omega\}$ with the class N of Nevanlinna. Howe it is important to have theorems of comparison for the classes U_{ω_1} and U_{ω_2} or $N\{\omega_1\}$ and $N\{\omega_2\}$ in the case when the functions $\omega_j(x) \in \Omega$ (j = 1, 2) generating these classes are arbitrary.

Solution of this problem is easily reduced to solvability of the Hausdorff moment problem:

$$\mu_n = \int_0^1 x^n \, d\alpha(x), \qquad (n = 0, 1, 2, \ldots)$$

where

$$\mu_0 = 1, \quad \mu_n = \left\{\int_0^1 \omega_1(x) x^{n-1} \, dx\right\}\left\{\int_0^1 \omega_2(x) x^{n-1} \, dx\right\}^{-1} \qquad (n = 1, 2, \ldots)$$

in the class of functions $\alpha(x)$ with the finite variation $\overset{1}{\underset{0}{\vee}} (\alpha) < +\infty$.

We suppose that the problem (1)–(2) must have a solution if the functions $\omega_j(x) \in \Omega$ (j = 1, 2) *are monotonic on* [0, 1); *also, the function* $\omega_1(x)/\omega_2(x)$ *decreases monotonically on* [(

It can be seen from Theorem 5.7 that our hypothesis is true at least for the special case when $\omega_1(x) \equiv 1$ and $\omega_2(x)$ is a function uniformly increasing on [0, 1). If this hypothesis is true, it will result in an improvement of the inclusion theorems noted above, i.e. it will be possib to establish that if $\omega_1(x)/\omega_2(x)$ is a function decreasing monotonically on [0, 1) then the inclusi $U_{\omega_2} \subset U_{\omega_1}$ and $N\{\omega_2\} \subset N\{\omega_1\}$ will hold.

Problem II. Part V refers to a work [6] which has not yet been described. This work contains an essential, though incomplete, result directed towards constructing a theory of factorization of functions meromorphic on the whole plane and which have an arbitrary growth of characteris The very first question arising while constructing the complete theory is formulated in the closin part of Ref. [6]. It runs as follows.

Let us denote by Ω^*_∞ the set of functions $\omega(x)$ continuous on the half-axis $[0, +\infty)$ and satisfying the conditions:

(1) $\omega(x)$ is positive, non-increasing on the half-axis $[0, +\infty)$, and $\omega(0) = 1$.

(2) The integrals

$$\Delta_k = k \int_0^\infty \omega(x) x^{k-1} \, dx$$

exist for k = 1, 2, … . Putting $\Delta_0 = 1$, we introduce the functions:

$$W_\omega^{(\infty)}(z;\zeta) = \int_{|\zeta|}^{\infty} \frac{\omega(x)}{x}\,dx - \sum_{k=1}^{\infty}\left\{\zeta^{-k}\int_0^{|\zeta|}\omega(x)x^{k-1}\,dx - \right.$$

$$\left. - \zeta^{-k}\int_{|\zeta|}^{\infty}\omega(x)x^{-k-1}\,dx\right\}\frac{z^k}{\Delta_k}, \quad |z| < \infty, \quad 0 < |\zeta| < \infty \tag{7.1}$$

$$A(z;\zeta) = \left(1 - \frac{z}{\zeta}\right)e^{-W_\omega(z;\zeta)} \tag{7.2}$$

Further, let $\{z_k\}_1^\infty$ $(0 < |z_k| \leqslant |z_{k+1}| < +\infty)$ be an arbitrary sequence of complex numbers for which

$$\sum_{k=1}^{\infty}\int_{|z_k|}^{\infty}\frac{\omega(x)}{x}\,dx < +\infty \tag{7.3}$$

The question is, whether any supplementary conditions besides (7.3) are needed to provide convergence of the product

$$\prod_{k=1}^{\infty} A_\omega^{(\infty)}(z;z_k) \tag{7.4}$$

on the whole complex plane.

We put forward a hypothesis that this must be the case at least when

$$\frac{d\log\omega(x)}{d\log x} \downarrow -\infty, \qquad \text{with } x\uparrow +\infty$$

This hypothesis is true in a special case when e.g.

$$\omega(x) = \frac{\rho\sigma^\mu}{\Gamma(\mu)}\int_x^{+\infty} e^{-\sigma t^\rho}\,t^{\mu\rho-1}\,dt$$

where ρ $(0 < \rho < +\infty)$, μ $(0 < \mu < +\infty)$ and σ $(0 < \sigma < +\infty)$ are arbitrary parameters.

Problem III. Let $\mu(t)$ be, in general, a complex-valued function on $[0, +\infty)$ for which

$$V_\mu(r) \equiv \int_0^\infty r^t|d\mu(t)| < +\infty, \qquad \forall r \in [0, +\infty) \tag{7.5}$$

Then it is obvious that the function

$$f_\mu(z) = \int_0^\infty z^t d\mu(t)$$

is regular on the whole Riemann surface.

$$G_\infty \equiv \{z; |\mathrm{Arg}\, z| < \infty, \qquad 0 < |z| < \infty\}$$

and it is easy to see that

$$\sup_{|\varphi|<+\infty} |f_\mu(re^{i\varphi})| \leqslant V_\mu(r); \quad \forall r \in [0, +\infty)$$

This remark brings us naturally to formulation of the next problem:

Let the function f(z) *be regular on* G_∞ *and satisfy the condition*

$$\widetilde{M}_f(r) \equiv \sup_{|\varphi|<+\infty} |f(re^{i\varphi})| < +\infty, \quad \forall r \in [0, +\infty) \tag{7.6}$$

Is there a function $\mu_f(t)$ *on* $[0, +\infty)$ *having the property (7.5) and such that*

$$f(z) = \int_0^\infty z^t d\mu_f(t), \qquad z \in G_\infty \quad ? \tag{7.7}$$

Note that our hypothesis is true in the special case when

$$f(re^{i(\varphi+2\pi)}) \equiv f(re^{i\varphi}), \quad \forall r \in [0, +\infty) \quad \text{and} \quad \forall \varphi \in (-\infty, +\infty)$$

Indeed, in this special case it is obvious that f(z) is an entire function and therefore can be expanded in the form

$$f(z) = \sum_{k=0}^\infty a_k z^k, \qquad |z| < +\infty$$

This means that the representation (7.7) exists with a certain measure $\mu(t)$ concentrated only at the points $\{k\}_0^\infty$ of the half-axis $[0, +\infty)$.

Problem IV. In connection with the main Theorem 6.3, we formulate the problem of finding its discrete analogue for the functions from the classes

$$L_{2,\mu}(0, \ell) \Leftrightarrow \int_0^\ell |g(y)|^2 y^{2(\mu-1)} dy < +\infty$$

where ℓ $(0 < \ell < +\infty)$ is arbitrary.

The problem is to construct the apparatus of series by the Mittag-Leffler-type functions representing the function $g(y) \in L_{2,\mu}(0, \ell)$ *for any* $\rho \geqslant 1/2$ *and possessing the property similar to statement (b) of this theorem in the case when* $\rho \geqslant 1$, *in the domain*

$$\mathscr{D}_\rho\{\ell\} = \left\{ z; \frac{\pi}{\rho} \leqslant |\arg z| \leqslant \pi, \quad 0 \leqslant |z| \leqslant \ell \right\}$$

Complete solution of this problem will result in a discrete analogue of our more general Theorem 6.4 for the functions g(z) from the class

$$\sum_{k=1}^{p} \int_0^{\ell} |g(re^{i\varphi_k})|^2 r^{2(\mu-1)} dr < +\infty$$

where

$$\frac{1}{2} < \mu < \frac{1}{2} + \frac{1}{\rho}, \qquad \rho \geqslant \omega = \max_{1 \leqslant k \leqslant p} \left\{ \frac{\pi}{\varphi_{k+1} - \varphi_k} \right\}$$

and $0 \leqslant \varphi_1 < \varphi_2 < \ldots < \varphi_p < \varphi_{p+1} = \varphi_1 + 2\pi$ are arbitrary.

Even in the simplest case of the finite cross type set,

$$E\{\ell\} = \{-\ell, \ell\} \cup \{-i\ell, i\ell\}$$

the positive solution of this problem (here it must certainly be $\rho \geqslant 2$) is of indisputable interest. Such a solution will provide an original Fourier-type series apparatus for the sets $E\{\ell\}$.

Problem V. Let

$$0 \leqslant \lambda_0 \leqslant \lambda_1 \leqslant \lambda_2 \leqslant \ldots \leqslant \lambda_n \leqslant \ldots$$

be an arbitrary sequence and $s_k \geqslant 1$ designate the multiplicity of occurrence for the number λ_k in the set $\{\lambda_1, \lambda_2, \ldots, \lambda_k\}$ Consider the family of quasipolynomials of the form

$$\Delta_n(x) = \sum_{k=0}^{n} a_k^{(n)} e^{-\lambda_k x} x^{s_k - 1} \qquad (n = 0, 1, 2, \ldots)$$

assuming that for the whole family $\{\Delta_n(x)\}$

$$\sup_{0 \leqslant x < +\infty} \{|\Delta_n(x)|\} \leqslant M \qquad (n = 0, 1, 2, \ldots)$$

Denote

$$\sup_{0 \leqslant x < +\infty} \{|\Delta_n'(x)|\} = \mu_n \qquad (n = 1, 2, \ldots)$$

and

$$\sup_{r\leqslant x\leqslant R} \{|\Delta_n'(x)|\} = \mu_n(r, R) \qquad (n = 1, 2, ...)$$

for any $0 < r < R < +\infty$.

The problem of estimating the numbers $\{\mu_n\}$ *and* $\{\mu_n(r, R)\}$ *from above is put forward. Consider it likely that the estimates of the form*

$$\mu_n \leqslant C_M \exp\left\{ 2\sum_{k=1}^{n} \frac{1}{\lambda_k}\right\} \qquad (n = 1, 2, ...)$$

$$\mu_n(r, R) \leqslant C_M(r, R) \exp\left\{ \sum_{k=1}^{n} \frac{1}{\lambda_k}\right\}$$

may be valid in which C_M *and* $C_M(r, R)$ *do not depend on* n.

In the case $\lambda_k = k$ $(k = 0, 1, 2, ...)$, our assumption holds by virtue of the well-known Markov-Bernstein theorem.

REFERENCES

Parts I and II

[1] DUREN, P., Theory of H^p Spaces, Ch.I, Academic Press, New York and London (1970).

Parts III and IV

[2] DJRBASHIAN, M.M., A generalized Riemann-Liouville operator and some of its applications, Math. USSR Izv. 2 (1968) 1027–1064.

Part V

[3] DJRBASHIAN, M.M., Theory of factorization of functions meromorphic in the disc, Math.USSR Sbornik 8 4 (1969) 493–591.

[4] DJRBASHIAN, M.M., ZAKHARIAN, V.S., Boundary properties of meromorphic functions of bounded form, Math. USSR Izv. 4 6 (1970) 1273–1354.

[5] ДЖРБАШЯН, М.М., ЗАХАРЯН, В.С., Граничные свойства подклассов мероморфных функций ограниченного вида, Изв. АН Арм.ССР, серия Математика, 6 2-3 (1971) 182-194.

[6] ДЖРБАШЯН, М.М., Факторизация функций, мероморфных в конечной плоскости, Изв. АН Арм.ССР "Математика" 5 6 (1970) 453-485.

Part VI

[7] DJRBASHIAN, M.M., "Harmonic analysis in the complex domain and its applications in the theory of analytical and infinitely differentiable functions", Lecture Notes in Mathematics **399**, Springer-Verlag, Berlin-Heidelberg-New York (1974) 94–118.

[8] ДЖРБАШЯН, М.М., Интегральные преобразования и представления функций в комплексной области, Москва, Наука (1966) главы III-VIII.

[9] DJRBASHIAN, M.M., An extension of the Denjoi-Carleman quasi-analytic classes, Am. Math. Soc. Transl. **107** 2 (1974).

[10] DJRBASHIAN, M.M., KOČARIAN, G.S., Uniqueness theorems for some classes of analytic functions, Math. USSR Izv. 7 1 (1973) 95–129.

A GENERALIZATION OF BEURLING'S ESTIMATE OF HARMONIC MEASURE

M. ESSÉN
Department of Mathematics,
Royal Institute of Technology,
Stockholm, Sweden

Abstract

A GENERALIZATION OF BEURLING'S ESTIMATE OF HARMONIC MEASURE.
In this note a more precise estimate is given of the harmonic measure of $\partial D \cap \{|z| = 1\}$ due to A. Beurling.

Let D *be an open connected subset of the open unit disc* Δ. *In 1933, A. Beurling gave an estimate of the harmonic measure of* $\partial D \cap \{|z| = 1\}$ *with respect to* D; *this estimate depends on the circular projection of the complement of* D *onto a ray from the origin. In this note a more precise estimate is given which also depends on the angular size of* $D \cap \{|z| = r\}$, $0 \leq r < 1$. *There are also results of this type in* $\underline{R}^d$, $d \geq 3$.

Let D be an open connected subset of the open unit disc Δ and set

$$\alpha = \partial D \cap \{|z| = 1\}, \qquad \beta = \partial D \cap \Delta$$

Let $\omega(z, \alpha, D)$ be the harmonic measure of α with respect to D, constructed by Perron's method with the boundary function which is 1 on α and 0 on β. Let

$$u(z) = \begin{cases} \omega(z, \alpha, D), & z \in D \\ 0, & z \in \Delta \setminus D \end{cases}$$

It is clear that u is a subharmonic function in Δ. We also introduce

$$E = \left\{ r \in (0, 1) : \inf_\theta u(re^{i\theta}) = 0 \right\}$$

$$F(r) = \{ \theta \in [-\pi, \pi] : \omega(re^{i\theta}, \alpha, D) > 0 \}$$

$$a(r) = mF(r), \; r \in E; \qquad a(r) = \infty, \; r \notin E$$

Beurling [2] has proved that

$$\operatorname{Max}_\theta \omega(re^{i\theta}, \alpha, D) \leq \text{const} \exp\left\{ -\frac{1}{2} \int_{E \cap (r,1)} t^{-1} \, dt \right\} \tag{1}$$

Let $A(r) = \sup_{r \leq t < 1} (2\pi/a(t))$.

Theorem. *Let* D *and* u *be as above. Then we have*

$$\left\{\int_{-\pi}^{\pi} u(re^{i\theta})^2\, d\theta\right\}^{1/2} \leqslant 2\sqrt{2\pi A(r)}\, \exp\left\{-\pi\int_r^1 \frac{dt}{ta(t)}\right\},\ r \leqslant 1 \qquad (2)$$

$$\operatorname*{Max}_{\theta}\, u(re^{i\theta}) \leqslant 6\pi\sqrt{2A(r)}\, \exp\left\{-\pi\int_{2r}^{1} \frac{dt}{ta(t)}\right\},\ r \leqslant 1/2 \qquad (3)$$

Remark 1. (1) and (3) are equivalent (modulo absolute constants) if $a(r) = 2\pi$, $r \in E$, i.e. the domain D is the unit disc Δ cut along segments along the negative real axis.

Remark 2. If the value A(r) is assumed on a small set in such a way that the exponential term does not compensate for the large term $\sqrt{A(r)}$, we first replace D by a larger domain D_1 so that A(r) decreases and $\omega(z, \alpha, D_1)$ majorizes $\omega(z, \alpha, D)$. Now, we can use estimate (2) or (3).

Remark 3

$$\int_r^1 (ta(t))^{-1}\, dt = \int_{E\cap(r,1)} (ta(t))^{-1}\, dt$$

is the "area" of that part of D which is in the "shadow" of E taken along concentric circles. If $D \cap \{r_1 \leqslant |z| \leqslant r_2\}$ is narrow, (2) or (3) imply that our estimate decreases considerably when we pass through the gap.

In the proof, we can either use results of M. Heins [5] or a theorem of A. Baernstein [1]. For reasons to be explained later, we prefer to use Baernstein's result.

With D and F(r) given as above, we define the circular symmetrization D* of D in the following way.

If $F(r) = [0, 2\pi]$, $D^* \cap \{|z| = r\} = \{|z| = r\}$

If this is not the case,

$$D^* \cap \{|z| = r\} = \{z = re^{i\varphi} : |\varphi| < a(r)/2\}$$

D* is also an open subset of Δ (cf. Hayman [4], 4.5.3).

Starting from D*, we define α^*, β^* and

$$v(z) = \begin{cases} \omega(z, \alpha^*, D^*), & z \in D^* \\ 0, & z \in \Delta\backslash D^* \end{cases}$$

v is subharmonic in Δ and it can be proved that $\varphi \to v(re^{i\varphi})$ is decreasing on $[0, a(r)/2)$. Baernstein's result (cf. Ref. [1], Theorem 7; a proof can also be found in Essén [3], p. 87) is as follows.

Theorem A. *Let* $\Phi : [0, \infty) \to R$ *be a convex non-decreasing function. Then*

$$\int_{-\pi}^{\pi} \Phi(u(re^{i\theta}))\, d\theta \leqslant \int_{-\pi}^{\pi} \Phi(v(re^{i\theta}))\, d\theta,\ 0 < r < 1$$

Two special cases:

(i) Choosing $\Phi(x) = x^p$, $p \geqslant 1$, taking p^{th} roots and letting $p \to \infty$, we obtain

$$\underset{\theta}{\text{Max}}\ u(re^{i\theta}) \leqslant \underset{\theta}{\text{Max}}\ v(re^{i\theta}),\ 0 < r < 1 \tag{4}$$

(ii) Choosing $\Phi(x) = x^2$, we obtain

$$\int_{-\pi}^{\pi} u(re^{i\theta})^2\, d\theta \leqslant \int_{-\pi}^{\pi} v(re^{i\theta})^2\, d\theta \overset{\Delta}{=} q(r)^2$$

We shall prove (2) by deducing an estimate of the Carleman mean q(r). Here the following result will be needed (cf. Essén [3], Theorem 5.2).

Theorem B. *Let* $p : (-\infty, 0] \to [0, \infty)$ *be a lower semicontinuous and locally bounded function. Assume that there exists a solution of the differential inequality*

$$\begin{cases} z''(t) - p(t)^2\, z(t) \geqslant 0, & -\infty < t \leqslant 0 \\ z(0) = 1, \ \lim_{t \to -\infty} z(t) = c \text{ exists}, & 0 \leqslant c < \infty \end{cases}$$

Let T *be given,* $T < 0$. *Let* p^* *be the measure-preserving, increasing rearrangement of* $p|_{[T,0]}$ *on* $(T, 0]$ *and let* $p^*(t) = \inf_{t \leqslant 0} p(t)$, $t \leqslant T$.

Then there exists a non-negative solution w *of the differential equation*

$$\begin{cases} w''(t) - (p^*(t))^2\, w(t) = 0,\ w(0) = 1,\ \lim_{t \to -\infty} w(t) = c^* \\ 0 \leqslant c^* < \infty \end{cases}$$

Furthermore

$$z(T) \leqslant w(T) \tag{5}$$

Remark. By a solution of a differential inequality or a differential equation, we mean a function z such that z and z′ are absolutely continuous, z″ is the a.e. existing derivative of z′ and the inequality or the equation is satisfied a.e.

The point of Theorem B is that it is fairly easy to find an estimate of w(T), and thus also of z(T).

The Baernstein *-function

Let u be subharmonic in Δ. Let r be given, $0 \leqslant r < 1$. We define

$$u^*(re^{i\theta}) = \sup_{|E| = 2\theta} \int_E u(re^{i\omega})d\omega, \quad 0 \leqslant \theta \leqslant \pi$$

where the supremum is taken over all measurable sets $E \subset [-\pi, \pi]$ of measure 2θ. A. Baernstein h proved that u* is subharmonic in the upper half-disc. (Proofs can be found in Baernstein [1] or in Essén [3], Ch. 9.)

Proof of the main result

From Theorem A we see that it suffices to consider the symmetrized region D* and the Carleman mean q(r). Without loss of generality, we can assume that $\beta^* = \partial D^* \cap \Delta$ is smooth. It known that q(r) is a convex function of log r (cf. e.g. Heins [6], Ex. 12, p. 82). In particular, q' exists a.e. and is an increasing function.

We introduce

$$Q(t) = q(e^t), \quad \alpha(t) = a(e^t), \; t \leqslant 0$$

$$\mathscr{E} = \{t : \inf_{\varphi} v(e^{t+i\varphi}) = 0\}$$

Then

$$Q(t)^2 = \int_{-\alpha/2}^{\alpha/2} v(e^{t+i\varphi})^2 \, d\varphi, \qquad \alpha = \alpha(t), \; t \in \mathscr{E}$$

Using an argument of Carleman (cf. e.g. Heins [6], p. 121–123), we see that in the interior $\mathscr{E}^0$ of $\mathscr{E}$,

$$Q''(t) - \left(\frac{\pi}{\alpha(t)}\right)^2 Q(t) \geqslant 0, \; t \in \mathscr{E}^0$$

Outside $\mathscr{E}^0$, we use the convexity of q(r) with respect to log r and obtain

$$Q''(t) \geqslant 0, \; t \leqslant 0$$

We do not know *a priori* that q' is absolutely continuous. Therefore, (7) holds only in the distributional sense, i.e. the left-hand member is a non-negative measure. Let

$$p(t) = \begin{cases} \dfrac{\pi}{\alpha(t)}, & t \in \mathscr{E}^0 \\ 0, & t \notin \mathscr{E}^0 \end{cases}$$

The function p is continuous in $\mathscr{E}^0$ and in the complement of $\mathscr{E}$. It is also lower semicontinuous. Thus

$$Q''(t) - p(t)^2 Q(t) \geqslant 0 \tag{8}$$

in the distributional sense on $(-\infty, 0)$, i.e.

$$Q'' - p^2 Q = d\mu = d\mu_a + d\mu_s$$

where $d\mu$ is a non-negative measure and $d\mu_a(t) = h(t)dt$ and $d\mu_s$ are the absolutely continuous and singular parts of $d\mu$, respectively. Consider now Q_1 which is the classical solution of the problem

$$Q_1''(t) - p(t)^2 Q_1(t) = h(t), \qquad Q_1(0) = Q(0), \quad \lim_{t \to -\infty} Q_1(t) \text{ exists} \tag{9}$$

If $Q_2 = Q - Q_1$, we have

$$Q_2'' - p^2 Q_2 \geqslant 0, \qquad Q_2(0) = 0, \quad \lim_{t \to -\infty} Q_2(t) \text{ exists} \tag{10}$$

in the distributional sense. It is easily seen that we must have $Q_2 \leqslant 0$, i.e. $Q \leqslant Q_1$.

We now apply Theorem B. If $T < 0$ is given, we rearrange p to obtain p^* and consider

$$w''(t) - p^*(t)^2 w(t) = 0 \qquad w(0) = Q(0), \quad \lim_{t \to -\infty} wt(t) \geqslant 0$$

Theorem B shows that $Q(T) \leqslant Q_1(T) \leqslant w(T)$.

To estimate $w(T)$, we note that p^* is increasing on $(-\infty, 0)$. It is now fairly easy to show that

$$w(T) \leqslant 2Q(0)\sqrt{A(e^T)} \exp\left\{-\int_T^0 p^*(\tau)d\tau\right\} \qquad \leqslant 2\sqrt{2\pi A(e^T)} \exp\left\{-\int_T^0 p(\tau)d\tau\right\}$$

Going back to our original notation, we obtain (2). Once we know that (2) holds, it is easy to prove (3).

Remark. If we want to use Carleman's original method to study the differential inequality (8), we must have

$$\inf_{T \leqslant t \leqslant 0} p(t) > 0 \text{ for each } T < 0$$

In the problem studied here, we must be able to handle situations when $p(t) = 0$ on certain intervals. This difficulty is solved by applying Theorem B: we obtain a differential equation $w'' - (p^*)^2 w = 0$ where the support of p^* is an interval $[d, 0]$. From this point on, we could also have used Carleman's original method (cf. e.g. Heins [5]).

Similar results are true in $\mathbb{R}^d$, $d \geqslant 2$. When $d \geqslant 3$, we use a cap symmetrization. If r is given, $F(r)$ is the subset of the unit sphere $\{|x| = 1\}$ where $\omega(rx, \alpha, D) > 0$, and $a(r)$ is the $(d-1)$-dimensional Lebesgue measure of $F(r)$. We define $D^* \cap \{|x| = r\}$ to be a spherical cap with the same (d − 1)-dimensional measure as the set $\{x = ry, y \in F(r)\}$. Theorem A, properly interpreted, also holds when $d \geqslant 3$: this is a result of C. Borell [7] and J. Sarvas. Borell's paper is contained in Essén [3], where further references are given (cf. Ref. [3], p. 52).

The reason we preferred to refer to a paper of A. Baernstein rather than to a paper of M. Heins in the earlier discussion is that the result of Baernstein is known to hold also in $\mathbb{R}^d$, $d \geqslant 3$.

The second step in our proof is to deduce a differential inequality for the Carleman mean. To do this, we have to investigate a singular boundary value problem.

REFERENCES

[1] BAERNSTEIN, A., Integral means, univalent functions and circular symmetrization, Acta Math. **133** (1974) 133–169.

[2] BEURLING, A., Etudes sur un problème de majoration, Thesis, Uppsala (1933).

[3] ESSÉN, M., The cos $\pi\lambda$ theorem, Springer Lecture Notes in Mathematics 467, Berlin-Heidelberg (1975).

[4] HAYMAN, W., Multivalent Functions, Cambridge Univ. Press (1967).

[5] HEINS, M., On a notion of convexity connected with a method of Carleman, J. Analyse Math. **7** (1960) 53-

[6] HEINS, M., Selected Topics in the Classical Theory of Functions of a Complex Variable, Holt, Rinehart and Winston, New York (1962).

[7] BORELL, C., "An inequality for a class of harmonic functions in n-space", Springer Lecture Notes in Mathematics 467 (1975) 99–112.

PSEUDOCONVEXITY AND THE PRINCIPLE OF MAXIMUM MODULUS

A.A. FADLALLA
Department of Mathematics,*
Cairo University,
Cairo, Egypt

Abstract

PSEUDOCONVEXITY AND THE PRINCIPLE OF MAXIMUM MODULUS.

Pseudoconvexity of a domain $G \subset \mathbb{C}^n$ and the existence of a function f holomorphic in $\bar{G}$ such that f assumes its maximum at a boundary point $P \in \partial G$ are closely related to each other. Such relationships are discussed.

Definition 1. A domain $G \subset \mathbb{C}^n$ is said to be pseudoconvex if (i) to every point $P \in \partial G$, there exist a neighbourhood U of P and a real-valued function $\varphi \in C^2(U)$ such that

$$Q \cap U = \{z : z \in U, \varphi(z) < 0\} \tag{1}$$

(ii) $d\varphi \neq 0$ in U ; (iii) the Hermitian form:

$$\sum \frac{\partial^2 \varphi}{\partial z_\mu \partial \bar{z}_\nu}\bigg|_P \lambda_\mu \bar{\lambda}_\nu \geqslant 0 \tag{2}$$

for all λ_μ ($\mu = 1, \ldots, n$) satisfying

$$\sum \frac{\partial \varphi}{\partial z_\mu}\bigg|_P \lambda_\mu = 0, \quad \sum \frac{\partial \varphi}{\partial \bar{z}_\mu}\bigg|_P \bar{\lambda}_\mu = 0 \tag{3}$$

If the Hermitian form (2) is >0, G is called a strictly pseudoconvex domain.

Definition 2. A hypersurface $H \subset \mathbb{C}^n$ is called pseudoconvex if (i) to every point $P \in H$, there exist a neighbourhood U of P and real-valued function $\varphi \in C^2(U)$, such that

$$H \cap U = \{z : z \in U, \varphi(z) = 0\} \tag{1'}$$

(ii) $d\varphi \neq 0$ in U; (iii) the Hermitian form:

$$\sum_{\mu,\nu=1}^{n} \frac{\partial^2 \varphi}{\partial z_\mu \partial \bar{z}_\nu}\bigg|_P \lambda_\mu \bar{\lambda}_\nu \tag{2'}$$

* Faculty of Science.

is semidefinite for all λ_μ ($\mu = 1, \ldots, n$) satisfying

$$\sum \frac{\partial\varphi}{\partial z_\mu}\bigg|_P \lambda_\mu = 0, \quad \sum \frac{\partial\varphi}{\partial \bar{z}_\mu}\bigg|_P \bar{\lambda}_\mu = 0 \tag{3'}$$

If the Hermitian form (2′) under conditions (3′) is definite, H is called strictly pseudoconvex. Now we give a short proof of

Theorem 1. Let G be a strictly pseudoconvex domain and $P \in \partial G$. Then there exists a function f holomorphic in a neighbourhood of $\bar{G}$ such that $|f(P)| = \text{Max}\ |f(G)|$ and $|f(Q)| < |f(P)|$ for all $Q \in \bar{G}$, $Q \neq P$.

(i) Let P be the origin, and ∂G in a neighbourhood of P be the hypersurface $\varphi = 0$. Furthermore let t be a real parameter and

$$a_\mu = \left(\frac{\partial\varphi}{\partial z_\mu}\right)_P \quad ; \quad a_{\mu\nu} = \frac{1}{2}\left(\frac{\partial^2\varphi}{\partial z_\mu \partial z_\nu}\right)_P$$

Then if $a > 0$ is sufficiently large, the analytic hypersurface

$$H: \sum_{\mu=1}^{n} a_\mu z_\mu + \sum_{\mu,\nu=1}^{n} a_{\mu\nu} z_\mu z_\nu - (t^2 + \frac{it}{a}) = 0, \qquad -\epsilon < t < \epsilon, \qquad \epsilon > 0$$

have the following property:

$$H \cap U \cap \bar{G} = \{P\}$$

where U is a sufficiently small neighbourhood of P and

$$S: \sum_{\mu=1}^{n} a_\mu z_\mu + \sum_{\mu,\nu=1}^{n} a_{\mu\nu} z_\mu z_\nu = 0$$

satisfies the relation $S \cap U \cap \bar{G} = \{P\}$, [Cf. Ref. [1]].

(ii) Grauert [2] proved that there exists a neighbourhood N of ∂G and a strictly plurisubharmonic function g defined in N, such that ∂G is the hypersurface $g = 0$ and $N \cap G = \{Q : Q \in N, g(Q) < 0\}$. Whence, for sufficiently small $e > 0$, the hypersurface $g - e = 0$ is the boundary of a strictly pseudoconvex domain G_1. Furthermore, ∂G_1 can be made as near to ∂G as we please. Let us now choose ∂G_1 so near to ∂G and U so small that:

(a) $\text{Max}\ |\varphi(u)| < b < \dfrac{1}{4a^2}$ where $\varphi = \displaystyle\sum_{\mu=1}^{n} a_\mu z_\mu + \sum_{\mu,\nu=1}^{n} a_{\mu\nu} z_\mu z_\nu$

(b) $\bar{H}' = \{H \cap \bar{G}_1 \cap U\}$ is a closed connected set (hypersurface).

Let $H' = \bar{H}' \cap G_1$

Now H' is a piece of the hypersurface $\varphi - (t^2 + it/a) = 0$. Solving this equation in t, we get

$$t = -\frac{i}{2a} \pm \sqrt{\varphi - \frac{1}{4a^2}}$$

According to (a), we can assume that

$$\frac{\pi}{2} < \arg\left[\varphi - \frac{1}{4a^2}\right] < \frac{3\pi}{2} \text{ in U}$$

Thus the function $\sqrt{\varphi - 1/4a^2}$ is one-valued in U, and H' is a piece of the hypersurface $t = -i/2a + \sqrt{\varphi - 1/4a^2}$. If $\Psi = -i/2a + \sqrt{\varphi - 1/4a^2}$ then Ψ is holomorphic in U, and H' is a piece of the analytic hypersurface $\nu = 0$, where $\nu = \text{Im}\,\Psi$. Now all points of $\{U\text{-}P\} \cap \overline{G}$ lie on one side of H'. Let it be the side $\nu > 0$.

Now H' is a piece of the analytic hypersurface $\nu = 0$, i.e. through any point $Q \in H'$ passes a complex analytic $(n-1)$-dimensional surface S_Q, which is a subset of H' (see (i)). It is obvious that S_P is the surface $\Psi = 0$ and that $S_P \cap \overline{G} = P$.

Now we define the open covering U_1 and U_2 of G_1 as follows. Let $U^* = \{Q : Q \in U, \nu(Q) \leq 0\}$, $U_1 = U \cap G_1$, $U_2 = G_1 - U^*$. Let f_2 be any function regular in U_2. We associate with U_1 and U_2 the meromorphic function $f_1 = 1/\Psi$ and f_2 respectively. It is clear that $f_1 - f_2$ is holomorphic in $U_1 \cap U_2$.

Since G_1 is a strongly pseudoconvex domain, it is a domain of holomorphy and, according to Oka [3], Cousin's first problem can be solved in G_1, and a function f_3 can be found which is equivalent – with regard to subtraction – to f_1 in U_1 and to f_2 in U_2. Thus f_3 has the points of S_P as poles and otherwise is regular in G_1. Whence f_3 is regular in $[\overline{G} - P]$ but singular in P.

(iii) We shall consider the function $f' = 1/f_3$ as a mapping of $\overline{G}$ in the compact complex plane $\overline{C}$. Let $U' \Subset U$ be a neighbourhood of P.

Since f_3 is regular in $\overline{G - U'}$, it is bounded in $\overline{G - U'}$. Now let $f_3 = h + f_1$ in U_1. Hence h is regular in U_1 and therefore it is bounded in $U' \cap \overline{G}$. Since all points of $\{U' \cap \overline{G} - [P]\}$ lie on the side $\nu > 0$, it follows that $\text{Im}\,(1/\Psi) \leq 0$ in $U' \cap \overline{G}$. Whence there exists a number k such that $\text{Im}\, f_3 < k$ in $\overline{G}$. Generally, we can assume that $k < 0$. Hence $w = f'$ can be considered as a mapping of $\overline{G}$ in the half plane: $\text{Im}\, w \geq 0$ of the complex plane of the variable w.

Let T be a conformal mapping of the half plane $\text{Im}\, w \geq 0$ on the unit circle $|z| \leq 1$, in the complex plane of the variable z. Then $f = T_0 f'$ possesses the required properties.

For pseudoconvex domains $G \subset \mathbb{C}^n$, such a theorem does not exist unless ∂G contains strictly pseudoconvex boundary points. For example, Rossi [4] proved that if such a function exists, then P should be the limiting point of strictly pseudoconvex boundary points, from which we draw the following conclusions.

Let G be a pseudoconvex domain and $P \in \partial G$. Suppose that there exists a function f holomorphic in a neighbourhood of $\overline{G}$ such that $|f(P)| = \text{Max}\, |f(G)|$. If there exists a neighbourhood U of P such that $U \cap \partial G$ contains no strictly pseudoconvex boundary points of G, then:

(1) In every neighbourhood of P, there exists points $Q \in \partial G$ such that $|f(Q)| = |f(P)|$.
(2) In every neighbourhood of P, there exists $Q \in \partial G$ such that $f(Q) = f(P)$.

Proof. If this were not true, we could find a neighbourhood V of P such that $f(Q) \neq f(P)$ for all $Q \in V \cap \overline{G}, Q \neq P$. Let $\Psi(z) = f(P) + f(z)$. Thus $|\Psi(z)|$ will assume its maximum in $V \cap \overline{G}$ only at P, which would be a contradiction to Rossi's theorem.

The Sommer [5] conditions

In the same direction Sommer proved the following. Let $G \subset \mathbb{C}^n$ be a pseudoconvex domain, $P \in \partial G$ and ∂G in a neighbourhood U of P be the hypersurface $\varphi = 0$, where φ is a real-valued function $\in C^4(U)$. Let the Hermitian form (2) under conditions (3) be of constant rank $k, 0 < k <$ in U. Then there exist a neighbourhood U′ of P and a co-ordinate system $(z_1, \ldots, z_n)$ in U′ such that

$$\partial G \cap U' = H' \times V_2, \quad G \cap U' = V_1 \times V_2, \quad H' \subset \partial V_1$$

where V_2 is a domain in the (n – k)-dimensional complex space $\mathbb{C}^{n-k}(z_{k+1}, \ldots, z_n)$ of the variables $z_{k+1}, \ldots, z_n$; V_1 is a domain in the k-dimensional complex space $\mathbb{C}^k(z_1, \ldots, z_k)$ of the variables $z_1, \ldots, z_k$; and H′ is a strictly pseudoconvex hypersurface in $\mathbb{C}^k(z_1, \ldots, z_k)$.

That is to say, the boundary of G in a neighbourhood of P is decomposable into (n – k)-dimensional analytic fibres; each fibre is of the form $P' \times V_2$, where $P' \in H'$. We denote the analytic fibre through P by $\mathscr{F}_P$.

Now if f is holomorphic in a neighbourhood of $\bar{G}$ and if $|f|$ assumes its maximum at P, it is obvious that $f(Q) = f(P)$, for all $Q \in \mathscr{F}_P$. Which is in agreement with Corollary 2 of Rossi's theorem. Now we prove

Theorem 2. Let f be a function holomorphic in G and continuous in $G \cup \{P\}$, i.e.

$$\lim_{m \to \infty} f(P_m)$$

exists for all sequences of points $\{P_m\} \subset G$ converging to P. Obviously this limit is unique and will be denoted by f(P). Furthermore let $|f(P)| = \text{Max}\,|f(G)|$. Then f can be extended continuously to $\mathscr{F}_P$ and $f(Q) = f(P)$ for all $Q \in \mathscr{F}_P$.

Proof. Let $P = P' \times P''$ where $P' \in H'$, $P'' \in V_2$. Now let $Q \in \mathscr{F}_P$; therefore $Q = P' \times Q''$, where $Q'' \in V_2$. Let $[Q_m] \subset U'$ be a sequence of points converging to Q and $Q_m = Q'_m \times Q''_m$, $Q'_m \in V_1$, $Q''_m \in V_2$. Thus $[Q'_m]$ converges to P′ and $[Q''_m]$ converges to Q″. Now let $P_m = Q'_m \times p''$ thus $[P_m]$ converges to P.

Let $\zeta = (z_{k+1}, \ldots, z_n) \in V_2$ and put $f_m(\zeta) = f(Q'_m, \zeta)$. Since $[f_m(\zeta)]$ is a bounded sequence of holomorphic functions in V_2, it is a normal family; it contains a uniformly convergent subsequence $\{f_{m'}(\zeta)\}$ to a holomorphic limit function $f_0(\zeta)$. Now $f_{m'}(P'') = f(Q'_{m'}, P'') = f(P_{m'}) \to f($ $= f_0(P'')$. Thus $|f_0(P'')| = |f(P)| = \text{Max}\,|f_0(V_2)|$; thus $f_0(\zeta)$ is constant and equals f(P) for all $\zeta \in V$

Now $f(Q_{m'}) = f(Q'_{m'}, Q''_{m'}) = f_{m'}(Q''_{m'}) \to f_0(Q'') = f(P)$, which proves the theorem.

REFERENCES

[1] BEHNKE, H., SOMMER, F., Über die Voraussetzungen des Kontinuitätssatzes, Math. Ann. **121** (1949/1950
[2] GRAUERT, H., Über Modifikationen und exzeptionelle analytische Mengen, Math. Ann. (1962).
[3] OKA, K., Domaines d'holomorphie, J. Hiroshima Univ. (1937).
[4] ROSSI, H., Ann. Math. **74** 3 (1961).
[5] SOMMER, F., Komplex-analytische Blätterung reeller Hyperflächen im C^n, Math. Ann. **137** (1959).

QUASICONFORMAL MAPPINGS

F.W. GEHRING
Department of Mathematics,
University of Michigan,
Ann Arbor, Michigan,
United States of America

Abstract

QUASICONFORMAL MAPPINGS.

1. Modulus of a curve family. 2. Conformal capacity of condensers. 3. Inner and outer dilatations. 4. Distortion and convergence. 5. One-quasiconformal mappings. 6. Mapping problems. 7. An existence theorem.

1. MODULUS OF A CURVE FAMILY

Given a family, Γ, of nonconstant curves γ in $\overline{R}^n$, we let $\mathrm{adm}(\Gamma)$ denote the family of Borel measurable functions $\rho : R^n \to [0, \infty]$ such that

$$\int_\gamma \rho\, ds \geq 1$$

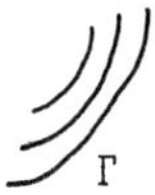

for all locally rectifiable $\gamma \in \Gamma$. We call

$$M(\Gamma) = \inf_{\rho \in \mathrm{adm}(\Gamma)} \int_{R^n} \rho^n dm , \quad \lambda(\Gamma) = M(\Gamma)^{\frac{1}{1-n}}$$

the <u>modulus</u> and <u>extremal length</u> of Γ, respectively.

When Γ is a family of arcs, we may think of $M(\Gamma)$ as the conductance and $\lambda(\Gamma)$ as the resistance of a system of homogeneous wires. $M(\Gamma)$ is big when the wires are plentiful or short, small when the wires are few or long.

Theorem 1. $M(\Gamma)$ is an outer measure on the collections of curve families Γ in $\overline{R}^n$. That is,

a) $M(\phi) = 0$,

b) $M(\Gamma_1) \leq M(\Gamma_2)$ when $\Gamma_1 \subset \Gamma_2$,

c) $M(\bigcup_j \Gamma_j) \leq \sum_j M(\Gamma_j)$.

Proof for c). We may assume $M(\Gamma_j) < \infty$ for all j . Then given $\varepsilon > 0$ we can choose for each j a $\rho_j \in \mathrm{adm}(\Gamma_j)$ such that

$$\int_{R^n} \rho_j^n dm \leq M(\Gamma_j) + 2^{-j}\varepsilon$$

Now set

$$\rho = \sup_j \rho_j \ , \ \Gamma = \bigcup_j \Gamma_j$$

Then $\rho : R^n \to [0, \infty]$ is Borel measurable. Moreover, if $\gamma \in \Gamma$ is locally rectifiable, then $\gamma \in \Gamma_j$ for some j ,

$$\int_\gamma \rho ds \geq \int_\gamma \rho_j ds \geq 1$$

and hence $\rho \in \mathrm{adm}(\Gamma)$. Thus

$$M(\Gamma) \leq \int_{R^n} \rho^n dm \leq \int_{R^n} \sum_j \rho_j^n dm \leq \sum_j M(\Gamma_j) + \varepsilon$$

Remark 1. If we apply the Carathéodory criterion to the outer measure M to define the notion of a measurable curve family, then we can show the following:

a) Γ is measurable if $M(\Gamma) = 0$,

b) Γ is not measurable if $0 < M(\Gamma) < \infty$,

c) Γ may or may not be measurable if $M(\Gamma) = \infty$.

Theorem 2. If each curve γ_1 in a family Γ_1 contains a subcurve γ_2 in a family Γ_2, then $M(\Gamma_1) \leq M(\Gamma_2)$.

Proof. Choose $\rho \in \mathrm{adm}(\Gamma_2)$ and suppose $\gamma_1 \in \Gamma_1$ is locally rectifiable. Then

$$\int_{\gamma_1} \rho ds \geq \int_{\gamma_2} \rho ds$$

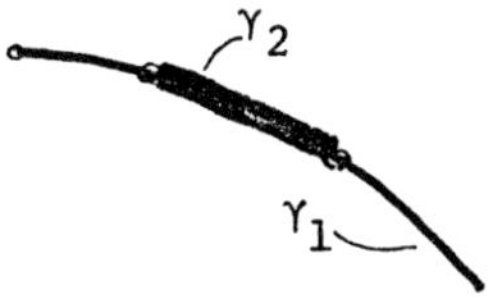

where γ_2 is the subcurve in Γ_2, and $\rho \in \mathrm{adm}(\Gamma_1)$.

Thus

$$M(\Gamma_1) \leq \int_{R^n} \rho^n dm$$

and taking the infimum over all such ρ yields

$$M(\Gamma_1) \leq M(\Gamma_2)$$

Theorem 3. $M(\Gamma)$ is additive on curve families in disjoint Borel sets. That is, if E_j are disjoint Borel sets and if the curves of Γ_j lie in E_j, then

$$M(\bigcup_j \Gamma_j) = \sum_j M(\Gamma_j)$$

Examples. a) Rectangular parallelopiped. If Γ is the family of curves γ joining two parallel faces of area A and distance h apart, then

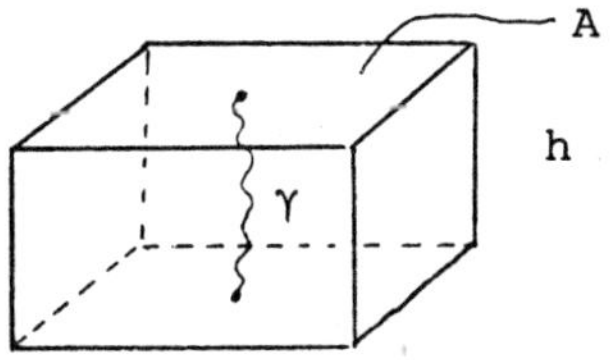

$$M(\Gamma) = \frac{A}{h^{n-1}}$$

b) Spherical ring. If Γ is the family of curves joining the sphere with center x_0 and radius a to the concentric sphere of radius b, then

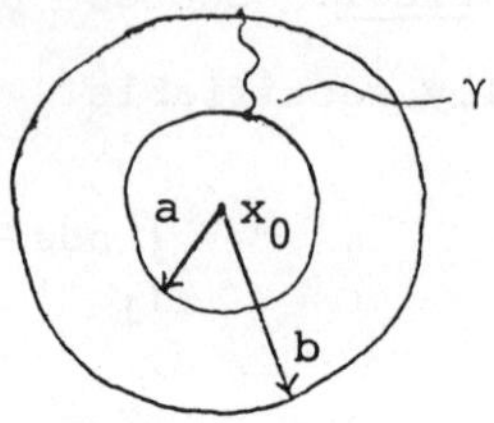

$$M(\Gamma) = \omega_{n-1}(\log\frac{b}{a})^{1-n}$$

where

$$0 \leq a < b \leq \infty , \ \omega_{n-1} = m_{n-1}(S)$$

Here $m_{n-1}(S)$ denotes the surface area of the unit sphere in

Proof for a). Choose $\rho \in \mathrm{adm}(\Gamma)$ and let γ_y be the vertical segment from y in the base E. Then $\gamma_y \in \Gamma$ and

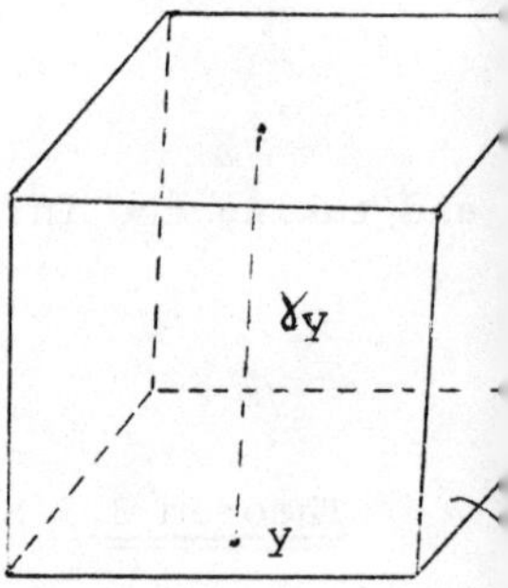

$$1 \leq \left(\int_\gamma \rho ds\right)^n \leq h^{n-1} \int_{\gamma_y} \rho^n ds$$

This holds for all such y and hence

$$\int_{R^n} \rho^n ds \geq \int_E \left(\int_{\gamma_y} \rho^n ds\right) dm_{n-1} \geq \frac{A}{h^{n-1}}$$

Since ρ is arbitrary,

$$M(\Gamma) \geq \frac{A}{h^{n-1}}$$

Next set $\rho = \frac{1}{h}$ inside the parallelopiped and $\rho = 0$ otherwise. Then $\rho \in \mathrm{adm}(\Gamma)$ and

$$M(\Gamma) \leq \int_{R^n} \rho^n dm = \frac{A}{h^{n-1}}$$

Theorem 4. If all the curves in a curve family Γ pass through a fixed point x_0, then $M(\Gamma) = 0$.

Proof. Suppose first that $x_0 \neq \infty$ and for each j let Γ_j denote the subfamily of $\gamma \in \Gamma$ which intersect x_0 and $S(x_0,1/j)$. Then each $\gamma \in \Gamma_j$ contains a subcurve γ' in the family of all curves joining x_0 to $S(x_0,1/j)$ in $B(x_0,1/j)$. Hence

$$M(\Gamma_j) \leq \omega_{n-1}\left(\log \frac{\frac{1}{j}}{0}\right)^{1-n} = 0$$

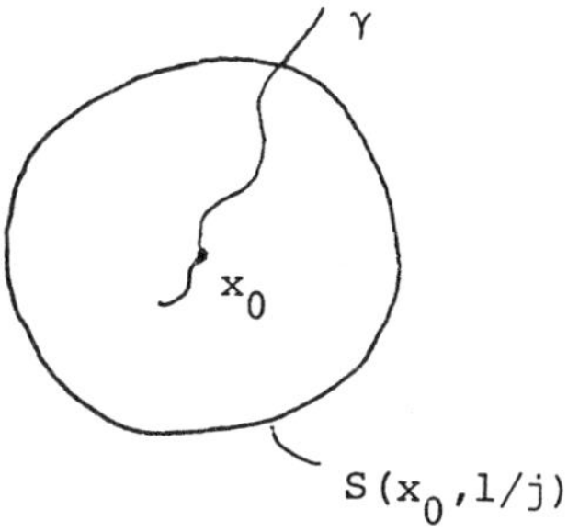

by example b) above. Since $\Gamma = \bigcup_j \Gamma_j$,

$$M(\Gamma) \leq \sum_j M(\Gamma_j) = 0$$

When $x_0 = \infty$, we argue as above with $S(x_0,1/j)$ replaced by $S(0,j)$.

Theorem 5. If $f : \overline{R}^n \to \overline{R}^n$ is a Möbius transformation, then

$$M(f(\Gamma)) = M(\Gamma)$$

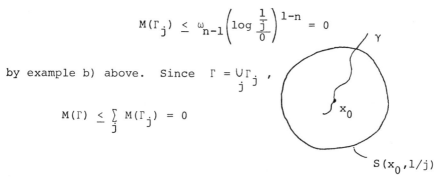

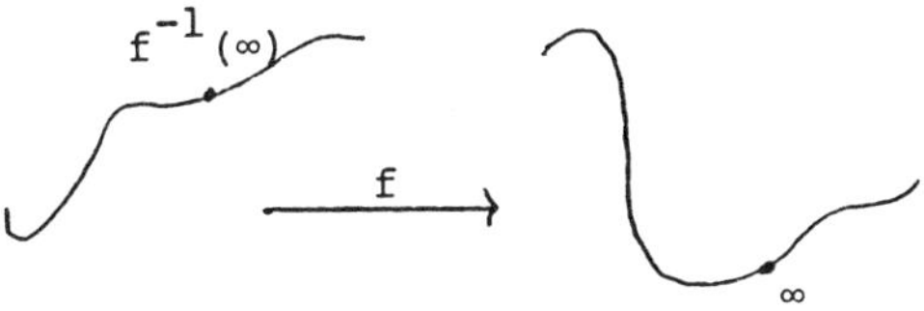

for all curve families Γ in $\overline{R}^n$.

Proof. Choose $\rho' \in \operatorname{adm} f(\Gamma)$, set

$$\rho(x) = \rho' \circ f(x)\,|f'(x)|$$

for $x \in R^n \sim \{f^{-1}(\infty)\}$, and let Γ_0 be the family of $\gamma \in \Gamma$ which pass through $f^{-1}(\infty)$. Then

$$M(\Gamma) = M(\Gamma \sim \Gamma_0)\ , \quad \rho \in \operatorname{adm}(\Gamma \sim \Gamma_0)$$

and hence

$$M(\Gamma) \le \int_{R^n} \rho^n dm = \int_{R^n} (\rho' \circ f)^n |f'|^n dm$$

$$= \int_{R^n} (\rho' \circ f)^n J(f) dm = \int_{R^n} (\rho')^n dm$$

Taking the infimum over every such ρ' gives $M(\Gamma) \le M(f(\Gamma))$. The result follows by repeating the preceding argument with f replaced by f^{-1}.

Theorem 6. If $f_j, f: R^n \to [0,\infty]$ are Borel measurable and if $f_j \to f$ in $L^n(R^n)$, then there exists a subsequence $\{j_k\}$ and a curve family Γ_0 with $M(\Gamma_0) = 0$ such that

$$\lim_{k\to\infty} \int_\gamma |f_{j_k} - f| ds = 0$$

for all locally rectifiable curves γ, $\gamma \notin \Gamma_0$.

Proof. Choose a subsequence $\{j_k\}$ so that

$$\int_{R^n} g_k^n \, dm < 2^{-(n+1)k}, \quad g_k = |f_{j_k} - f|$$

and let Γ_0 be the family of all locally rectifiable γ in $\overline{R}^n$ such that

$$\limsup_{k\to\infty} \int_\gamma g_k ds > 0$$

We want to show that $M(\Gamma_0) = 0$.

Let Γ_k be the family of all locally rectifiable curves in R^n for which

$$\int_\gamma g_k ds \ge 2^{-k}$$

Then $\rho = 2^k g_k \in \operatorname{adm}(\Gamma_k)$ and

$$M(\Gamma_k) \leq \int_{R^n} \rho^n dm \leq 2^{nk} \int_{R^n} g_k^n \, dm < 2^{-k}$$

Now $\gamma \in \Gamma_0$ implies $\gamma \in \Gamma_k$ for infinitely many k. Thus, for each ℓ,

$$\Gamma_0 \subset \bigcup_{k=\ell}^{\infty} \Gamma_k \, , \; M(\Gamma_0) \leq \sum_{k=\ell}^{\infty} M(\Gamma_k) < 2^{-\ell+1}$$

and hence $M(\Gamma_0) = 0$.

Theorem 7. If $\{\Gamma_j\}$ is an increasing sequence of curve families, then

$$M(\bigcup_j \Gamma_j) = \lim_{j \to \infty} M(\Gamma_j)$$

Idea of proof. Let $\Gamma = \bigcup_j \Gamma_j$. Then by monotonicity of the modulus,

$$M(\Gamma) \geq \lim_{j \to \infty} M(\Gamma_j)$$

For the reverse inequality, we may assume that the limit is finite. Then since $L(R^n)$ is uniformly convex, we can choose $\rho_j \in \operatorname{adm}(\Gamma_j)$ so that $\rho_j \to \rho$ in $L^n(R^n)$ and so that

$$\int_{R^n} \rho^n dm = \lim_{j \to \infty} \int_{R^n} \rho_j^n dm = \lim_{j \to \infty} M(\Gamma_j)$$

By Theorem 6 there exists a subsequence $\{j_k\}$ and a family Γ_0 with $M(\Gamma_0) = 0$ such that

$$\int_\gamma \rho ds = \lim_{k \to \infty} \int_\gamma \rho_{j_k} ds$$

for all locally rectifiable $\gamma \in \Gamma \sim \Gamma_0$. Since each such γ lies in Γ_{j_k} for large k ,

$$\int_\gamma \rho ds \geq 1$$

Thus $\rho \in \mathrm{adm}(\Gamma \sim \Gamma_0)$, and we conclude that

$$M(\Gamma) = M(\Gamma \sim \Gamma_0) \leq \int_{R^n} \rho^n dm = \lim_{j \to \infty} M(\Gamma_j)$$

Remark 2. Since the Γ_j are increasing in Theorem 7,

$$\Gamma = \bigcup_j \Gamma_j = \lim_{j \to \infty} \Gamma_j$$

in the set theoretic sense and so we see that the conclusion of Theorem 7 is a continuity property for the modulus. Unfortunate no such result holds for decreasing families Γ_j with

$$\Gamma = \bigcap \Gamma_j = \lim_{j \to \infty} \Gamma_j$$

2. CONFORMAL CAPACITY OF CONDENSERS

We let $q(x,y)$ denote the chordal distance between points $x,y \in \bar{R}^n$. That is

$$q(x,y) = \begin{cases} \dfrac{|x-y|}{\sqrt{1+|x|^2}\sqrt{1+|y|^2}} & \text{if } x,y \neq \infty \\[2ex] \dfrac{1}{\sqrt{1+|x|^2}} & \text{if } x \neq \infty,\ y = \infty \end{cases}$$

A condenser is a domain $R \subset \bar{R}^n$ whose complement is the union of two distinguished disjoint compact sets C_0 and C_1.

For convenience we write

$$R = R(C_0, C_1)$$

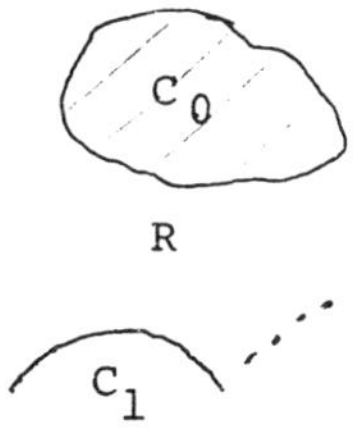

A <u>ring</u> is a condenser $R = R(C_0, C_1)$ where C_0 and C_1 are continua. We call C_0 and C_1 the <u>complementary components</u> of R.

Given a condenser $R = R(C_0, C_1)$ with $R \subset R^n$, we let adm (R) denote the class of functions $u : \bar{R}^n \to R^1$ with the following properties:

a) u is continuous in $\bar{R}^n$,

b) u has distribution derivatives in R,

c) $u = 0$ on C_0 and $u = 1$ on C_1.

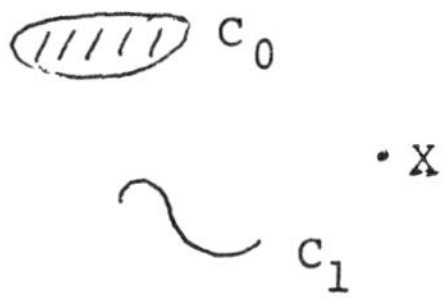

Note that

$$u(x) = \min\left(\frac{q(x, C_0)}{q(C_1, C_0)}, 1\right) \in \text{adm}(R)$$

and hence adm $(R) \neq \phi$. We call

$$\text{cap}(R) = \inf_{u \in \text{adm}(R)} \int_R |\nabla u|^n \, dm$$

$$\text{mod}(R) = \left(\frac{\omega_{n-1}}{\text{cap}(R)}\right)^{\frac{1}{n-1}}$$

the <u>conformal</u> <u>capacity</u> and <u>modulus</u> of R, respectively.

<u>Example</u>. If R is the ring in R^n bounded by concentric spheres of radii a and b, $0 \leq a < b \leq +\infty$ then

$$\text{cap }(R) = \omega_{n-1} \left(\log \frac{b}{a}\right)^{1-n}$$

$$\text{mod }(R) = \log \frac{b}{a}$$

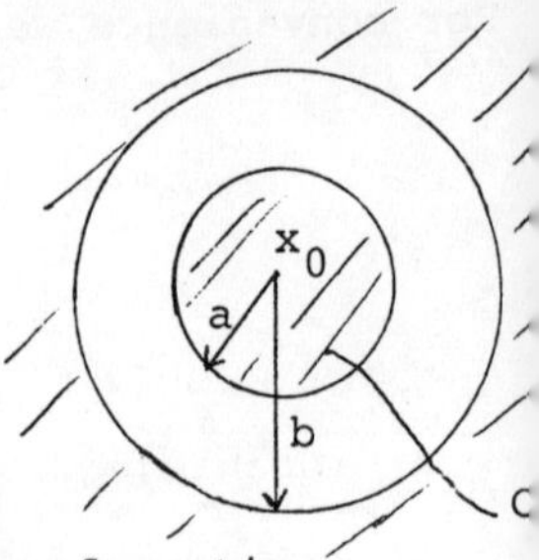

Remark. If $f : \bar{R}^n \to \bar{R}^n$ is a Möbius transformation, then $\text{cap }(f(R)) = \text{cap }(R)$ for all condensers R with $R, f(R) \subset R^n$. Hence we can use this fact to define cap (R) for all condensers in $\bar{R}^n$ which contain ∞ as an interior point.

Given $E, F, G \subset \bar{R}^n$, we let $\Delta(E,F;G)$ denote the family of all curves γ with

a) one endpoint in $\bar{E}$ and the other in $\bar{F}$,

b) interior in G.

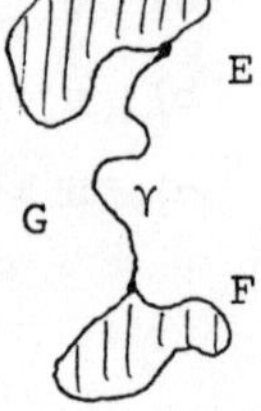

Theorem 1. If $R = R(C_0, C_1)$ is a ring and if $\Gamma = \Delta(C_0, C_1; R)$, then cap $(R) = M(\Gamma)$.

Outline of proof. By performing a preliminary Möbius transformation we may assume that $\infty \in C_1$. Choose a locally Lipschitzian function $u \in \text{adm }(R)$ and set

$$\rho(x) = \begin{cases} |\nabla u(x)| & x \in R \\ 0 & x \in C_0 \cup C_1 \end{cases}$$

If $\gamma : [a,b] \to \bar{R}^n$ is a locally rectifiable curve in Γ, then

$$\int_\gamma \rho \, ds \geq \left|\int_\gamma \nabla u \, ds\right| = |u(\gamma(b)) - u(\gamma(a))| \geq 1$$

Hence $\rho \in \operatorname{adm}(\Gamma)$ and

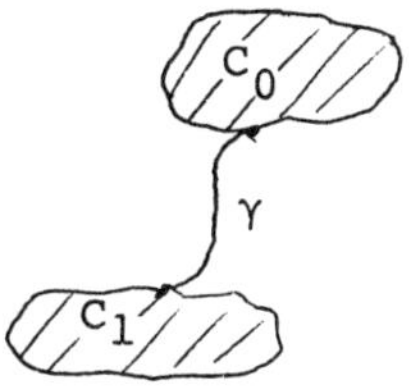

$$M(\Gamma) \leq \int_R |\nabla u|^n \, dm$$

By a smoothing argument, we can show that taking the infimum of the right-hand side over all such u gives $\operatorname{cap}(R)$. Thus

$$M(\Gamma) \leq \operatorname{cap}(R)$$

Choose a bounded continuous $\rho \in \operatorname{adm}(\Gamma)$ and set

$$u(x) = \min\left(1, \inf_\gamma \int_\gamma \rho \, ds\right)$$

for $x \in R$, where the infimum is taken over all locally rectifiable γ joining C_0 to x in R. Then u has distribution derivatives and

$$\lim_{x \to C_0} u(x) = 0 \quad , \quad \lim_{x \to C_1} u(x) = 1$$

Hence we can extend u to $\bar{R}^n$ so that $u \in \operatorname{adm}(R)$. Then since $|\nabla u| = \rho$ in R,

$$\operatorname{cap}(R) \leq \int_R \rho^n \, dm \leq \int_{R^n} \rho^n \, dm$$

Another smoothing argument shows the infimum over such ρ gives $M(\Gamma)$. Thus $\operatorname{cap}(R) \leq M(\Gamma)$.

Given a ray L from x_0 to ∞ and a compact set $E \subset \bar{R}^n$, we define the <u>spherical symmetrization of E in L</u> as the set E^* satisfying the following conditions:

a) $x_0 \in E^*$ iff $x_0 \in E$,

b) $\infty \in E^*$ iff $\infty \in E$,

c) For $r \in (0,\infty)$, $E^* \cap S(x_0,r) \neq \phi$ iff $E \cap S(x_0,r) \neq \phi$ in which case $E^* \cap S(x_0,r)$ is a closed spherical cap centered on L with the same m_{n-1} measure as $E \cap S(x_0,r)$.

We see that E^* is compact and that E^* is connected (Fig if E is .

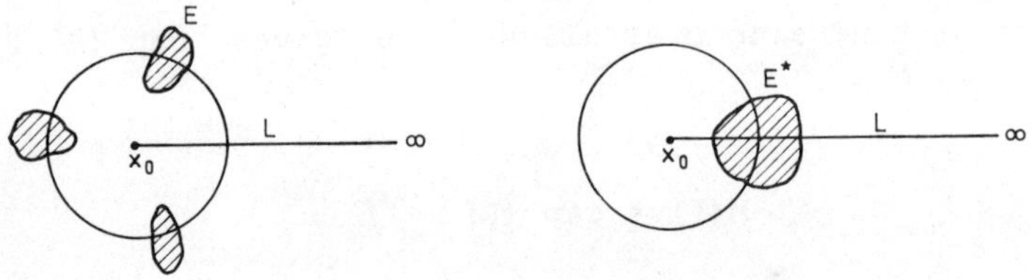

FIG.1. Spherical symmetrization of a set.

Theorem 2. If E^* is the spherical symmetrization of E in a ray L , then

a) $m_n(E^*) = m_n(E)$,

b) $m_{n-1}(\partial E^*) \leq m_{n-1}(\partial E)$.

Outline of proof. To prove a) we apply Fubini's theorem and obtain

$$m_n(E^*) = \int_0^\infty m_{n-1}(E^* \cap S(x_0,r))\, dr = \int_0^\infty m_{n-1}(E \cap S(x_0,r))\, d$$

$$= m_n(E)$$

For b), assume first that E is a polyhedron. Then for $r \in (0,\infty)$ the Brunn-Minkowski inequality implies that

$$E^*(r) = \{x : \text{dist}\,(x,E^*) \leq r\} \subset \{x : \text{dist}\,(x,E) \leq r\}^* = E\,(r)$$

and hence that

$$m_{n-1}(\partial E^*) \leq \limsup_{r \to 0} \frac{m_n(E^*(r)) - m_n(E^*)}{2r}$$

$$\leq \limsup_{r \to 0} \frac{m_n(E(r)) - m_n(E)}{2r} = m_{n-1}(\partial E)$$

The general result follows by a limiting argument.

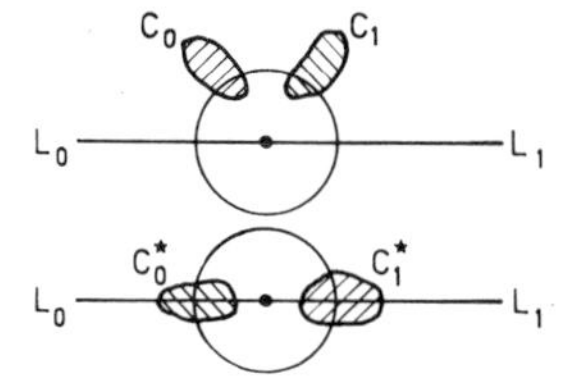

FIG.2. Spherical symmetrization of a ring.

Theorem 3. If $R = R(C_0, C_1)$ is a condenser and if C_0^* and C_1^* are the spherical symmetrizations of C_0 and C_1 in opposite rays L_0 and L_1, then $R^* = R(C_0^*, C_1^*)$ is a condenser with $\operatorname{cap}(R^*) \le \operatorname{cap}(R)$ (see Fig. 2).

Idea of proof. Choose a locally Lipschitzian $u \in \operatorname{adm}(R)$ and define u^* so that $\{x : u^*(x) \le t\} = \{x : u(x) \le t\}^*$. Then $u^* \in \operatorname{adm}(R^*)$ and Theorem 2 allows one to show that

$$\operatorname{cap}(R^*) \le \int_{R^n} |\nabla u^*|^n \, dm \le \int_{R^n} |\nabla u|^n \, dm$$

Taking the infimum over all such u yields the result.

Let $e_1, e_2, \ldots, e_n$ denote the basis vectors in R^n. For $t \in (0,\infty)$ let $R_T(t)$ denote the ring domain in R^n whose complement consists of the ray from te_1 to ∞ and the segment from $-e_1$ to 0. $R_T(t)$ is called the Teichmüller ring. The following properties for its modulus can be established:

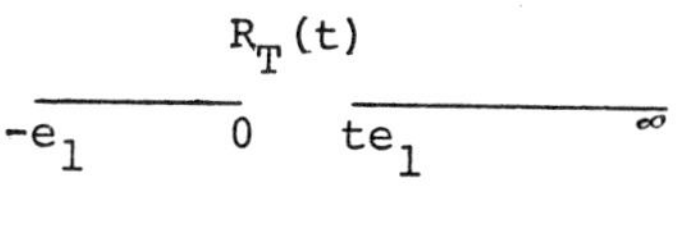

a) $\operatorname{mod} R_T(t) - \log(t+1)$ is nondecreasing in $(0,\infty)$,

b) $\lim_{t\to 0} \operatorname{mod} R_T(t) = 0$,

c) $\lim_{t\to\infty} \left(\operatorname{mod} R_T - \log(t+1)\right) = \log \lambda_n < \infty$,

d) $\lambda_2 = 16$ and $\lim_{n\to\infty} \lambda_n^{1/n} = e^2$.

Thus

e) $\operatorname{mod} R_T(t)$ is strictly increasing in $(0,\infty)$,

f) $\log(t+1) \leq \operatorname{mod} R_T(t) \leq \log \lambda_n(t+1)$.

Theorem 4. If $R = R(C_0, C_1)$ is a ring with $a, b \in C_0$ and $c, \infty \in C_1$, then

$$\operatorname{mod} R \leq \operatorname{mod} R_T\left(\frac{|c-a|}{|b-a|}\right)$$

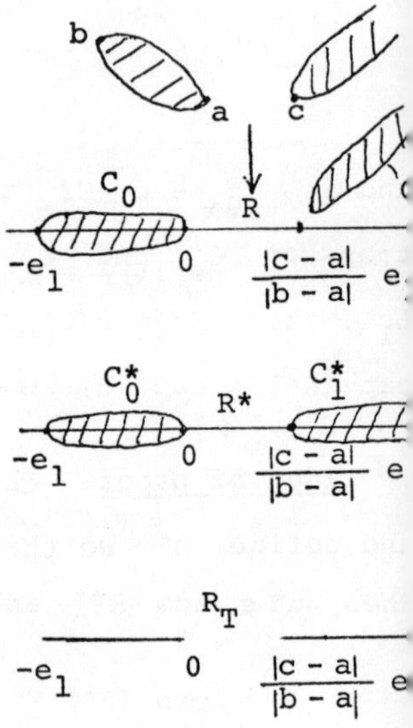

Proof. By performing a preliminary similarity mapping, we may assume that $a = 0$, $b = -e_1$. Then the spherical symmetrizations C_0^*, C_1^* of C_0, C_1 in the negative and positive halves of the x_1-axis contain the complementary components of $R_T\left(\frac{|c-a|}{|b-a|}\right)$. Thus

$$\operatorname{cap} R_T\left(\frac{|c-a|}{|b-a|}\right) \leq \operatorname{cap} R^* \leq \operatorname{cap} R$$

as desired.

Corollary 1. If $R = R(C_0, C_1)$ is a ring with $a, b \in C_0$ and $c, d \in C_1$, then

$$\operatorname{mod} R \leq \operatorname{mod} R_T\left(\frac{q(a,c)\,q(b,d)}{q(a,b)\,q(c,d)}\right)$$

Proof. By performing a preliminary chordal isometry we may assume that $d = \infty$. Then

$$\frac{|c-a|}{|b-a|} = \frac{q(a,c)\sqrt{|c|^2+1}}{q(a,b)\sqrt{|b|^2+1}} = \frac{q(a,c)\,q(b,d)}{q(a,b)\,q(c,d)}$$

and we can apply Theorem 4.

Corollary 2. If $R = R(C_0, C_1)$ is a ring, then

a) $\mod R \le \mod R_T\left(\frac{1}{q(C_0)q(C_1)}\right)$

b) $\mod R \le \mod R_T\left(\frac{4q(C_0,C_1)}{q(C_0)q(C_1)}\right)$.

Proof. For a), choose $a,b \in C_0$ and $c,d \in C_1$ so that

$$q(a,b) = q(C_0) \quad , \quad q(c,d) = q(C_1)$$

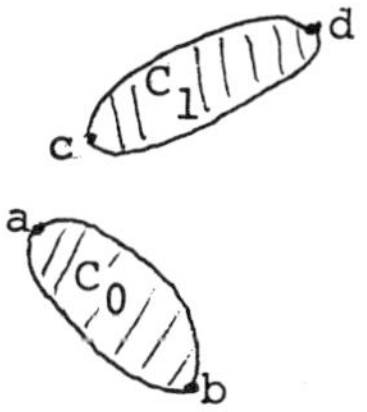

Then

$$\frac{q(a,c)q(b,d)}{q(a,b)q(c,d)} \le \frac{1}{q(C_0)q(C_1)}$$

and we can apply Corollary 1. For b) choose $a \in C_0$ and $c \in C_1$ so that

$$q(a,c) = q(C_0,C_1)$$

Next pick $b \in C_0$, and $d \in C_1$ so that

$$q(a,b) \geq \tfrac{1}{2}\, q(C_0) \quad , \quad q(c,d) \geq \tfrac{1}{2}\, q(C_1)$$

Then

$$\frac{q(a,c)q(b,d)}{q(a,b)q(c,d)} \le \frac{4q(C_0,C_1)}{q(C_0)q(C_1)}$$

and we again apply Corollary 1.

Convergence of sets. We say that a sequence of sets E_j converges uniformly to a set E if for each $\varepsilon > 0$ there exists a j_0 such that

$$\sup_{x \in E_j} q(x,E) < \varepsilon \quad , \quad \sup_{x \in E} q(x,E_j) < \varepsilon$$

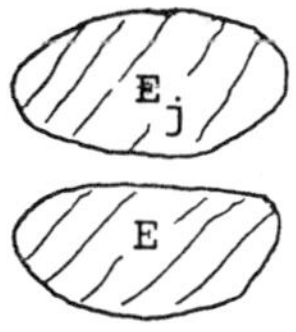

for $j \ge j_0$.

Theorem 5. If the complementary components of a sequence of rings R_j converge uniformly to the corresponding complementary components of a ring R, then

$$\operatorname{cap}(R) = \lim_{j\to\infty} \operatorname{cap}(R_j)$$

3. INNER AND OUTER DILATATIONS

Suppose D, D' are domains in $\bar{R}^n$ and that $f : D \to D'$ is a homeomorphism. We call

$$K_I(f) = \sup_{\Gamma} \frac{M(f(\Gamma))}{M(\Gamma)} \quad , \quad K_0(f) = \sup_{\Gamma} \frac{M(\Gamma)}{M(f(\Gamma))}$$

the inner and outer dilatations of f, where the suprema are taken over all curve families in D for which $M(\Gamma)$ and $M(f(\Gamma))$ are not simultaneously 0 or ∞. Similarly we call

$$K_I^*(f) = \sup_{R} \frac{\operatorname{cap}(f(R))}{\operatorname{cap}(R)} \quad , \quad K_0^*(f) = \sup_{R} \frac{\operatorname{cap}(R)}{\operatorname{cap}(f(R))}$$

the inner and outer ring dilatations of f, where the suprema are taken over all rings R with $\bar{R} \subset D$. Obviously

$$K_I^*(f) \le K_I(f) \quad , \quad K_0^*(f) \le K_0(f)$$

We say that f is K-quasiconformal if

$$\max(K_I(f), K_0(f)) \le K < \infty$$

Suppose that $f : R^n \to R^n$ is a linear bijection. We call

$$H_I(f) = \frac{J(f)}{\ell(f)^n} \quad , \quad H_0(f) = \frac{|f|^n}{J(f)}$$

the inner and outer analytic dilatations of f , where

$$|f| = \sup_{|x|=1} |f(x)| \ , \quad \ell(f) = \inf_{|x|=1} |f(x)| \ , \quad J(f) = |\det f|$$

If E is the image of the unit ball B under f , then

$$H_I(f) = \frac{m(E)}{m(B_I)} \ , \qquad H_0(f) = \frac{m(B_0)}{m(E)}$$

where B_I and B_0 are the largest inscribed and smallest circumscribed ball about E .

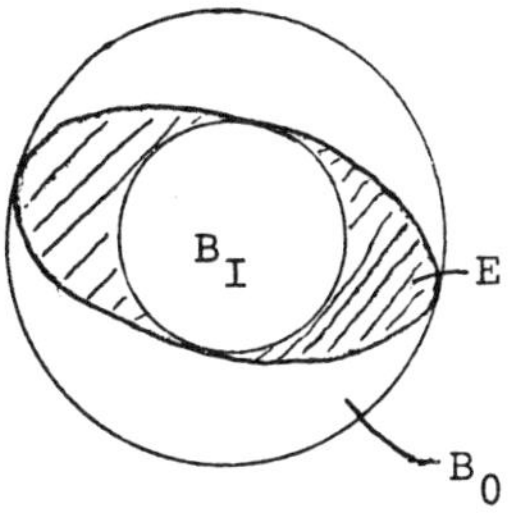

Theorem 1. If $f : D \to D'$ is a diffeomorphism, then

$$K_I(f) \leq \sup_{x\in D} H_I(f'(x)) \ , \qquad K_0(f) \leq \sup_{x\in D} H_0(f'(x))$$

Proof. For the second part choose a curve family Γ in D . Let $\rho' \in \text{adm}\,(f(\Gamma))$ and set

$$\rho(x) = \rho' \circ f(x)\,|f'(x)|\,\chi_D(x)$$

If $\gamma \in \Gamma$ is locally rectifiable, then

$$\int_\gamma \rho(x)\,ds = \int_\gamma \rho' \circ f(x)\,|f'(x)|\,ds \geq \int_{f(\gamma)} \rho'(x)\,ds \geq 1$$

and hence $\rho \in \text{adm}\,(\Gamma)$. Thus

$$M(\Gamma) \leq \int_D \rho(x)^n\,dm \leq \sup_{x\in D} H_0(f'(x)) \int_D (\rho' \circ f(x))^n\, J(f'(x))\,dm$$

Hence

$$M(\Gamma) \leq \sup_{x \in D} H_0(f'(x)) \int_{D'} \rho'(x)^n \, dm$$

and taking infimum over all such ρ' yields

$$M(\Gamma) \leq \sup_{x \in D} H_0(f'(x))\, M(f(\Gamma))$$

as desired.

<u>Theorem 2.</u> If $f : R^n \to R^n$ is a linear bijection, then

$$H_I(f) \leq K_I^*(f), \qquad H_0(f) \leq K_0^*(f)$$

<u>Proof of first inequality.</u> By performing preliminary orthogonal mappings we may assume that

$$f(e_i) = \lambda_i e_i \quad , \quad \lambda_1 \geq \cdots \geq \lambda_n > 0$$

for $i = 1, \cdots, n$. Then

$$H_I(f) = \frac{J(f)}{\ell(f)^n} = \frac{\lambda_1 \cdots \lambda_n}{\lambda_n^n}$$

and we want to prove that

$$(*) \qquad \frac{\lambda_1 \cdots \lambda_{n-1}}{(\lambda_n)^{n-1}} \leq K_I^*(f)$$

For $r \in (0, \infty)$ let R be the ring $R(F, C(G))$, where

$$F = \{x : |x_i| \leq 1,\ x_n = 0\} \ , \quad G = \{x : |x_i| < 1 + r,\ |x_n| <$$

Then we see that (Fig. 3)

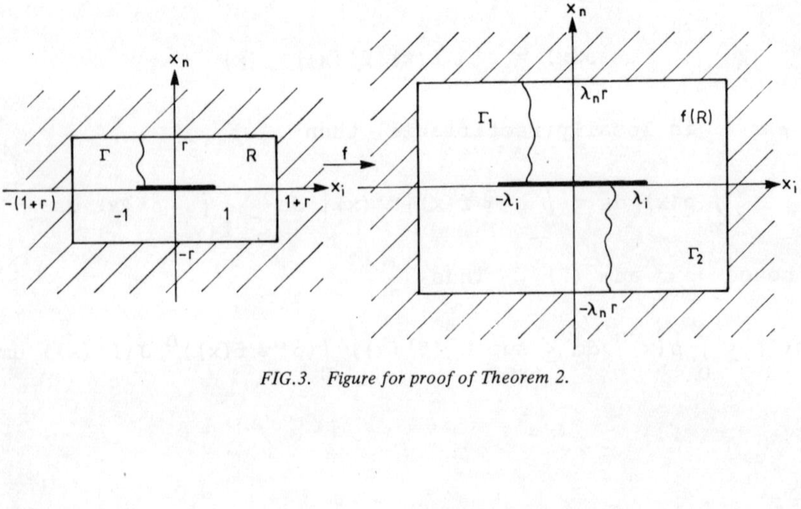

FIG.3. Figure for proof of Theorem 2.

$$\rho = \frac{1}{r} \chi_R \in \text{adm}(\Gamma)$$

where Γ is the family of curves joining F and $C(G)$ in R . Hence

$$\text{cap}(R) = M(\Gamma) \leq \int_{R^n} \rho^n \, dm = \left(\frac{1}{r}\right)^n 2(1+r)\cdots 2(1+r) \cdot 2r$$
$$= 2^n \left(\frac{1+r}{r}\right)^{n-1}$$

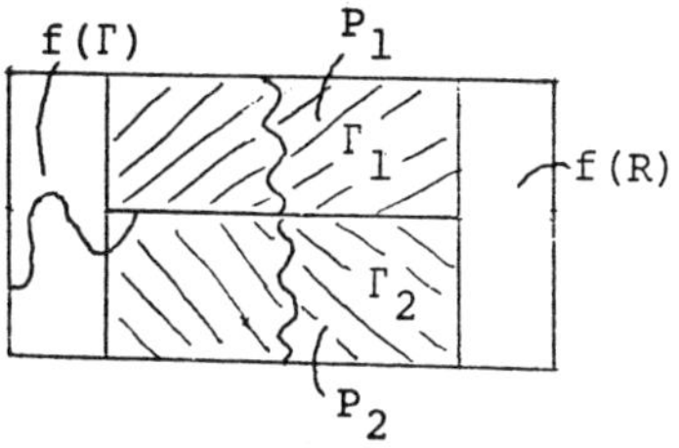

Next if Γ_1 and Γ_2 are the curve families joining the bases of the indicated parallelopipeds P_1 and P_2 in $f(R)$, then

$$M(\Gamma_1) = M(\Gamma_2) = \frac{A}{h^{n-1}} = \frac{2\lambda_1 \cdots 2\lambda_{n-1}}{(\lambda_n r)^{n-1}} = 2^{n-1} \frac{\lambda_1 \cdots \lambda_{n-1}}{(\lambda_n r)^{n-1}}$$

Finally, since Γ_1 and Γ_2 lie in disjoint Borel sets

$$\text{cap}\, f(R) = M(f(\Gamma)) \geq M(\Gamma_1) + M(\Gamma_2) = 2^n \frac{\lambda_1 \cdots \lambda_{n-1}}{(\lambda_n r)^{n-1}}$$

and hence

$$K_I^*(f) \geq \frac{\text{cap}\, f(R)}{\text{cap}\, R} \geq \frac{\lambda_1 \cdots \lambda_{n-1}}{(\lambda_n)^{n-1}} (1+r)^{1-n}$$

Letting $r \to 0$ yields (*) .

Theorem 3. If $f_j : D_j \to D_j'$ and $f : D \to D'$ are homeomorphisms and if

$$f = \lim_{j\to\infty} f_j \quad , \quad f^{-1} = \lim_{j\to\infty} f_j^{-1}$$

uniformly on compact sets in D and D', respectively, then

$$K_I^*(f) \le \liminf_{j \to \infty} K_I^*(f_j) \quad , \quad K_0^*(f) \le \liminf_{j \to \infty} K_0^*(f_j)$$

<u>Proof</u>. For first part choose the ring R with $\bar{R} \subset D$. Then $\bar{R} \subset D_j$ for $j \ge j_0$ and we see that the complementary components of $f_j(R)$ converge uniformly to the corresponding complementary components of $f(R)$. Thus by Theorem 5 of Chapter 2, we have

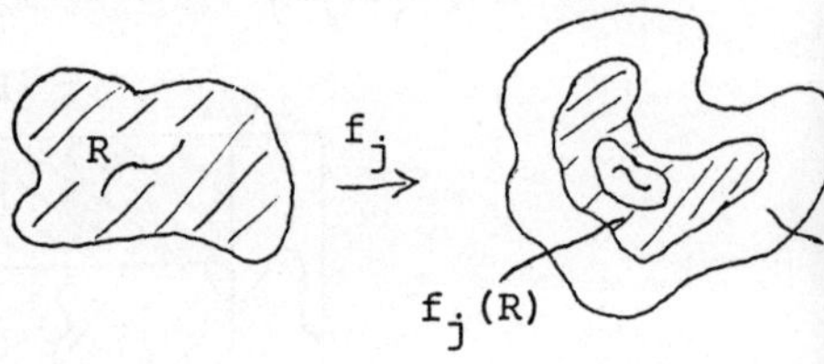

$$\operatorname{cap} f(R) = \lim_{j\to\infty} \operatorname{cap} f_j(R) \le \liminf_{j \to \infty} K_I^*(f_j) \operatorname{cap} R$$

whence

$$\frac{\operatorname{cap} f(R)}{\operatorname{cap} R} \le \liminf_{j \to \infty} K_I^*(f_j)$$

Taking the supremum over all such R yields the result.

<u>Theorem 4</u>. If f is a homeomorphism which is differentia[ble] at $x_0 \in R^n$ with $J(f'(x_0)) \neq 0$, then

$$H_I(f'(x_0)) \le K_I^*(f) \quad , \quad H_0(f'(x_0)) \le K_0^*(f)$$

<u>Proof</u>. We may assume $x_0 = 0$ and $f(x_0) = 0$. For each j let

$$f_j(x) = j f(\tfrac{x}{j})$$

Then $f'(0) : R^n \to R^n$ is a linear bijection and

$$f'(0) = \lim_{j\to\infty} f_j \quad , \quad f'(0)^{-1} = \lim_{j\to\infty} f_j^{-1}$$

uniformly on compact sets in R^n. Thus by Theorems 2 and 3,

$$H_I(f'(0)) \le K_I^*(f'(0)) \le \liminf_{j\to\infty} K_I(f_j) = K_I(f)$$

$$H_0(f'(0)) \le K_0^*(f'(0)) \le \liminf_{j\to\infty} K_0(f_j) = K_0(f)$$

as desired.

Theorem 5. If $f : D \to D'$ is a diffeomorphism, then

$$K_I(f) = K_I^*(f) = \sup_{x\in D} H_I(f'(x))$$

$$K_0(f) = K_0^*(f) = \sup_{x\in D} H_0(f'(x))$$

Proof. By Theorems 1 and 4,

$$K_I(f) \le \sup_{x\in D} H_I(f'(x)) \le K_I^*(f) \le K_I(f)$$

$$K_0(f) \le \sup_{x\in D} H_0(f'(x)) \le K_0^*(f) \le K_0(f)$$

Remarks. Theorem 5 implies that a diffeomorphism $f : D \to D'$ is K-quasiconformal if and only if

$$H_I(f'(x)) \le K \quad , \quad H_0(f'(x)) \le K$$

for all $x \in D$. One can prove that a homeomorphism $f : D \to D'$ is K-quasiconformal if and only if

a) f has locally L^n-integrable distribution derivatives,

b) $H_I(f'(x)) \le K$ and $H_0(f'(x)) \le K$ a.e. in $D \sim \{\infty, f^{-1}(\infty)\}$.

4. DISTORTION AND CONVERGENCE

Theorem 1. If $f : D \to D'$ is K-quasiconformal and if $C(D) \ne \phi$, then

$$q(f(x),f(y))\,q(C(D')) \le a\left(\frac{q(x,y)}{q(x,C(D))}\right)^b$$

for $x,y \in D$, where $a = 2\lambda_n$ and $b = K^{\frac{1}{1-n}}$.

Proof. Fix $x,y \in D$ with $x \neq y$ and suppose that

$$q(x,y) < q(x,C(D))$$

By performing a preliminary chordal isometry we may assume that $x = 0$. Next choose r and s so that

$$q(0,y) < r = \frac{s}{\sqrt{1+s^2}} < q(0,C(D))$$

and let R denote the spherical ring

$$R = \{z : q(0,y) < q(0,z) < r\} = \{z : |y| < |z| < s\}$$

Then $\bar{R} \subset D$ and R separates the points 0, y from $C(D)$. Also, it is easy to see that

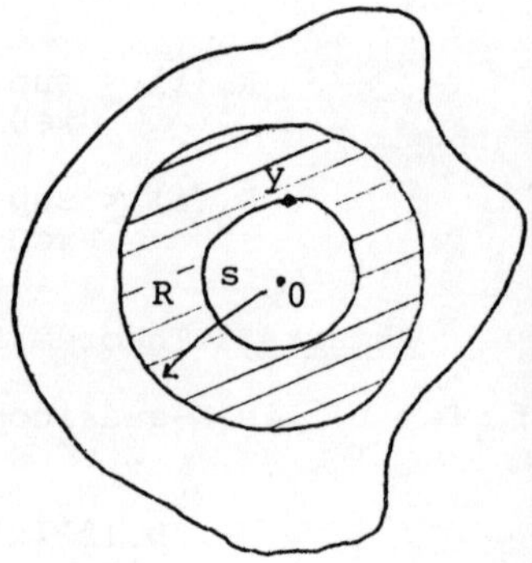

$$\frac{s}{|y|} > \frac{r}{q(0,y)}$$

and hence that

$$\text{mod } R = \log \frac{s}{|y|} > \log \frac{r}{q(0,y)}$$

Let C_0 and C_1 denote the complementary components of $f(R)$, labeled so that

$$f(0),\ f(y) \in C_0\ ,\ C(D') \subset C_1$$

Then by the first estimate in Corollary 2 of Chapter 2,

$$\begin{aligned} \text{mod } f(R) &\leq \text{mod } R_T\left(\frac{1}{q(C_0)\, q(C_1)}\right) \\ &\leq \log \lambda_n\left(\frac{1}{q(C_0)q(C_1)} + 1\right) \\ &\leq \log \frac{2\lambda_n}{q(f(0),f(y))q(C(D'))} \end{aligned}$$

Since f is K-quasiconformal,

$$\operatorname{cap} f(R) \leq K \operatorname{cap} R$$

$$\operatorname{mod} f(R) \geq K^{\frac{1}{1-n}} \operatorname{mod} R$$

and combining the above inequalities yields

$$q(f(0),f(y))q(C(D')) \leq a\left(\frac{q(0,y)}{r}\right)^{b}$$

with $a = 2\lambda_n$ and $b = K^{\frac{1}{1-n}}$. Letting $r \to q(0,C(D))$ completes the proof in case $q(x,y) < q(x,C(D))$.

Suppose next that

$$q(x,y) \geq q(x,C(D))$$

Then since $a = 2\lambda_n \geq 32$,

$$q(f(x),f(y))q(C(D')) \leq 1 < a\left(\frac{q(x,y)}{q(x,C(D)}\right)^{b}$$

Corollary 1. If $f : D \to D'$ is K-quasiconformal, then f is $K^{\frac{1}{1-n}}$ - Hölder continuous with respect to the chordal metric in D and with respect to the euclidean metric in $D \sim \{\infty, f^{-1}(\infty)\}$.

Proof. Choose domains D_1 and D_2 such that

$$D = D_1 \cup D_2 \ , \ q(C(D_i')) > 0$$

where $D_i' = f(D_i)$. Then Theorem 1 implies the desired result in D_1, in D_2 and hence in D.

Remark 1. If $f : B \to B$ is K-quasiconformal with $f(0) = 0$, then one can show by a similar argument that

$$|f(x) - f(y)| \leq a|x - y|^b$$

for $x,y \in B$, where $a = 4\lambda_n$ and $b = K^{\frac{1}{1-n}}$.

Remark 2. If $b \in (0,1]$, the mapping

$$f(x) = |x|^{b-1}x$$

is K-quasiconformal, where $K = b^{1-n}$. Thus the exponent for Hölder continuity in Corollary 1 is sharp.

Corollary 2. If $r > 0$ and if F is a family of K-quasiconformal mappings $f : D \to D'_f$ such that $q(C(D'_f)) \geq r$ for all $f \in F$, then F is equicontinuous.

Proof. If $f \in F$, then $C(D'_f) \neq \phi$ implies that $C(D) \neq \phi$ and hence by Theorem 1,

$$q(f(x),f(y)) \leq \frac{1}{r}q(f(x),f(y))q(C(D'_f)) \leq \frac{a}{r}\left(\frac{q(x,y)}{q(x,C(D))}\right)^b$$

for all $x,y \in D$.

Corollary 3. If F is a family of K-quasiconformal mappings $f : D \to D'_f$ which is not equicontinuous in D, then there exists an $x_0 \in D$ and a sequence $f_j \in F$ such that $\{f_j$ is equicontinuous in $D \sim \{x_0\}$.

Proof. By hypothesis, there exists an $x_0 \in D$ and an $r \in (0,1)$ such that the statement

$$x \in D,\ q(x,x_0) < \frac{1}{j} \text{ implies } q(f(x),f(x_0)) < r$$

fails for infinitely many $f \in F$. Thus we can pick a sequence of $x_j \in D \sim \{x_0\}$ and distinct $f_j \in F$ such that

$$q(x_j,x_0) < \frac{1}{j}\ , \ q(f_j(x_j),f_j(x_0)) \geq r$$

for all j . For each k let

$$D_k = D \sim \{x_0, x_k, x_{k+1}, \ldots\}$$

If $k \leq j < \infty$, then

$$q(C(f_j(D_k))) \geq q(f_j(x_j),f_j(x_0)) \geq r$$

while if $1 \leq j < k$, then

$$q(C(f_j(D_k))) \geq \min_{1 \leq j < k} q(f_j(x_k),f_j(x_0)) = s > 0$$

Hence the mappings f_j are equicontinuous in D_k by Corollary 1 and therefore at each point of

$$D \sim \{x_0\} = \bigcup_1^{\infty} D_k$$

<u>Theorem 2</u>. Suppose $f_j : D \to D'_j$ are K-quasiconformal and that $f_j \to f$ pointwise in D . Then one of the following is true:

a) f is a homeomorphism and convergence is uniform on compact sets ,

b) f assumes only two values, one only at one point,

c) f is constant.

<u>Outline of proof</u>. Suppose f assumes three values at x_1, x_2, x_3 and let

$$D_1 = D \sim \{x_2,x_3\}\ ,\ D_2 = D \sim \{x_3,x_1\}\ ,\ D_3 = D \sim \{x_1,x_2\}$$

Since $f(x_2) \neq f(x_3)$,

$$\inf_j q(C(f_j(D_1))) \geq \inf_j q(f_j(x_2),f_j(x_3)) = r_1 > 0$$

and the f_j are equicontinuous in D_1. Permuting the roles of x_1, x_2, x_3 shows the f_j are also equicontinuous in D_2, in D_3 and hence in D. Thus the convergence is uniform on compact sets in D and f is continuous. We now show f is univalent in D.

Fix $x \in D$. By the equicontinuity of the f_j, we can choose a chordal ball U about x such that

$$\bar{U} \subset D \ , \ q(U_j') \leq \frac{1}{2} \quad \text{whence} \quad q(C(U_j')) \geq \frac{1}{2}$$

where $U_j' = f_j(U)$. We claim f is either univalent or constant in U. Otherwise we can choose $a,b,c \in U$ such that

$$f(a) \neq f(b) = f(c)$$

Let R be the indicated ring and let $R_j' = f_j(R)$. Then

$$\text{mod } R > 0$$

On the other hand, by the second estimate in Corollary 2 of Chapter 2,

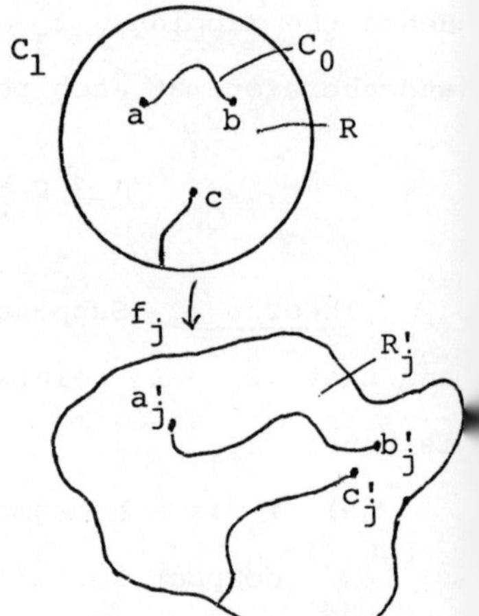

$$\text{mod } R_j' \leq \text{mod } R_T\left(\frac{4q(b_j',c_j')}{q(a_j',b_j')q(C(U_j'))}\right)$$

$$\leq \text{mod } R_T\left(8\,\frac{q(b_j',c_j')}{q(a_j',b_j')}\right) \to 0$$

where $a_j' = f_j(a)$, $b_j' = f_j(b)$, $c_j' = f_j(c)$, since

$$q(a_j',b_j') \to q(f(a),f(b)) > 0 \ , \ q(b_j',c_j') \to q(f(b),f(c)) = 0$$

Finally the fact that the f_j are K-quasiconformal implies that

$$\text{mod } R \leq K^{\frac{1}{1-n}} \text{ mod } R_j' \to 0$$

and we have a contradiction.

Now let G_1, G_2 be the sets of points $x \in D$ which have a neighborhood in which f is univalent or constant, respectively. Then G_1, G_2 are open with

$$D = G_1 \cup G_2 \ , \ G_1 \cap G_2 = \phi$$

If f is not univalent we can show that $G_2 \neq \phi$ and hence $D = G_2$. By continuity f is then constant in D, yielding a contradiction.

Theorem 3. Suppose $C(D)$ contains at least two points, that $f_j : D \to D'$ is K-quasiconformal, and that $f_j \to f$ pointwise in D. Then the convergence is uniform on compact sets and

a) $f : D \to D'$ is K-quasiconformal,

or

b) f is a constant $c \in \partial D$.

Homogeneous domains. We say that D is K-homogeneous if for each $a,b \in D$ there exists a K-quasiconformal $f : D \to D$ such that

a) $f(a) = b$,

b) f is homotopic to the identity in D.

We say D is quasiconformally-homogeneous if it is K-homogeneous for some K.

<u>Theorem 4</u>. If D is quasiconformally-homogeneous, then either

a) $C(D)$ is connected,

or

b) $C(D)$ contains exactly two points.

<u>Proof</u>. Assume otherwise. Then there exist distinct points $a,b,c \in \partial D$ such that a and b do not lie in the same component of $C(D)$. Choose a compact set $E \subset D$ separating a and b and let U be a neighborhood of c such that a and b can be joined in $C(U)$.

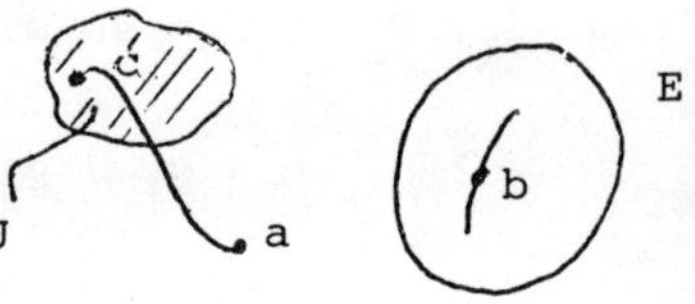

Choose $c_0, c_j \in D$ so that $c_j \to c$. By hypothesis we can choose for each j a K-quasiconformal $f_j : D \to D$ such that $f_j(c_0) = c_j$, f_j homotopic to the identity in D.

Since $a,b,c \notin D$,

$$q(C(f_j(D))) = q(C(D)) \geq q(a,b) > 0$$

and the f_j are equicontinuous in D by Corollary 2. Hence we can choose a subsequence $\{j_k\}$ such that $f_{j_k} \to f$ uniforml on compact sets in D. Then

$$f(c_0) = \lim_{k\to\infty} f_{j_k}(c_0) = \lim_{k\to\infty} c_{j_k} = c \in \partial D$$

Thus $f \equiv c$ in D by Theorem 3, and by the uniform convergenc we can choose an integer j so that

$$f_j(E) \subset U$$

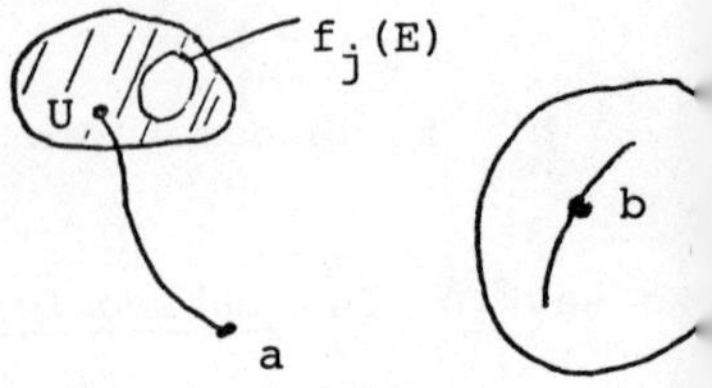

The homotopy condition on f_j implies that $f_j(E)$ still separates a and b. But this contradicts the way that the neighborhood U was chosen. We conclude that either C(D) is connected or that C(D) contains exactly two points.

Corollary 4. When $n = 2$, D is quasiconformally-homogeneous if and only if it is conformally equivalent to one of the following domains:

$$\bar{R}^2, \; R^2, \; R^2 \sim \{0\}, \; B$$

5. ONE-QUASICONFORMAL MAPPINGS

Lemma 1. If $f : \bar{R}^n \to \bar{R}^n$ is 1-quasiconformal and if $f(0) = 0$ and $f(\infty) = \infty$, then $f|R^n$ is linear.

Proof. Suppose that x, y, z are finite, ordered points on a line L. We begin by showing that

$$(*) \qquad \frac{|f(y) - f(x)|}{|y - x|} = \frac{|f(z) - f(y)|}{|z - y|}$$

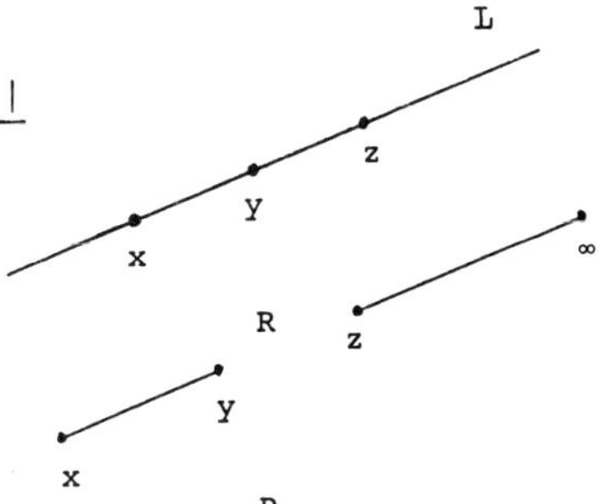

For this let R denote the ring whose complement consists of the segments of L joining x to y and z to ∞. Then R is similar to the Teichmüller ring $R_T\left(\frac{|z-y|}{|y-x|}\right)$ and

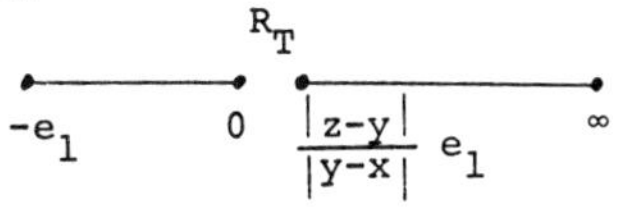

$$\text{mod } R = \text{mod } R_T\left(\frac{|z-y|}{|y-x|}\right)$$

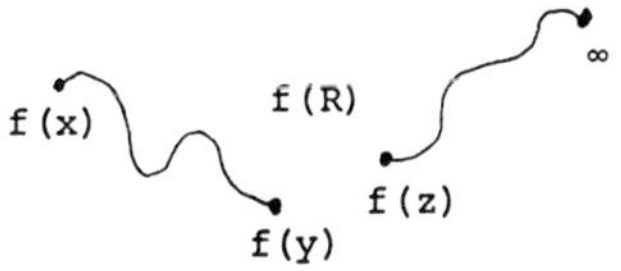

Next by the estimate in Theorem 4 of Chapter 2,

$$\text{mod } f(R) \le \text{mod } R_T\left(\frac{|f(z) - f(y)|}{|f(y) - f(x)|}\right)$$

Finally, since f is 1-quasiconformal, $\mathrm{mod}\ f(R) = \mathrm{mod}\ R$, and hence

$$\mathrm{mod}\ R_T\left(\frac{|z-y|}{|y-x|}\right) \le \mathrm{mod}\ R_T\left(\frac{|f(z)-f(y)|}{|f(y)-f(x)|}\right)$$

Since $\mathrm{mod}\ R_T(t)$ is strictly increasing in t,

$$\frac{|z-y|}{|y-x|} \le \frac{|f(z)-f(y)|}{|f(y)-f(x)|}$$

and hence

$$\frac{|f(y)-f(x)|}{|y-x|} \le \frac{|f(z)-f(y)|}{|z-y|}$$

Interchanging the roles of x and z reverses this inequality giving (*).

We show next that for each $x, y \in R^n$,

$$(**) \qquad f(x+y) = f(x)+f(y)$$

For this let L and z be as indicated. Then applying inequality (*) to the following triples of points:

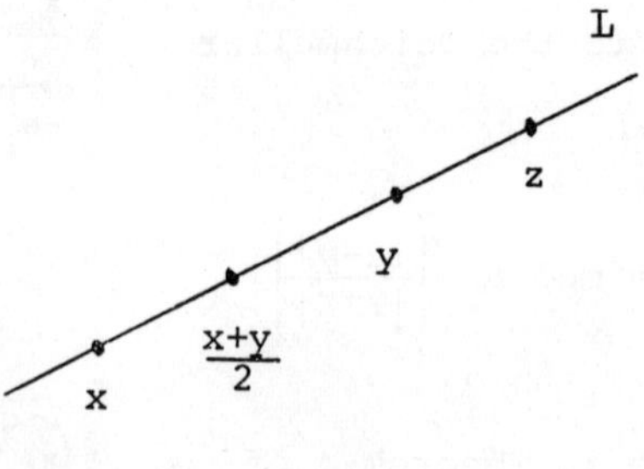

$$(x, \tfrac{x+y}{2}, y), \ (\tfrac{x+y}{2}, y, z), \ (x, y, z)$$

yields

$$2\frac{|f(\frac{x+y}{2})-f(x)|}{|x-y|} = 2\frac{|f(y)-f(\frac{x+y}{2})|}{|x-y|} = \frac{|f(z)-f(y)|}{|z-y|} = \frac{|f(y)-f(x)|}{|y-x|}$$

or simply

$$|f(\tfrac{x+y}{2})-f(x)| = |f(y)-f(\tfrac{x+y}{2})| = \tfrac{1}{2}|f(y)-f(x)|$$

The triangle inequality implies that

$$f(\tfrac{x+y}{2}) = \tfrac{1}{2}\big(f(x)+f(y)\big)$$

This applied to the points $x+y$ and 0 gives

$$f(\tfrac{x+y}{2}) = \tfrac{1}{2}\big(f(x+y)+f(0)\big) = \tfrac{1}{2}f(x+y)$$

and hence (**) follows.

Finally (**) and the fact that f is continuous imply that

$$f(ax) = af(x)$$

for all $a \in R^1$ and $x \in R^n$. Hence f is linear.

Theorem 1. A homeomorphism $f : \bar{R}^n \to \bar{R}^n$ is 1-quasiconformal if and only if it is a Möbius transformation.

Proof. The sufficiency is clear by Theorem 5 of Chapter 1. For the necessity suppose $f : \bar{R}^n \to \bar{R}^n$ is 1-quasiconformal and choose a Möbius transformation g which maps $f(0)$, $f(e_1)$, $f(\infty)$ onto 0, e_1, ∞. Set $h = g \circ f$. Then h is 1-quasiconformal and Lemma 1 implies that h is a linear bijection. Hence by

Theorem 5 of Chapter 3

$$\frac{J(h)}{\ell(h)^n} = H_I(h) = K_I(h) \leq 1$$

$$\frac{|h|^n}{J(h)} = H_0(h) = K_0(h) \leq 1$$

Hence

$$\inf_{|x|=1}|h(x)| = \ell(h) = |h| = \sup_{|x|=1}|h(x)|$$

and since $h(e_1) = 1$,

$$|h(x)| = |x|$$

for all $x \in R^n$. Thus h is an orthogonal mapping and $f = g^{-1} \circ h$ is a Möbius transformation.

Remark 1. Theorem 1 holds for $n \geq 2$. However when $n > 2$ we get a much stronger result using the following classical result due to Liouville.

Theorem 2. Suppose that D, D' are domains in R^n where $n > 2$. If $f : D \to D'$ is a C^4 homeomorphism and if

$$|f'(x)|^n = J\big(f'(x)\big)$$

in D, then $f = g|D$ where g is a Möbius transformation.

Theorem 3. If $f : D \to D'$ is 1-quasiconformal where $n > 2$ then $f=g|D$ where g is a Möbius transformation.

Before indicating a proof of Theorem 3, we need to make some comments concerning the extremal function for a ring domain

Suppose that $R = R(C_0, C_1)$ is a ring in R^n. Then

$$\operatorname{cap}(R) = \inf_{\operatorname{adm}(R)} \int_R |\nabla u|^n dm$$

Suppose next that the complementary components C_0 and C_1 are both nondegenerate, i.e. contain at least two points. Then using the fact that $L^n(R)$ is uniformly convex and an old argument on the Dirichlet problem due to Lebesgue, one can show there exists a unique extremal $u \in \operatorname{adm}(R)$ with

$$\operatorname{cap}(R) = \int_R |\nabla u|^n dm$$

This function then satisfies the variational condition

$$\int_R |\nabla u|^{n-2} \nabla u \cdot \nabla w \, dm = 0$$

for all $w \in C^1(R)$ with compact support in R, i.e. u is a weak solution of the equation

$$\operatorname{div} (|\nabla u|^{n-2} \nabla u) = 0$$

When $n = 2$, this is the Laplace equation. When $n > 2$, the equation is no longer linear, but rather quasilinear.

When $n = 2$, Weyl's Lemma implies that u is C^2 and hence harmonic. This also follows from exhibiting an explicit expression for u by means of conformal mapping. When $n > 2$, the situation is more complicated. If we know that for each compact set $E \subset R$ there exists a constant $M \in (0, \infty)$ such that

(***) $$\frac{1}{M} \leq |\nabla u(x)| \leq M$$

a. e. in E, then we can apply regularity theorems due to de Giorgi,

Morrey and Hopf to conclude that u is real analytic in R. A recent result of Uraltseva implies $u \in C^1(R)$ without (***)

Examples. a) Suppose that R is the spherical ring $R = R(C_0, C_1)$, where

$$C_0 = \overline{B(x_0, a)}, \quad G = C(B(x_0, b))$$

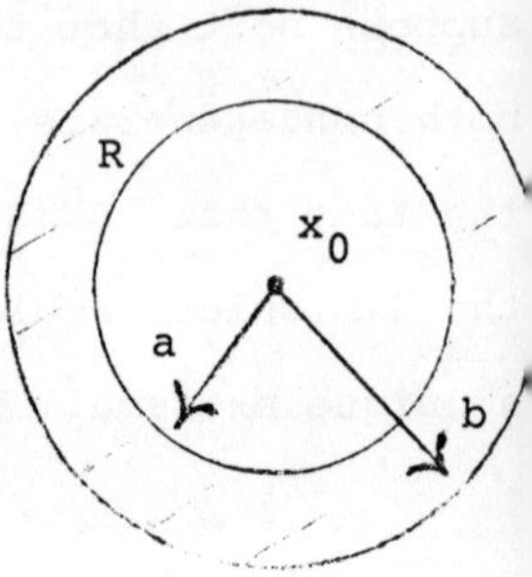

and $x_0 \in R^n$, $0 < a < b < \infty$. Then it is easy to verify that

$$u(x) = \frac{\log\left|\frac{x - x_0}{a}\right|}{\log \frac{b}{a}}, \quad x \in R$$

is the extremal function for R. Note that

$$|\nabla u(x)| = \frac{1}{\log \frac{b}{a}} \quad \frac{1}{|x - x_0|}$$

in R, and hence u satisfies (***).

b) Suppose next that $n = 3$ and that $R = R(C_0, C_1)$ is the ring where

$$C_0 = S(a) \cap R^2, \quad C_1 = C(B(b))$$

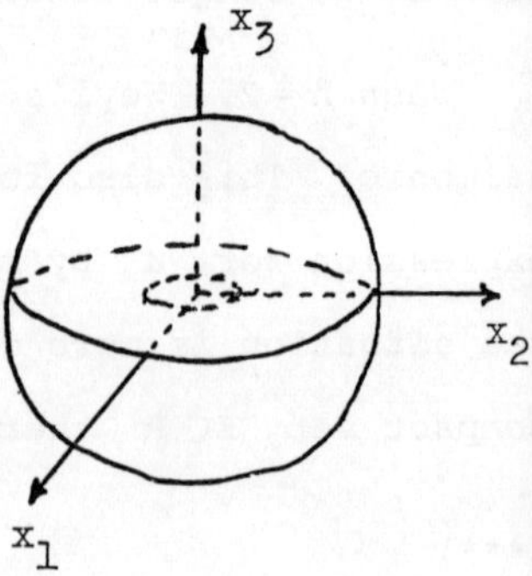

and $0 < a < b < \infty$. Then the extremal function u for R is C^1 by Uraltseva's result and symmetric in the x_1x_2 - plane and the x_3 - axis. Hence $\nabla u(0) = (0,0,0)$, $|\nabla u(0)| = 0$, and (***) does not hold. By slightly modifying R we can find a ring R' with extremal function u' such that R' is homeomorphic to a spherical ring and such that $|\nabla u'|$ vanishes at a point of R'.

<u>Sketch of proof of Theorem 3</u>

We may assume that $D, D' \subset R^n$. Then by Theorem 2 it is sufficient to show that each point $x_0 \in D$ has a neighborhood in which f is real analytic. Fix x_0 and choose $b \in (0,\infty)$ so that the closure of

$$U' = \{y : 0 < |y - f(x_0)| < b\}$$

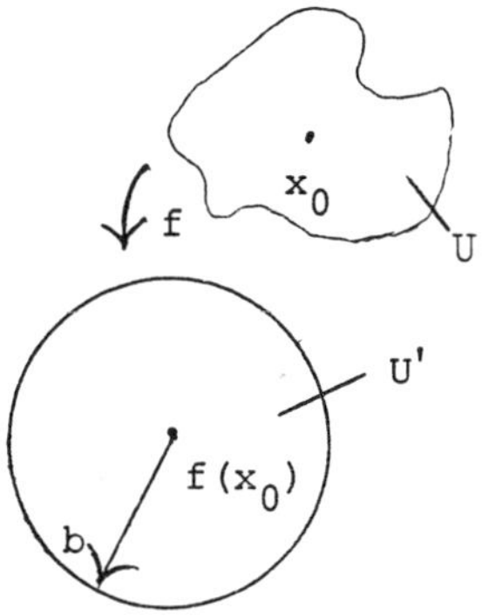

lies in D'. Let $U = f^{-1}(U')$. We will show that $|f(x) - f(x_0)|$ is real analytic for $x \in U$, and this will essentially establish the desired result

Choose $a \in (0,b)$ and let $R = f^{-1}(R')$, where R' is the spherical ring

$$R' = \{y : a < |y - f(x_0)| < b\}$$

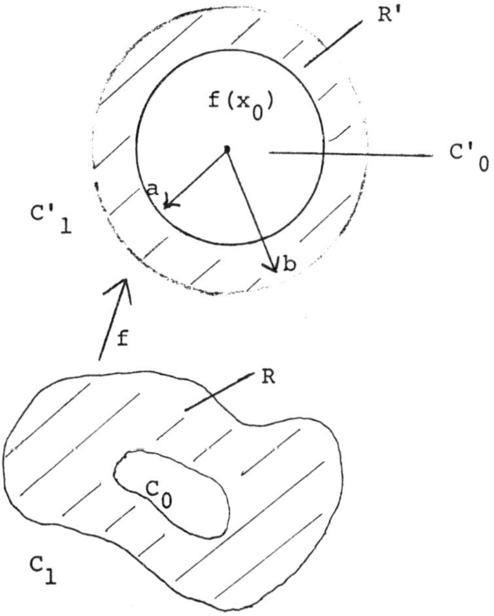

Then as we saw above,

$$u'(y) = \frac{\log \left| \frac{y - f(x_0)}{a} \right|}{\log \frac{b}{a}}$$

is equal in R' to the extremal for R'. Set

$$u(x) = u' \circ f(x), \quad x \in R$$

u has distribution derivatives in R and boundary values 0 on C_0 and 1 on C_1. Because f is 1-quasiconformal

$$\begin{aligned}\operatorname{cap} R' = \operatorname{cap} R &\le \int_R |\nabla u(x)|^n \, dm \\ &\le \int_R |\nabla u' \circ f(x)|^n |f'(x)|^n \, dm \\ &= \int_R |\nabla u' \circ f(x)|^n J(f'(x)) \, dm \\ &= \int_{R'} |\nabla u'(x)|^n \, dm = \operatorname{cap} R'\end{aligned}$$

We conclude that u is equal in R to the unique extremal function for R. Next with Theorem 1 in Chapter 4, it is not difficult to show that u satisfies the condition (***) on each compact set $E \subset R$. Thus

$$u(x) = u' \circ f(x) = \frac{\log \frac{|f(x)-f(x_0)|}{a}}{\log \frac{b}{a}}$$

is real analytic in R, whence

$$|f(x)-f(x_0)| = a\left(\frac{b}{a}\right)^{u(x)} = b^{u(x)} a^{1-u(x)}$$

is real analytic in R. We then let $a \to 0$ to conclude $|f(x)-f(x_0)|$ real analytic in U.

Given a homeomorphism $f : D \to D'$ we set

$$H(x,f) = \limsup_{r \to 0} \frac{L(x,f,r)}{\ell(x,f,r)}$$

for each $x \in D \sim \{\infty, f^{-1}(\infty)\}$, where for small $r > 0$

$$L(x,f\ r) = \sup_{|x-y|=r} |f(y)-f(x)|$$

$$\ell(x,f\ r) = \inf_{|x-y|=r} |f(y)-f(x)|$$

The function $H(x,f)$ measures how much f distorts the shape of small sphere at x and thus measures how far f is from being conformal at x. We can extend the definition of $H(x,f)$ to all $x \in D$ by using auxiliary Möbius transformations.

Theorem 4. If $f: D \to D'$ is a homeomorphism, if $H(x,f)$ is bounded in D, and if

$$(*) \qquad H(x,f) \leq K^{\frac{1}{n-1}}$$

a.e. in D, then f is a K-quasiconformal mapping.

Idea of proof. A length area argument allows one to prove that f has locally L^n-integrable distribution derivatives in D and that f is differentiable a.e. in D. At each such point of differentiability x_0, inequality (*) implies that

$$H_I(f'(x_0)) \leq K, \quad H_0(f'(x_0)) \leq K$$

and the result follows from the last Remark in Chapter 3.

Corollary. Suppose that $n > 2$. If $f: D \to D'$ is a homeomorphism and if $H(x,f) = 1$ for all $x \in D$, then $f = g|D$ where g is a Möbius transformation.

6. MAPPING PROBLEMS

Suppose that D and D' are domains in $\bar{R}^n$. How can we decide if D and D' are quasiconformally equivalent? That is, when does there exist a quasiconformal mapping $f: D \to D'$? The general problem is extremely difficult even when $n = 2$. If we restrict ourselves to the case where D' is the unit ball B, we can give some results.

Theorem 1. Suppose that $n = 2$. Then D is quasiconformally equivalent to B if and only if $C(D)$ is a nondegenerate continuum.

Proof. If $C(D)$ is a nondegenerate continuum, then D is a simply connected domain in the extended complex plane and $C(D)$ contains at least two points. Hence there exists a conformal mapping $f : D \to B$ by the Riemann mapping theorem.

If there exists a K-quasiconformal $f : D \to B$, the fact $C(B)$ is connected implies $C(D)$ is connected. Suppose $C(D)$ consists of a single point x_0, choose $a \in (0,1)$ and let $R = f^{-1}(R')$, where

$$R' = \{z : a < |z| < 1\}$$

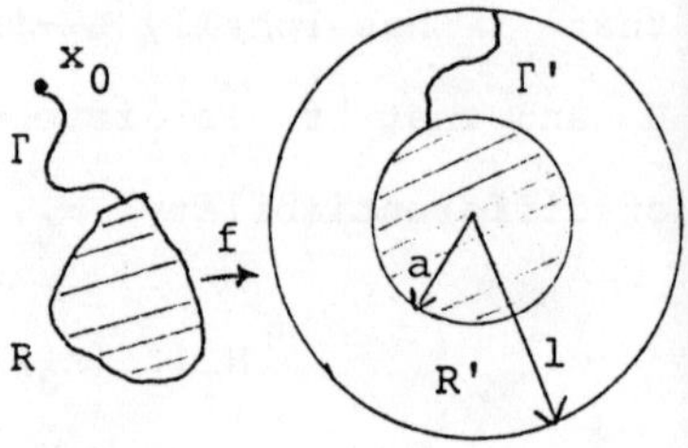

Next let $\Gamma = \Delta(C_0, C_1; R)$, let $\Gamma' = f(\Gamma)$, and let Γ_0 denote the family of all curves which meet x_0. Then $\Gamma \subset \Gamma_0$ and

$$M(\Gamma) \le M(\Gamma_0) = 0$$

by Theorem 4 in Chapter 1. On the other hand,

$$M(\Gamma') = 2\pi(\log \frac{1}{a})^{-1} > 0$$

while the fact that f is K-quasiconformal implies that

$$M(\Gamma') \le KM(\Gamma)$$

Hence we have a contradiction, and it follows that $C(D)$ must contain at least two points.

Lemma 1. Suppose that $n \geq 3$, that D is an open ball or half space in R^n, that $E \subset \bar{D}$ and $F \subset \partial D$ are disjoint continua, and that E joins $a, b \in \partial D$ while F separates a, b in ∂D. If $\Gamma = \Delta(E, F; D)$, then

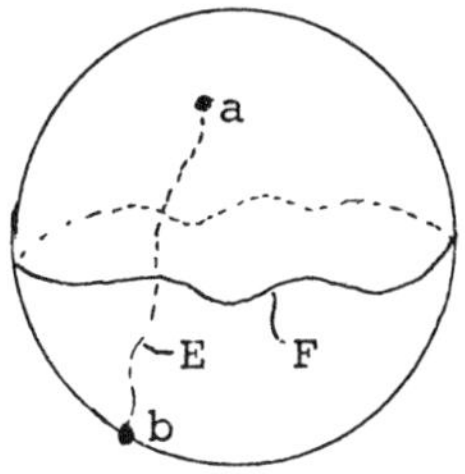

$$M(\Gamma) \geq \mu_n > 0$$

where $\mu_n = \frac{1}{2} \text{cap } R_T(1)$

Proof. By performing a preliminary Möbius transformation we may assume that D is a half space and that $\infty \in F$. Let E^* denote the symmetric image of E in ∂D and $R = R(C_0, C_1)$ the ring where

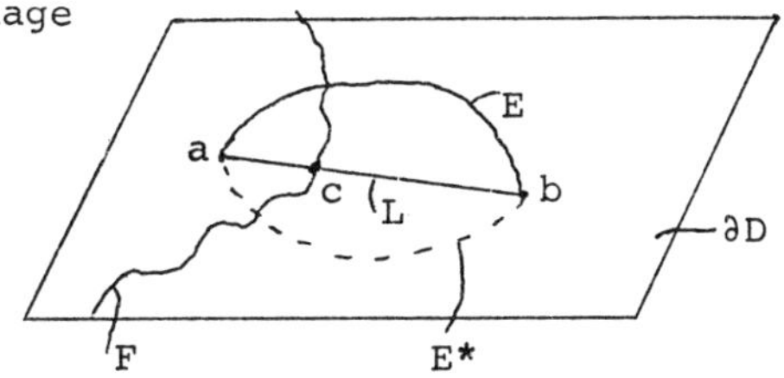

$$C_0 = E \cup E^* \quad , \quad C_1 = F$$

Next let L be the line segment joining a and b in ∂D. Then $F \cap L \neq \phi$ and we can choose a point $c \in F \cap L$. Since

$$|c - a| \leq |b - a| \quad , \quad a, b \in C_0 \quad , \quad c, \infty \in C_1$$

Theorem 4 of Chapter 2 implies that

$$\text{mod } R \leq \text{mod } R_T\left(\frac{|c - a|}{|b - a|}\right) \leq \text{mod } R_T(1)$$

Next if $\Gamma' = \Delta(C_0, C_1; R)$, then it is not difficult to show that

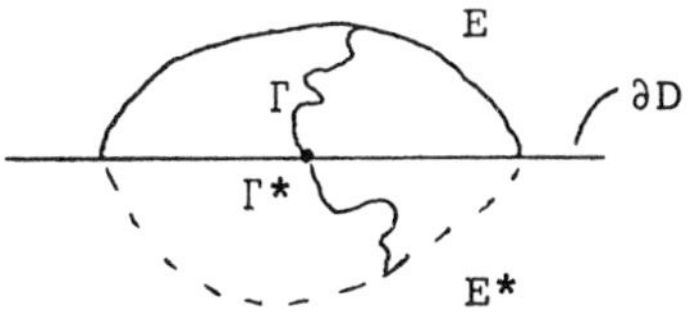

$$M(\Gamma') \overset{1}{=} M(\Gamma) + M(\Gamma^*)$$

where Γ^* is the symmetric image of Γ in ∂D, and hence that

$$M(\Gamma) = \frac{1}{2} M(\Gamma') = \frac{1}{2} \operatorname{cap} R \geq \frac{1}{2} \operatorname{cap} R_T(1)$$

Theorem 2. Suppose that $n \geq 3$, that $x_0 \in R^n$ and that $0 < a < b < \infty$. If $f : D \to B$ is K-quasiconformal and if

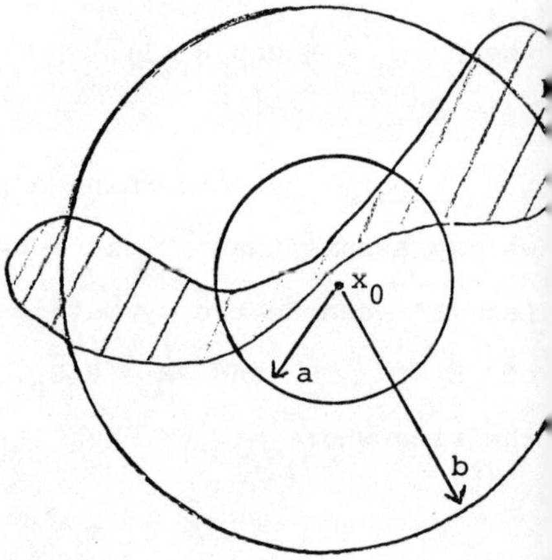

$C(D) \cap \overline{B(x_0,b)}$ has at least two components which meet $\overline{B(x_0,a)}$, then

$$K \geq c\left(\log \frac{b}{a}\right)^{n-1}$$

where c is a positive constant which depends only on n.

Proof. By an elementary limiting argument, we may assume that f has a homeomorphic extension which maps $\overline{D}$ onto $\overline{B}$.

By hypothesis there exist points $x_1, x_2 \in \overline{B(x_0,a)}$ which belong to different components of

$$C(D) \cap \overline{B(x_0,b)}$$

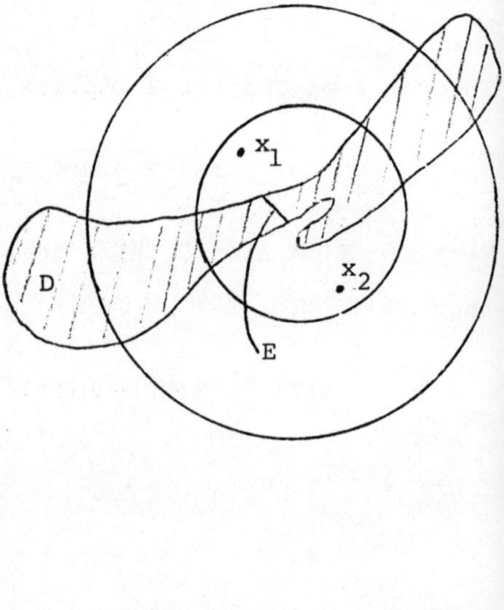

This then implies there exists a segment E with endpoints $y_1, y_2 \in \partial D$ such that

$$E \sim \{y_1, y_2\} \subset D \ , \ E \subset \overline{B(x_0, a)}$$

and such that y_1, y_2 do not belong to the same component of

$$\partial D \cap \overline{B(x_0, b)}$$

Since ∂D is homeomorphic to ∂B, there exists a continuum F separating y_1, y_2 in ∂D such that

$$F \subset \partial D \cap C(B(x_0, b)) \ .$$

Let $\Gamma = \Delta(E, F, D)$ and let Γ_0 be the curve family associated with the spherical ring $R_0 = \{x : a < |x - x_0| < b\}$. Then each $\gamma \in \Gamma$ contains a subcurve $\gamma_0 \in \Gamma_0$ and hence

$$M(\Gamma) \leq M(\Gamma_0) = \omega_{n-1}\left(\log \frac{b}{a}\right)^{1-n}$$

Next let $E', F', \Gamma', z_1, z_2$ denote the images of E, F, Γ, y_1, y_2 under f. Then F' separates z_1, z_2 in ∂B and Lemma 1 implies that $M(\Gamma') \geq \mu_n$.

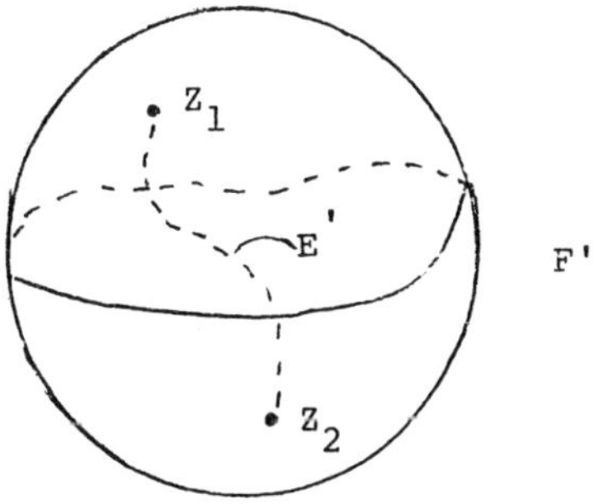

Finally since f is K-quasiconformal,

$$M(\Gamma') \leq K M(\Gamma)$$

and the above inequalities yield

$$K \geq c\left(\log \frac{b}{a}\right)^{n-1}$$

where $c = \dfrac{\mu_n}{\omega_{n-1}} > 0$.

<u>Theorem 3</u>. Suppose that $n \geq 3$, that $x_0 \in R^n$ and that $0 < a < b < \infty$. If $f : D \to B$ is K-quasiconformal and if $C(D) \cap C(B(x_0,a))$ has at least two components which meet $C(B(x_0,b))$, then

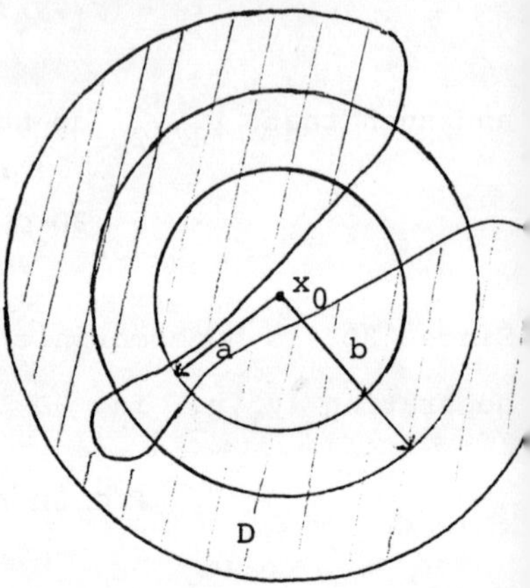

$$K \geq c\left(\log \frac{b}{a}\right)^{n-1}$$

where c is the constant in Theorem 1.

<u>Proof</u>. Let g denote inversion in $S(x_0,1)$ and set

$$h = f \circ g \; , \; D' = g(D) \; .$$

Then $h : D' \to B$ is K-quasiconformal,

$$C(D') \cap \overline{B(x_0,1/a)} = g(C(D) \cap C(B(x_0,a)))$$

has at least two components which meet $\overline{B(x_0,1/b)} = g(C(B(x_0,b)))$ and hence Theorem 2 yields $K \geq c\left(\log \frac{1/a}{1/b}\right)^{n-1} = c\left(\log \frac{b}{a}\right)^{n-1}$.

<u>Remark 2</u>. Though the lower bounds for K in Theorems 2 and 3 are not sharp for any values of a and b, they are of the right order as $b/a \to \infty$.

<u>Examples in $\overline{R}^3$</u>.

a) The infinite plate domain

$$D = \{x : x_1^2 + x_2^2 < \infty, |x_3| < 1\}$$

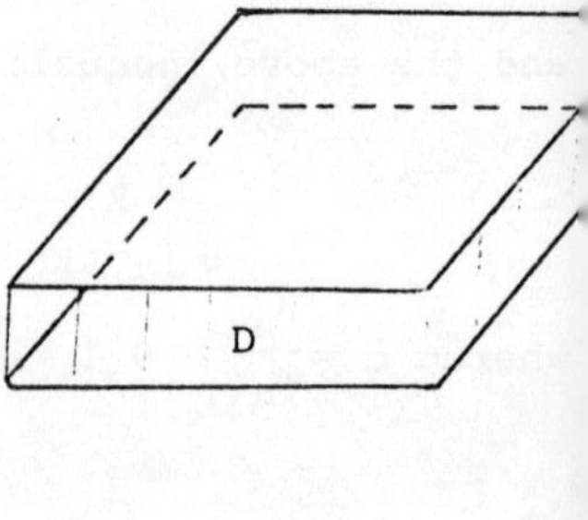

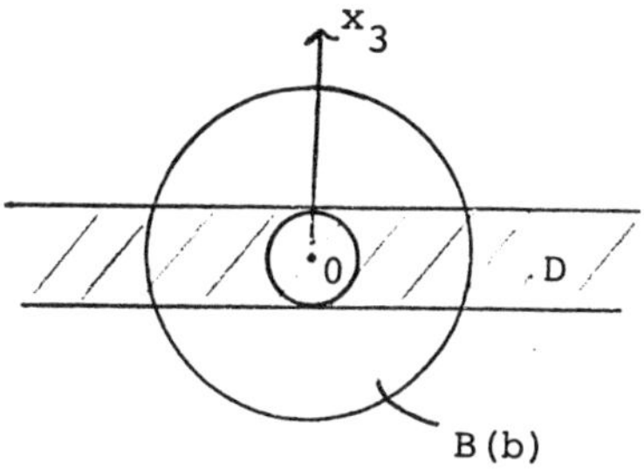

is not quasiconformally equivalent to B . For suppose there exists a K-quasiconformal $f : D \to B$. Then for each $b \in (1,\infty)$, $C(D) \cap \overline{B(b)}$ has two components which meet $\overline{B(1)}$. Thus

$$K \geq c(\log b)^2$$

and letting $b \to \infty$ yields a contradiction.

b) The infinite cylinder

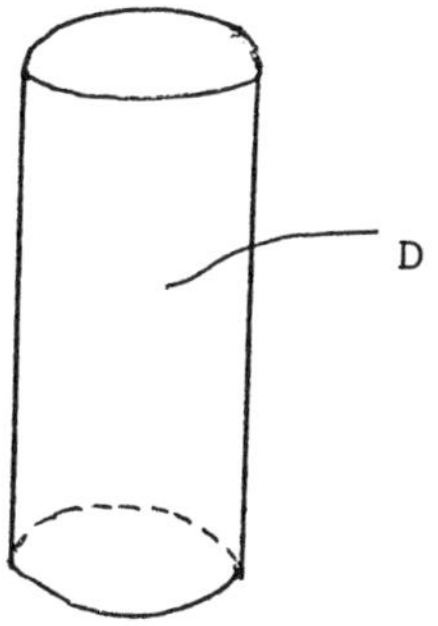

$$D = \{x : x_1^2 + x_2^2 < 1 , |x_3| < \infty\}$$

is quasiconformally equivalent to B . For if we let (r,θ,x_3) and (ρ,θ,ϕ) denote cylindrical and spherical coordinates in R^3 , then the function $f(r,\theta,x_3) = (\rho,\theta,\phi)$ given by

$$\rho = e^{x_3} , \phi = \frac{\pi}{2} r$$

maps D quasiconformally onto the half space $D' = \{x : x_3 > 0\}$ which is itself conformally equivalent to D .

c) If γ is an arc, then the slit domain

$$D = C(\gamma)$$

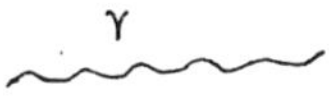

is not quasiconformally equivalent to B . For suppose there exists a K-quasiconformal $f : D \to B$. Let x_0 be an interior

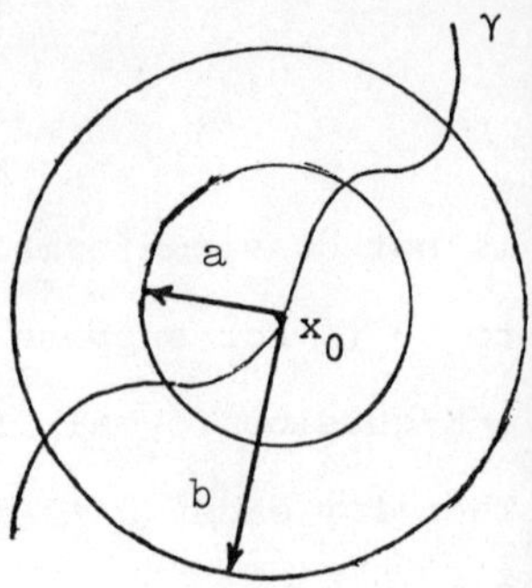

point of γ, and choose $b \in (0,\infty)$ so that both endpoints of γ lie in $C(B(x_0,b))$. Then for each $a \in (0,b)$ $C(D) \cap C(B(x_0,a))$ has two components which meet $C(B(x_0,b))$. Hence

$$K \geq c\left(\log \frac{b}{a}\right)^2$$

by Theorem 3, and letting $a \to 0$ yields a contradiction.

d) If σ is a half plane, the slit domain

$$D = C(\sigma)$$

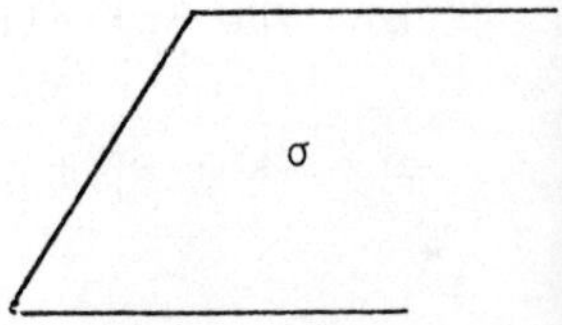

can be folded quasiconformally onto a half space D', and hence D is quasiconformally equivalent to B.

7. AN EXISTENCE THEOREM

We say that a domain $D \subset \overline{R}^n$ is a Jordan domain if ∂D is homeomorphic to the unit sphere S.

<u>Example.</u> There exist complementary Jordan domains D_1 an D_2 in $\overline{R}^3$ such that

a) D_1 is not quasiconformally equivalent to B,

b) D_2 is quasiconformally equivalent to B,

c) $\partial D_1 = \partial D_2$.

<u>Proof.</u> Let D_1 be the semi-infinite plate domain

$$D_1 = \{x : x_1 > 0 \, , \, |x_2| < 1\}$$

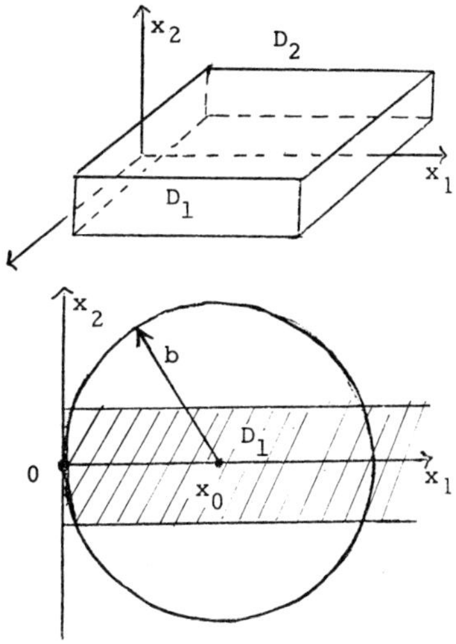

and let $D_2 = C(\bar{D}_1)$. Then it is easy to see that $\partial D_1 = \partial D_2$ is homeomorphic to S .

Now suppose there exists a K-quasiconformal $f : D_1 \to B$ and for each $b \in (1,\infty)$ let $x_0 = be_1$. Then $C(D) \cap \bar{B}(x_0,b)$ has two components which meet $\bar{B}(x_0,1)$, and hence

$$K \geq c(\log b)^2$$

by Theorem 2 of Chapter 6. Letting $b \to \infty$ yields a contradiction and hence D_1 is not quasiconformally equivalent to B .

To see that D_2 is quasiconformally equivalent to B let H_1, H_2 be the half spaces

$$H_1 = \{x : x_2 > 1\} , \; H_2 = \{x : x_2 < -1\}$$

let H_1', H_2' be the quarter spaces

$$H_1' = \{x : x_1 < 0, x_2 > 1\} , \; H_2' = \{x : x_1 < 0, x_2 < -1\}$$

and let

$$H_3 = \{x : x_1 < 0, |x_2| \leq 1\} , \; D_3 = \{x : x_1 < 0\}$$

For $i = 1,2$ there exists a quasiconformal folding $f_i : H_i \to H_i'$ with

$$f_i \mid \bar{H}_i \cap H_3 = \text{identity}$$

Then $D_2 = H_1 \cup H_2 \cup H_3$, $D_3 = H_1' \cup H_2' \cup H_3$ and

$$f(x) = \begin{cases} f_i(x) & x \in H_i \quad i = 1,2 \\ x & x \in H_3 \end{cases}$$

defines a quasiconformal mapping of D_2 onto D_3 . Since D_3 is conformally equivalent to B , D_2 is quasiconformally equivalent to B .

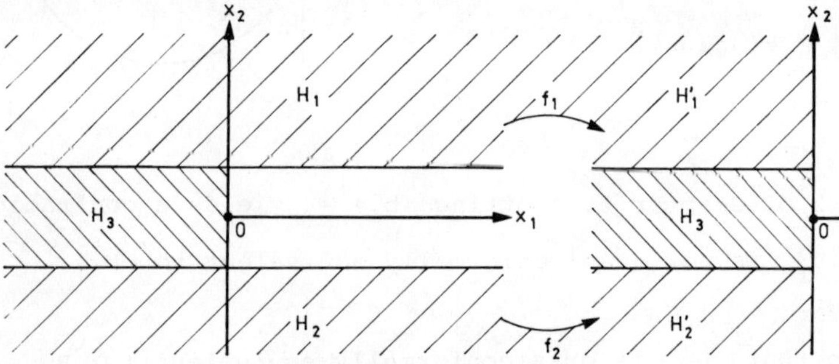

FIG.4. Construction of the mapping of D_2 onto D_3.

Remark 1. When $n = 2$, we see from Theorem 1 of Chapter that a domain D is quasiconformally equivalent to B if and if ∂D is a nondegenerate continuum. The example in Fig. 4 show that when $n > 2$, it is no longer possible to characterize the domains D which are quasiconformally equivalent to B by loo only at their boundaries ∂D . We shall prove, however, that i the part or D near ∂D is quasiconformally equivalent to the part of B near ∂B , then D is quasiconformally equivalent to B . The proof of

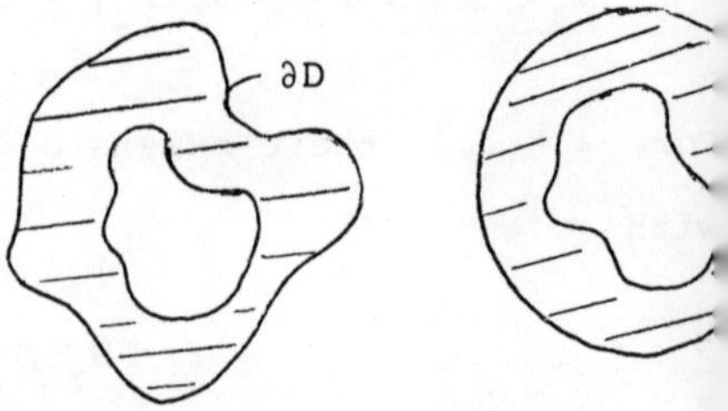

this is quite long and rather involved. We begin with a result on Jordan domains from which the general theorem readily follows.

Theorem 1. Suppose that D_1, D_2 are Jordan domains in $\overline{R}^n$ with $\overline{D}_1 \cap \overline{D}_2 = \phi$ and that B_1, B_2 are open balls in R^n with $\overline{B}_1 \cap \overline{B}_2 = \phi$. Suppose also that $f : C(D_1 \cup D_2) \to C(B_1 \cup B_2)$ is a homeomorphism such that $f \mid C(\overline{D}_1 \cup \overline{D}_2)$ is quasiconformal and $f(\partial D_i) = \partial B_i$ for $i = 1,2$. Then there exists a homeomorphism $f^* : C(D_2) \to C(B_2)$ such that $f^* \mid C(\overline{D}_2)$ is quasiconformal and

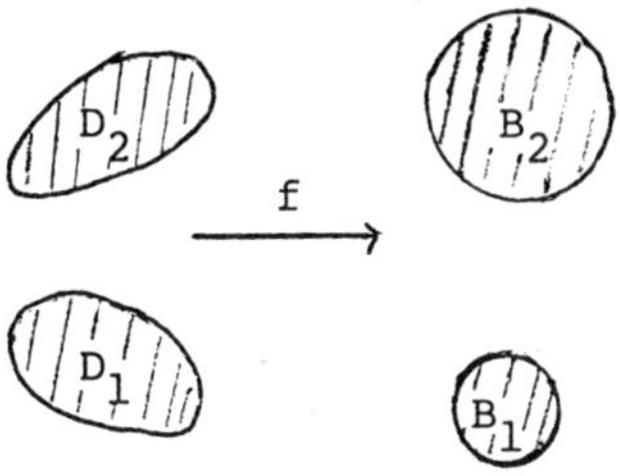

$$f^* \mid \partial D_2 = f \mid \partial D_2$$

Remark 2. Theorem 1 says that the boundary correspondence

$$f \mid \partial D_2 : \partial D_2 \to \partial B_2$$

has a quasiconformal extension

$$f^* : C(\overline{D}_2) \to C(\overline{B}_2)$$

The proof is somewhat tricky. We consider first two special cases.

Lemma 1. Theorem 1 holds under the following additional hypotheses:

a) $\overline{D}_1$, $\overline{D}_2$, $\overline{B}_1$, $\overline{B}_2$ lie in the unit ball B ,

b) $f(x) = x$ in a neighborhood of $C(B)$.

Proof. By performing a preliminary orthogonal mapping, we may assume there exist numbers $-1 < a < b < 1$ such that

$$B_1 \subset H_1 = \{x : x_n < a\}$$

$$B_2 \subset H_2 = \{x : x_n > b\}$$

Here we have made essential use of the geometrical character of B_1 and B_2, e.g. that they are not linked in R^n and, in fact, can be separated by an (n-1)-dimensional hyperplane.

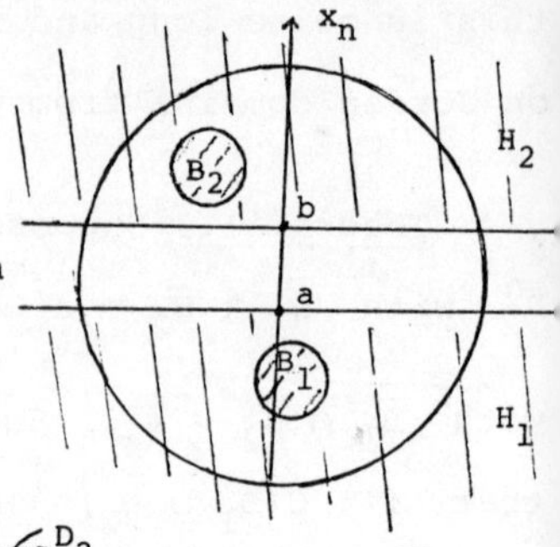

Next let

$$E = \bigcup_{j=0}^{\infty} (D_1 \cup D_2) + 3je_1$$

$$E' = \bigcup_{j=0}^{\infty} (B_1 \cup B_2) + 3je_1$$

$$F = \bigcup_{j=0}^{\infty} \overline{B} + 3je_1$$

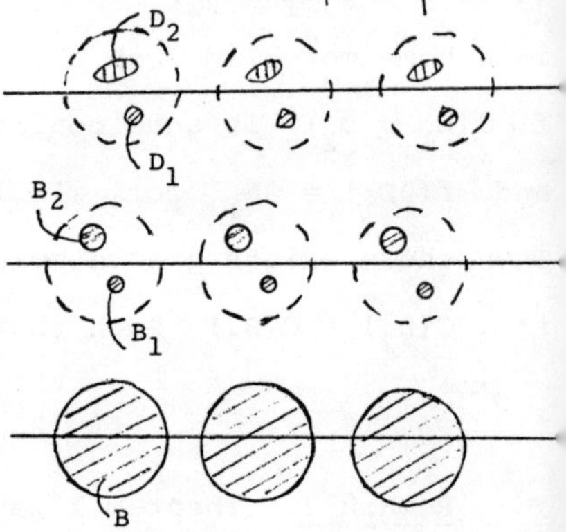

Then it is clear that $E, E' \subset F$ and that

$$F \sim E = \bigcup_{j=0}^{\infty} (\overline{B} \sim (D_1 \cup D_2)) + 3je_1$$

$$F \sim E' = \bigcup_{j=0}^{\infty} (\overline{B} \sim (B_1 \cup B_2)) + 3je_1$$

Now set

$$g(x) = \begin{cases} f(x - 3je_1) + 3je_1 & \text{if } x \in (\overline{B} \sim (D_1 \cup D_2)) + 3je_1 \\ x & \text{if } x \in C(F) \end{cases}$$

Then g is a homeomorphism of $C(E)$ onto $C(E')$. Moreover, each $x \in C(\overline{E})$ has a neighborhood U such that $g\,|\,U$ is either the identity or the composition of f with at most two translations. Hence $g : C(\overline{E}) \to C(\overline{E}')$ is also quasiconformal.

Set $r = \frac{b-a}{3} > 0$ and define $k : R^1 \to R^1$ as follows:

$$k(t) = \begin{cases} 0 & \text{if } -\infty < t \leq a + r \\ \frac{a+r-t}{r} & \text{if } a + r < t \leq b - r \\ -1 & \text{if } b - r < t < \infty \end{cases}$$

Next define $h : \bar{R}^n \to \bar{R}^n$ by

$$h(x) = \begin{cases} x + 3k(x_n)e_1 & \text{if } x \in R^n \\ \infty & \text{if } x = \infty \end{cases}$$

Then h is a piecewise linear, and hence quasiconformal, mapping of $\bar{R}^n$ onto $\bar{R}^n$ (see Fig. 5).

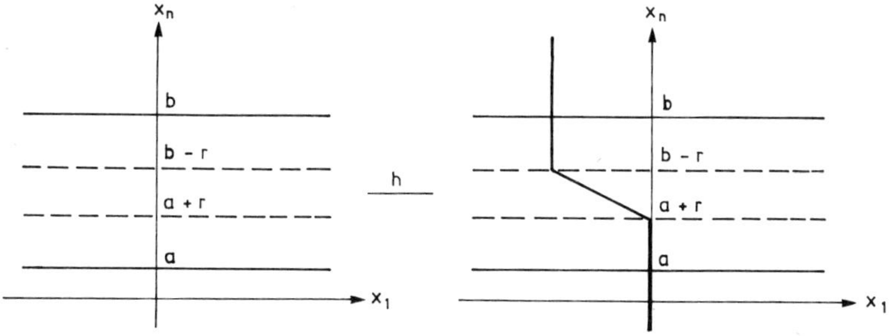

FIG.5. Mapping h in Lemma 1.

Finally let

$$f^*(x) = \begin{cases} g^{-1} \circ h \circ g(x) + 3e_1 & \text{if } x \in R^n \sim E \\ x + 3e_1 & \text{if } x \in D_1 + 3_j e_1 \ , \ j \geq 0 \\ x & \text{if } x \in D_2 + 3_j e_1 \ , \ j \geq 1 \\ \infty & \text{if } x = \infty \end{cases}$$

We shall show that f^* is the desired mapping. It is clear from Fig. 6 that f is a univalent mapping of $C(D_2)$ onto $C(B_2)$.

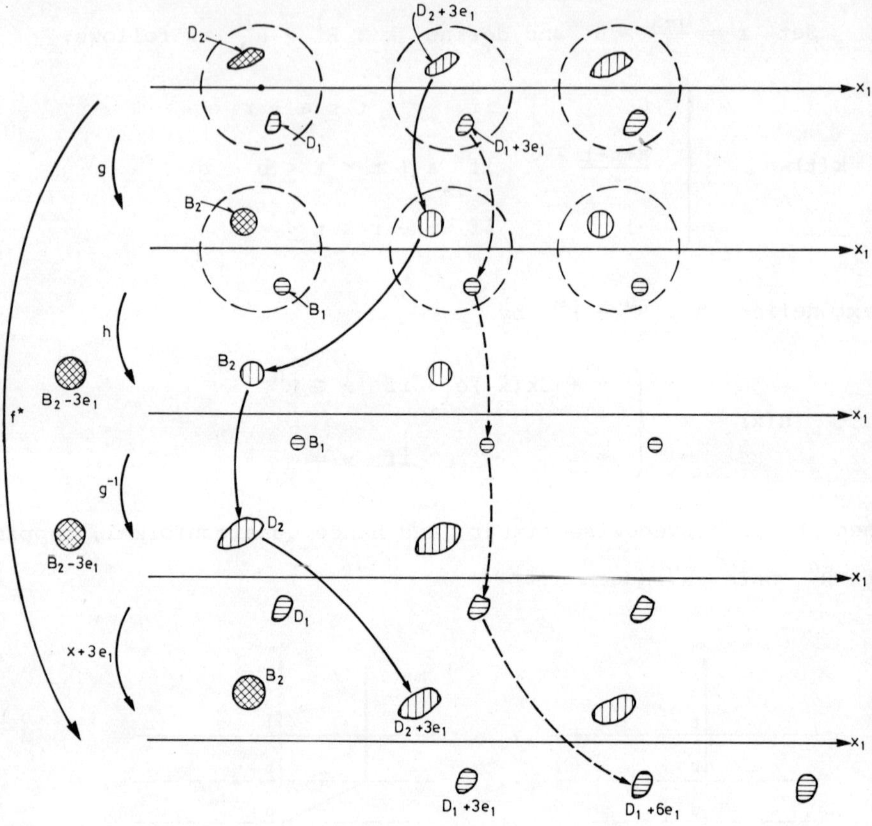

FIG.6. Construction of the mapping f in Lemma 1.*

We also see that f is continuous except possibly at points of

$$H = \left(\bigcup_{j=0}^{\infty} \partial D_1 + 3je_1 \right) \cup \left(\bigcup_{j=1}^{\infty} \partial D_2 + 3je_1 \right) \cup \{\infty\}$$

Choose a point $x_0 \in \partial D_1 + 3je_1$, where $j \geq 0$. Then x_0 has a neighborhood U such that

$$g(U \sim E) \subset (B + 3je_1) \cap \{x : x_n < a + r\}$$

and it follows that $g^{-1} \circ h \circ g(x) = x$, $f^*(x) = x + 3e_1$

for $x \in U \sim E$. We conclude that

(*) $$f^*(x) = x + 3e_1$$

for $x \in U$. Similarly if $x_0 \in \partial D_2 + 3je_1$ where $j \geq 1$, then x_0 has a neighborhood U such that

$$g(U \sim E) \subset (B + 3je_1) \cap \{x : x_n > b - r\}$$

and we have $g^{-1} \circ h \circ g(x) = x - 3e_1$, $f^*(x) = x$ for $x \in U \sim E$. Thus

(**) $$f^*(x) = x$$

for $x \in U$. Finally it is easy to verify that

$$|f^*(x) - x| < 5$$

for all $x \in R^n$. From the above it follows that f^* is continuous at each point of H.

Now let $x \in \partial D_2$. Then

$$\begin{aligned} f^*(x) &= g^{-1} \circ h \circ g(x) + 3e_1 \\ &= g^{-1}(f(x) - 3e_1) + 3e_1 \\ &= (f(x) - 3e_1) + 3e_1 = f(x) \end{aligned}$$

and thus $f^* \mid \partial D_2 = f \mid \partial D_2$. Finally since h is quasiconformal, it is not difficult to see from (*) and (**) that f^* maps $C(\bar{D}_2)$ quasiconformally onto $C(\bar{B}_2)$. This completes the proof of Lemma 1.

<u>Lemma 2.</u> Theorem 1 holds under the following additional hypotheses:

a) $\bar{D}_1, \bar{D}_2, \bar{B}_1, \bar{B}_2$ lie in the unit ball B,

b) $0 \in D_2'$,

c) $C(B) \subset f(C(\bar{B}))$.

Proof. By hypothesis we can choose $0 < a < b < 1$ so tha

$$B(a) \subset D_2 \ , \ D_1 \cup D_2 \subset B(b)$$

Next define $g : \overline{R}^n \to \overline{R}^n$ by

$$g(x) = \begin{cases} |x|^{c-1}x & \text{if } x \in B \\ x & \text{if } x \in C(B) \end{cases}$$

where $c = \dfrac{\log b}{\log a}$. Then g is quasiconformal,

$$g(B(a)) = B(b)$$

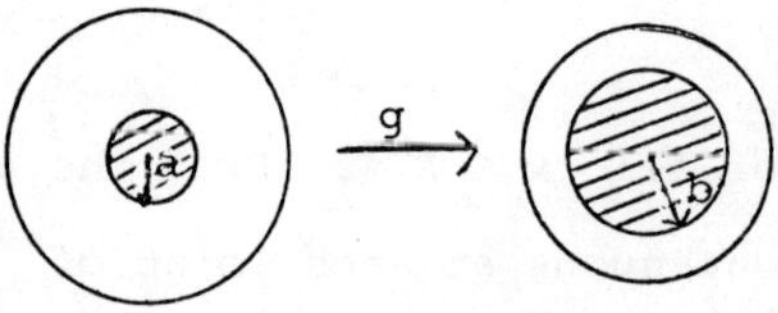

and hence

$$g(C(D_2)) \subset C(D_1 \cup D_2).$$

We conclude that $f \circ g$ is a homeomorphism of $C(D_2)$ into $C(B_1 \cup B_2)$.

Now let

$$D_1' = f \circ g(D_1), \quad D_2' = C(f \circ g(C(D_2)))$$

Then D_1', D_2' are Jordan domains with

$$D_1' \cap D_2' = \phi \ ,$$

$$\overline{D}_1' \cup \overline{D}_2' \subset C(f \circ g(C(B))) \subset C(f(C(\overline{B}))) \subset B.$$

Next let $U = f(C(\overline{B}))$, $h = f \circ (f \circ g)^{-1}$.

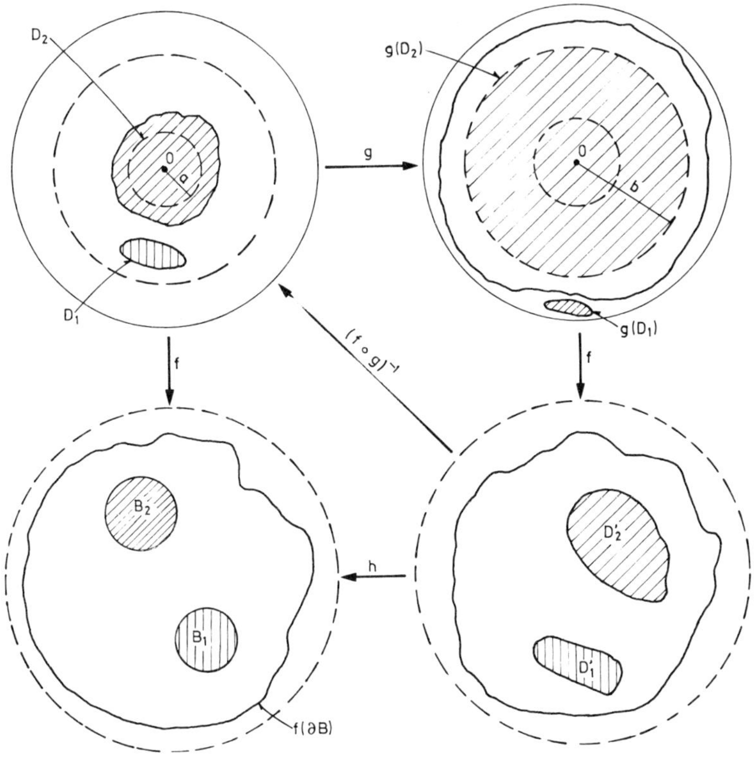

FIG.7. Construction of the mapping f in Lemma 2.*

Then U is a neighborhood of $C(B)$, h is a homeomorphism of $C(D_1' \cup D_2')$ onto $C(B_1 \cup B_2)$, h is quasiconformal in $C(\bar{D}_1' \cup \bar{D}_2')$ and

$$h(\partial D_i') = f(\partial D_i) = \partial B_i$$

for $i = 1, 2$ (see Fig. 7).

Also if $x \in U$, then $f^{-1}(x) \in C(B)$ and

$$h(x) = f \circ g^{-1} \circ f^{-1}(x) = f \circ f^{-1}(x) = x$$

Hence by Lemma 1 we get a homeomorphism $h^*: C(D_2') \to C(B_2)$ such that $h^* \mid C(\bar{D}_2')$ is quasiconformal and

$$h^* \mid \partial D_2' = h \mid \partial D_2$$

Now set

$$f^* = h^* \circ (f \circ g)$$

Then $h^*: C(D_2) \to C(B_2)$ is a homeomorphism, $f^* \mid C(\bar{D}_2)$ is quasiconformal, and

$$f^* \mid \partial D_2 = h^* \mid \partial D_2' = h \mid \partial D_2' = f \mid \partial D_2$$

This completes the proof of Lemma 2.

Proof of Theorem 1. Obviously $\infty \in C(\bar{B}_1 \cup \bar{B}_2)$. By performing a preliminary Möbius transformation we may assume that

$$\infty = f^{-1}(\infty) \in C(\bar{D}_1 \cup \bar{D}_2)$$

Then there exists $x_0 \in D_2$ and $0 < a < b < \infty$ such that

$$\bar{D}_1 \cup \bar{D}_2 \subset B(x_0, a), \quad \bar{B}_1 \cup \bar{B}_2 \subset B(x_0, b)$$

$$f(C(B(x_0, a))) \supset C(B(x_0, b))$$

Let g_1, g_2 be the similarity mappings which carry $B(x_0, a)$, $B(x_0, b)$ onto B and set

$$h = g_2 \circ f \circ g_1^{-1}$$

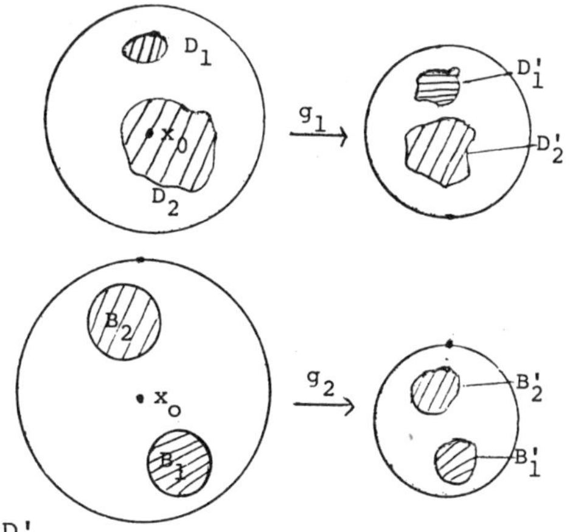

Then h satisfies the hypotheses of Lemma 2 and there exists a homeomorphism $h^*: C(D_2') \to C(B_2')$ which is quasiconformal in $C(\bar{D}_2')$ with

$$h^* \mid \partial D_2' = h \mid \partial D_2'$$

where $D_i' = g_1(D_i)$ and $B_i' = g_2(B_i)$. Hence

$$f^* = g_2^{-1} \circ h^* \circ g_1$$

is the desired extension.

<u>Theorem 2</u>. Suppose that D is a domain in $\bar{R}^n$, that U is a neighborhood of ∂D, and that f is a quasiconformal mapping of $D \cap U$ into B such that

$$\lim_{y \to x} |f(y)| = 1$$

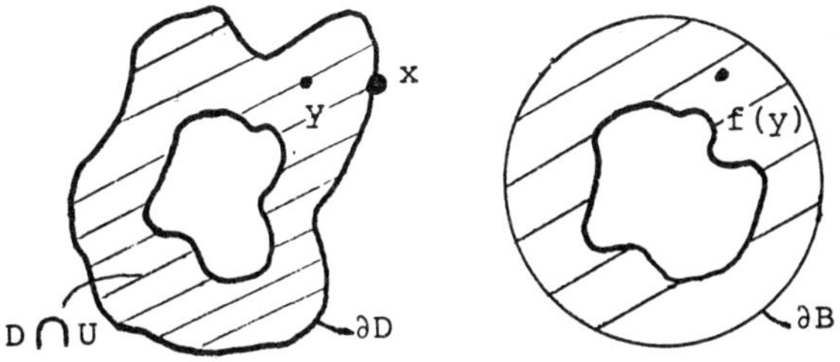

for each $x \in \partial D$. Then there exists a neighborhood U^* of ∂D and a quasiconformal mapping $f^*: D \to B$ such that

$$f^* \mid D \cap U^* = f \mid D \cap U^*$$

Idea of proof. An elementary topological argument and two preliminary Möbius transformations allow one to deduce this result from Theorem 1.

IAEA-SMR-18/102

GEOMETRIC THEORY OF DIFFERENTIAL EQUATIONS IN THE COMPLEX DOMAIN

R. GERARD
Centre d'équations différentielles,
Institut de recherche mathématique avancée,
Département de mathématique,
Université de Strasbourg,
Strasbourg, France

Abstract

GEOMETRIC THEORY OF DIFFERENTIAL EQUATIONS IN THE COMPLEX DOMAIN.
It is well known how useful the notion of foliation is in the study of differential equations in the real case but it is not an indispensable tool, although fairly useful. This paper shows how the notion of foliation can be used in the theory of complex differential equations and complex pfaffian equations and how the analytic behaviour of solutions can be derived from the properties of the foliations. The use of foliations allows us to give geometric interpretations of some classical results on differential equations in the complex domain, to generalize these results to other situations, and to solve new problems in the theory. The paper is divided into three parts: I. Complex analytic foliations on a complex analytic manifold: Definitions, elementary properties; Examples; Applications. II. Painlevé's foliations on a fibration: Definitions, examples; Local properties; The main theorem on Painlevé's foliation; Applications. III. Algebraic pfaffian equations on a projective space; Non-existence of compact leaves; The structure of the set of algebraic solutions.

Part I

COMPLEX ANALYTIC FOLIATIONS

The following definitions can be found in the real case in [1] and [2].

1. FOLIATED STRUCTURE ON A COMPLEX MANIFOLD

a) The trivial foliated structure of codimension p on C^n.

Look at the affine space C^n as the product $C^p \times C^{n-p}$ and let $x = (x_1, x_2, \ldots, x_p)$ the coordinates in C^p and $y = (y_1, y_2, \ldots, y_{n-p})$ the coordinates in C^{n-p}. Then the simplest foliation in codimension p of C^n is the foliation in which the leaves are the planes $x = Cte$. Denote by F_p this foliation. A local automorphism of F_p is by definition a local isomorphism of C^n preserving locally the leaves; that means that in the neighbourhood of each point on which the isomorphism is well defined it can be written in the follo

form :

$$\begin{cases} X = h_1(x) \\ Y = h_2(x,y) \end{cases}$$

where h_1 is a local isomorphism of $\mathbb{C}^p$.

If U is an open subset of $\mathbb{C}^n$, a "plaque" in U is a connected component

the trace of a leaf on U.

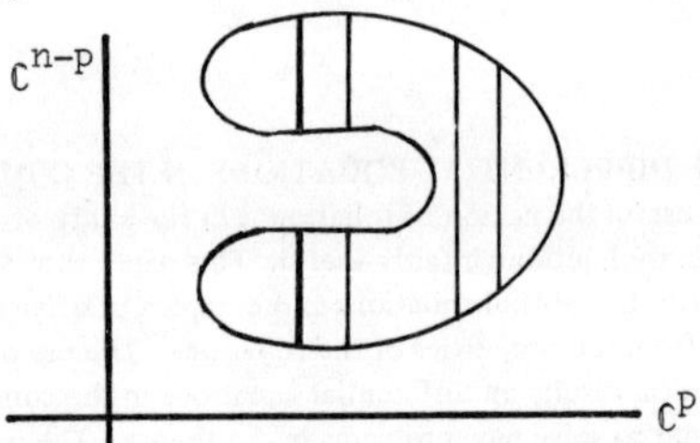

b) Foliated structure on a complex analytic manifold

A foliated structure of codimension p on a complex manifold V_n

dimension n with countable basis is a maximal atlas $\mathcal{A} = (U_i, f_i)_{i \in I}$ such

for all (i,j), h_{ij} is a local automorphism for the foliation F_p on $\mathbb{C}^n$

A "plaque" in U_i for this foliated structure is by definition the inverse

image by f_i of a "plaque" in $f_i(U_i)$ for the trivial foliated structure o

codimension p on $\mathbb{C}^n$.

An open subset U of V_n is called distinguished if it is isomorphic to a

product of two disks $P_{n-p} \subset \mathbb{C}^{n-p}$, $P_p \subset \mathbb{C}^p$ such that the inverse images of t

"plaques" in $P_{n-p} \times P_p$ are the "plaques" in U.

It is easy to see that :

1) all distinguished open subsets of V_n containing a point m for

a fundamental system of neighbourhoods of the point m ;

2) for all (i,j), h_{ij} induces an isomorphism from each "plaque"

$f_i(U_i \cap U_j)$ onto a "plaque" of $f_j(U_i \cap U_j)$;

3) the intersection of a finite number of "plaques" is a union of

"plaques".

The last property enablesus to define a new complex analytic structure on V_n ; the open subsets for this new structure are the "plaques". It is clear that this new structure is finer than the old one; with this structure V_n is a complex analytic manifold of dimension $n-p$.

A leaf for the foliated structure of codimension p on V_n is by definition a connected component of V_n for the fine structure and with this structure a leaf is an analytic submanifold of dimension $n-p$ of V_n .

A leaf is called proper if the topology on the leaf defined by the "plaques" is the same as the topology induced by the natural topology of the manifold V_n .

Let $\mathfrak{F}$ be an analytic foliation of codimension p on the complex analytic manifold V_n and $\mathcal{G} = \{U_i, f_i\}_{i \in I}$ the atlas defining this foliation. A map $g : U \subset V_n \to \mathbb{C}^p$ (U open subset) is called distinguished if it is locally of the form $g = \pi \circ f_i$ where π is the canonical projection of $\mathbb{C}^n$ onto $\mathbb{C}^p$.

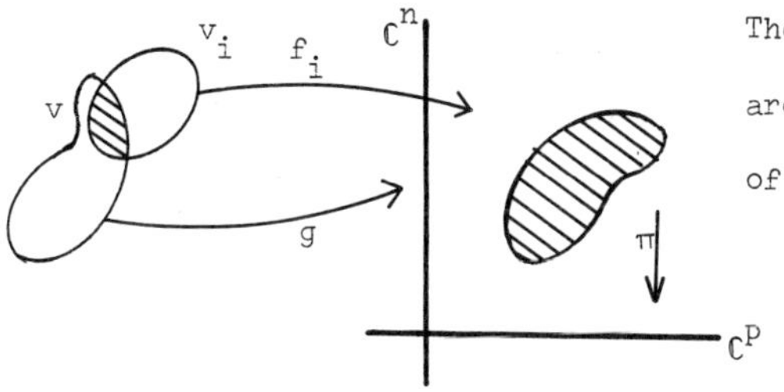

The "plaques" in $U \cap U_i$ (for all $i \in I$) are the inverse images of the points of $\mathbb{C}^p$.

If f and g are two distinguished maps having a common part in their range, there exists a local isomorphism h of $\mathbb{C}^p$ such that $f = h \circ g$.

Now it is easy to see that a foliation $\mathfrak{F}$ of codimension p on V_n is also given by a family $\{V_i, f_i\}_{i \in J}$ such that :

1) $\{V_i\}_{i \in J}$ is a covering of V_n ;
2) $f_i : V_i \to \mathbb{C}^p$ are submersions;
3) the family $\{V_i, f_i\}_{i \subset J}$ is complete in the following sense: if $f : U \to \mathbb{C}^p$ (U open) is a submersion then for all i such that $U \cap U_i \neq \emptyset$ and for all $x \in U \cap U_i$ there exists a local isomorphism h of $\mathbb{C}^p$ such that $f = h \circ f_i$ in the neighbourhood of x .

This definition of foliations is the natural one arising in the study of differential equations in the complex domain.

Now let us recall some elementary properties of the leaves which wi be used later:

1) Continuity theorem:

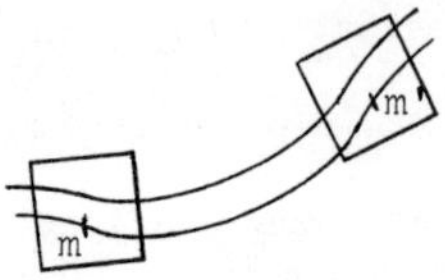

If m and m' are two points belonging to the same leaf then for distinguished neighbourhood U' of there exists a distinguished neighbc U of m such that each leaf meeting U goes through U' .

2) The equivalence relation defined by the leaves is open.

3) The closure in V_n of a union of leaves is a union of leaves.

The leaves' space is the quotient space of V_n by the equivalenc relation associated to the foliation. In general this space is v complicated and exceptionally Hausdorff.

In the study of this space, the following result is known :

THEOREM. If F is an open equivalence relation on a reduced analytic space then X/R is an analytic (reduced) space if and only if the graph of R in X x X is an analytic set.

For the proof and other results, see [3] and [4].

2. EXAMPLES

2.1. Foliation defined by a holomorphic differential system

Let $\frac{dy}{dx} = F(x,y)$ be a holomorphic differential system on $\mathbb{C} \times \mathbb{C}^n$; means that F is a holomorphic map from $\mathbb{C} \times \mathbb{C}^n$ into $\mathbb{C}^n$.

If (x^o,y^o) is a point of $\mathbb{C} \times \mathbb{C}^n$ then the Cauchy theorem states that there exists a unique solution $y = \varphi(x,y^o,x^o)$ which takes the value y^o at the x^o . And we have also $y^o = \varphi(x^o,y,x)$; in other words there exists a disk U centered at y^o such that the local solution are the inverse images of the points of $\mathbb{C}^n$ by the map

$$U \times V \to \mathbb{C}^n$$

$$(x,y) \to \varphi(x^o,y,x)$$

The collection $(U \times V, \varphi)$ gives us the distinguished maps for the foliation defined by the given differential system.

We can also look at this foliation in the following way : take a point (x^o,y^o) in $\mathbb{C} \times \mathbb{C}^n$ and the local solution at this point, $y = \psi(x,y^o,x^o)$ the analytic

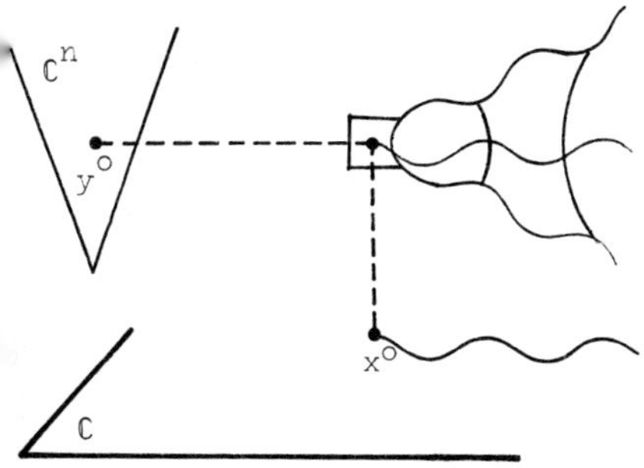

continuation of ψ, is a solution of our system and the graph of this analytic continuation in $\mathbb{C} \times \mathbb{C}^n$ gives us a leaf of the associated foliation.

This introduction to foliation can be found in Painlevé's lectures at Stockholm (1894) (see [5]) and in <u>this sense Painlevé was the first foliater</u> (G. Reeb dixit!)!

2.2. Trajectories of a holomorphic vector field on $\mathbb{P}_2(\mathbb{C})$

It is easy to see that a holomorphic vector field in the complex projective plane $\mathbb{P}_2(\mathbb{C})$ is induced by a linear vector field in $\mathbb{C}^3$. A generic vector field on $\mathbb{P}_2(\mathbb{C})$ has three singular points a,b,c. By the choice of the coordinates we may assume that $a = (1,0,0), b = (0,1,0), c = (0,0,1)$. This vector field defines in $\mathbb{P}_2(\mathbb{C}) - \{a,b,c\}$ a foliation. The leaves of this foliation are trajectories of the group $\mathbb{C}$ and the only one parameters connected groups are : $\mathbb{C}$, cylinder, torus, point. In our case it is easy to give the explicit parametrization of the trajectories :

$$x = x_o\, e^{t}$$

$$y = y_o\, e^{\lambda t}$$

in some coordinate neighbourhood.

Now it is clear that no leaf is a torus. This admits the following generaliza

as we shall see later :

<u>If $\omega = 0$ is a completely integrable Pfaffian equation on</u> $\mathbb{P}_n(\mathbb{C}$

S <u>the set of singularities of</u> ω <u>then the foliation defined by</u> $\omega = 0$ <u>in</u>

$P_n(C) - S$ <u>has no compact leaf</u>.

Now let us give the description of the foliation given by a generic holomorp

vector field on $\mathbb{P}_2(\mathbb{C})$.

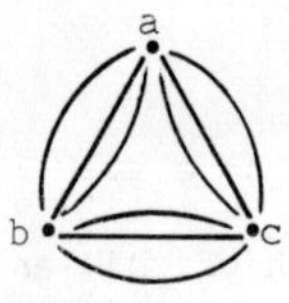

1) If λ is a rational number then al leaves are cylinders whose end $B(\gamma) = \bar{\bar{\gamma}}$ (γ is a leaf) is the union of two sing points.

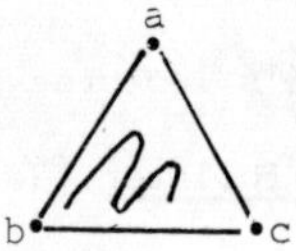

2) If λ is real but not rational t are three particular leaves which are cylinders whose end is composed of two s lar points. The other leaves are comple planes; the end of such a leaf contains three-dimensional manifold $(T^2 \times \mathbb{R})$ an two singular points (for details see [6

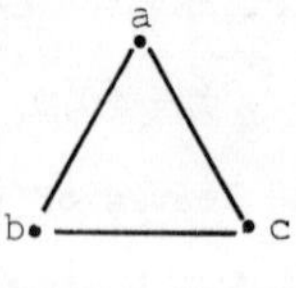

3) If λ is complex but not real there three particular leaves which are cylin ending at two singular points; the othe leaves are proper complex planes.

2.3. Foliation defined by a completely integrable pfaffian system on a complex analytic manifold

For more details look at [7] (Frobenius theorem).

A foliation $\mathfrak{F}$ of dimension p on a complex analytic manifold V_n $(n \geq p)$ gives us an integrable subbundle of rank p of the holomorphic t bundle of V_n and conversely an integrable holomorphic subbundle of the holomorphic tangent bundle gives an analytic foliation on the complex analytic manifold V_n.

4) Other examples are given in the following paragraph.

3. APPLICATIONS

3.1. The holomorphic Riccati equation

A holomorphic Riccati equation is an equation of the following type:

$$y' = a(x)y^2 + b(x)y + c(x) \tag{1}$$

where a,b,c are entire functions.

This equation can also be written in the following form :

$$dy - (a(x)y^2 + b(x)y + c(x))\, dx = 0 \tag{2}$$

Equations (1) and (2) define the same foliation in $\mathbb{C}^2$.

Now we extend equation (2) to an equation on $\mathbb{P}_1(\mathbb{C}) \times \mathbb{C}$, if (y,z) are homogenous coordinates on $\mathbb{P}_1(\mathbb{C})$ this extended equation is

$$z\, dy - y\, dz - (a(x)y^2 + b(x)yz) + c(x)\, z^2)\, dx = 0 \tag{3}$$

The manifold $\mathbb{P}_1(\mathbb{C}) \times \mathbb{C}$ is covered by two coordinate neighbourhoods :

$$U_1 : (y,x) \quad (z \neq 0)$$

$$U_2 : (z,x) \quad (y \neq 0)$$

Equation (3) is also given by two equations :

$$dy - (a(x)y^2 + b(x)y + c(x))dx = 0 \quad \text{in } U_1$$

$$-dz - (a(x) + b(x)z + c(x)z^2)dx = 0 \quad \text{in } U_2$$

Now it is clear that each leaf of the foliation defined by the equation (3) i transversal to the fibers $x = Cte$. This means that our foliation is transve to the fibration $\mathbb{P}_1(\mathbb{C}) \times \mathbb{C} \to \mathbb{C}$ which has compact fibers, then by a theorem o Ehresmann [8] each leaf is a covering of the basis. It follows that each leaf isomorphic to the basis $\mathbb{C}$ by the canonical projection. This result implies that every solution of a holomorphic Riccati equation is a uniform functior This is a well known classical result but everywhere proved by analytical met

3.2. Differential equations of the form $y' = \dfrac{P(x,y)}{Q(x,y)}$

Let us suppose that P and Q are polynomials in y with analyt coefficients and moreover that P and Q are relatively prime.

We recall that for such an equation there are two types of singularities for solution (a pole is not considered as a singularity).

1) Fixed singularities: These singularities can be determined expl citly on the equation. They are :

a) the point $\xi \in \mathbb{C}$ such that there exists an η satisfying

$$P(\xi,\eta) = 0 \text{ and } Q(\xi,\eta) = 0 .$$

The points (ξ,η) are singularities of the vector field defined by the equation in $\mathbb{C}^2$.

b) the points $\tilde{\xi}$ such that $Q(\tilde{\xi},y) = 0$ identically in y .

2) Movable singularities: These singularities depend on the choic of the solution and cannot be given explicitly before knowing explicitly th considered solution.

Now write the given equation in the following form :

$$\omega = Q(x,y)\,dy - P(x,y)\,dx = 0$$

and extend this equation to an equation $\tilde{\omega} = 0$ on $\mathbb{P}_1(\mathbb{C} \times \mathbb{C})$.

Denote by Σ the set of the singularities of $\tilde{\omega}$. Then $\tilde{\omega} = 0$ defines a foliation on $\mathbb{P}_1(\mathbb{C}) \times \mathbb{C} - \Sigma$. This foliation may have vertical leaves (for the projection onto $\mathbb{C}$). These leaves are parts of the fibers $x = \tilde{\xi}$ for which $Q(\tilde{\xi}, y) = 0$ identically in y. On such a fiber there is always a singular point of $\tilde{\omega}$.

Let Ξ be the subset of $\mathbb{C}$ made up by the points ξ which are the projection of the points of Σ. The set Ξ has only isolated points. Denote by $\tilde{\Xi}$ the subset of $\mathbb{C}$ which is the union of the points of $\mathbb{C}$ such that the fiber over such a point contains a leaf of the foliation associated to the equation $\tilde{\omega} = 0$. The point of $\Xi \cup \tilde{\Xi}$ are the fixed singularities of our equation in the sense given before. We shall see later that:

a) a point of Ξ is at most a transcendental singular point for the solution and never an essential singular point ;

b) a point of $\tilde{\Xi}$ can be an essential singular point for a solution.

A movable singularity is a point at which the leaf has a vertical tangent. As we shall see later, such a singularity is at most algebraic.

Now let us determine the equations of the above form without movable singularities. This means that for such an equation the associated foliation in $\mathbb{P}_1(\mathbb{C}) \times \mathbb{C}$ has no leaf with vertical tangent plane ; and this implies that Q has to be independent of y and the equation has the following form :

$$\frac{dy}{dx} = \frac{P(x,y)}{Q(x)}$$

The same thing has to be true at infinity. Near infinity in y the equation has the following form (change of variables $z = \frac{1}{y}$)

$$-\frac{dz}{dx} = \frac{P(x,z)}{z^{n-2}Q(x)} \qquad n = \text{degree of } P \ (\text{in } y)$$

To have no vertical tangent for the leaves it is necessary that $n - 2 \leq 0$.

All this means that our equation has to be a Riccati equation of the follo

form :

$$y' = \frac{a(x)y^2 + b(x)y + c(x)}{d(x)}$$

where a,b,c,d are holomorphic functions.

It is easy to see that such a Riccati equation has no movable singular poin

3.3. Uniformity of the elliptic functions

The classical elliptic functions are solutions of the differential equation

$$y'^2 = (1-y^2)(1-k^2y^2)$$

We are going to prove in a geometric way the uniformity of the solutions of elliptic equation. First we are trying to do the same as for the Riccati equation. Look at the equation $dy - zdx$ on the surface with equation $z^2 = (1-y^2).(1-k^2y^2)$ in $\mathbb{C}^3$, now extend the situation to $\mathbb{P}_2(\mathbb{C})(y,z,t) \times \mathbb{C}$ into

$$tdy - ydt - ztdx = 0$$

on

$$t^2z^2 = (t^2-y^2)(t^2-k^2y^2)$$

The first difficulty arising now is the existence of singularities for our pfaffian equation on the surface. To clarify the situation we make another compactification.

Look at the following diagram :

$$\mathbb{P}T(\mathbb{P}_1(\mathbb{C})) \times \mathbb{C}(x) \to T(\mathbb{P}_1(\mathbb{C})) \times \mathbb{C}(x) \to \mathbb{P}_1(\mathbb{C}) \times \mathbb{C}(x) \to \mathbb{C}(x)$$

where :

$\mathbb{P}_1(\mathbb{C}) \times \mathbb{C}(x)$ is obtained from $\mathbb{C}(y) \times \mathbb{C}(x)$ by compactificatior the y-plane;

$T(\mathbb{P}_1(\mathbb{C}))$ is the holomorphic tangent bundle to $\mathbb{P}_1(\mathbb{C})$;

$\mathbb{P}T(\mathbb{P}_1(\mathbb{C}))$ is the projective tangent bundle to $\mathbb{P}_1(\mathbb{C})$.

The last bundle is covered by the following coordination neighbourhoods :

$$O_1 : (x,y,s) \ , \ s = \frac{dy}{dx}$$

$$O_2 : (x,Y,s) \ , \ yY = 1$$

$$O_3 : (x,Y,S) \ , \ S = \frac{dY}{dx}$$

$$O_4 : (x,Y,T) \ , \ TS = 1$$

Our surface is given in $\mathbb{P}T(\mathbb{P}_1(\mathbb{C})) \times \mathbb{C}$ by the equations

$$s^2 = (1-y^2)(1-k^2y^2) \quad \text{in } O_1$$

and

$$S^2 = (Y^2-1)(Y^2-k^2) \quad \text{in } O_3$$

and is contained in $O_1 \cup O_3$. The equation on this surface is now given by

$$dy - s dx = 0 \quad \text{in } O_1 \quad \text{and} \quad dY - S dx = 0 \quad \text{in } O_3$$

Now we see that there are no singularities for the equation on our surface but the foliation defined by our equation is not transversal to the fibration. Put

$$f(y,s) = -s^2 + (1-y^2)(1-k^2y^2)$$

$$F(Y,S) = -S^2 + (Y^2-1)(Y^2-k^2)$$

and look at

$$(dy - sdx) \wedge df = s\,\tilde{\omega}_1$$

$$(dY - Sdx) \wedge dF = S\,\tilde{\omega}_2$$

$$\tilde{\omega}_1 = -2dy \wedge ds + 2s dx \wedge ds + 2y[(1-k^2y^2) + k^2(1-y^2)]dx \wedge dy$$

$$\tilde{\omega}_2 = -2dY \wedge dS + 2S\, dx \wedge dS - 2Y(2Y^2 - k^2 - 1)dx \wedge dY$$

Now it is easy to see that

a) $\tilde{\omega}_1 = 0$, $\tilde{\omega}_2 = 0$ define our foliation (we have lost, $s = 0$ and $S = 0$);

b) $\tilde{\omega}_1$, $\tilde{\omega}_2$ have no singularities on the surface.

Further more

$$\tilde{\omega}_1 \wedge dx = -2dy\, ds \neq 0$$

$$\tilde{\omega}_2 \wedge dx \neq 0$$

So our foliation is now transverse to a fibration with compact fiber, and Ehre theorem implies the uniformity of the elleptic functions.

3.4. Equations $F(x,y,y') = 0$ without movable singular points

For an analytic study see [5].

3.5. Uniformity of the hyperelliptic functions

Exercise.

3.6. Geometric study of Briot's and Bouquet's equations (see [9])

One of the problems is the following : determine all the algebraic different equations of the form $F(y,y') = 0$ with only meromorphic solutions. It is a possible to give in a geometric way the classical results of Briot and Bouq concerning the periodicity of the solutions of $F(y,y') = 0$.

3.7. On abelian functions

Problem : Determine by using geometry the polynomials F_1, F_2, F_3, F_4 such tha the pfaffian system :

$$F_1(z,t,p,q,p_1,q_1) = 0 \quad , \quad F_2(z,t,p,q,p_1,q_1) = 0$$

$$F_3(z,t,p,q,p_1,q_1) = 0 \quad , \quad F_4(z,t,p,q,p_1,q_1) = 0$$

$$dz = p\,dx + q\,dy \quad , \quad dt = p_1 dx + q_1 dy$$

is completely integrable and has only uniform solutions.

3.8. The first Painlevé equation

It would be very interesting to have a geometric proof of the uniformity of transcendental function of Paul Painlevé which is solution of the different equation:

$$y'' = 6y^2 + x \quad \text{(first Painlevé equation)}$$

This equation can be replaced by the following pfaffian system :

$$\begin{cases} dz = (6y^2 + x)dx \\ dy - zdx = 0 \end{cases} \quad \text{on} \quad \mathbb{C}^2(y,z) \times \mathbb{C}(x)$$

Now the classical compactifications of $\mathbb{C}^2$ into $\mathbb{P}_2(\mathbb{C})$ or $\mathbb{P}_1(\mathbb{C}) \times \mathbb{P}_1(\mathbb{C})$ or an analogous compactification as those used for the elliptic functions does not lead to a nice geometric situation allowing the use of Ehresmann's theorem. It would be very interesting to give a geometric interpretation of the transformation used by P. Painlevé in the proof of the uniformity of the solutions of the equation $y'' = 6y^2 + x$ (see [10]).

Part II

PAINLEVE'S FOLIATIONS ON A COMPLEX MANIFOLD

1. DEFINITIONS AND EXAMPLES

Let E and B be two connected topological manifolds of finite dimension and $\pi : E \to B$ a continous projection.

Definition 1. A foliation $\mathfrak{F}$ in E is called simple for the projection if for each point m E there exists a distinguished neighbourhood of m such that the "plaque" of m and the fiber of m $(\pi^{-1}(\pi(m))$ cut themself only in the point m.

Examples

1) If $\pi : E \to B$ is a fibration and $\mathfrak{F}$ a foliation which is transversal to the fibration then $\mathfrak{F}$ is simple for the projection π.

2) If $\pi : E \to B$ is a covering then any foliation in E is simple for π.

3) Let

$$\omega = P(x,y)dx + Q(x,y)dy = 0 \quad \text{a pfaffian equation in} \quad C^2$$

where $P(x,y)$ and $Q(x,y)$ are polynomials relatively prime. Let S be the of singular points of ω and suppose that $\omega = 0$ has no integral curves in every fiber $x = Cte$. Then $\omega = 0$ defines in $C^2 - S$ a foliation which is simple for the projection on the x-plane if we are restricted to the fibrat

$$\pi : C^2 - \pi^{-1}(\pi(S)) \to C(x) = \pi(S)$$

4) Other examples can be given by completely integrable pfaffian systems.

Definition 2. A foliation $\mathcal{F}$ in E is a Painlevé foliation of the first type for the projection π if for each path $\{\ell,[0,1]\}$ in B and each point $m \in \pi^{-1}(\ell(0))$ this path can be lifted into the leaf of m into a path start at m. An important consequence of this definition is that if $\mathcal{F}$ is a Painlevé foliation of the first type for $\pi : E \to B$ then the restriction of π to each leaf is surjective.

Examples

1) Each foliation transversal to a fibration with compact fiber and of the sar dimension as the basis is a Painlevé foliation of the first type.

2) In R^2 : $\frac{d\rho}{d\theta} = \rho(1-\rho^2)$ ρ,θ, polar coordinates in R^2 define a Painlevé foliation of the first type for the radial projection of $R^2 - \{0\}$ onto S^1.

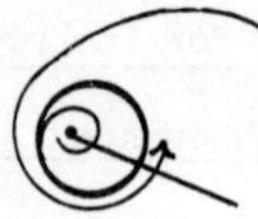

3) In R^2 the foliation represented in the picture is not a Painlevé foliat of the first type for the canonical projection on the x-plane and also not for the canonical projection on the y-plane

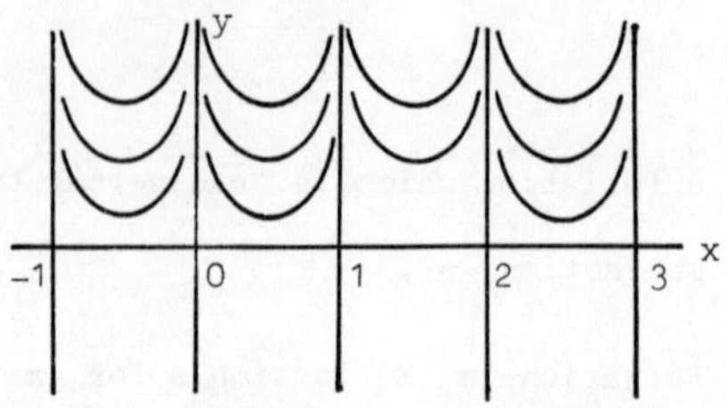

4) As we shall see later, each foliation defined by a differential equation in the complex domain is nearly a Painlevé foliation for a good projection.

Now we are going to define Painlevé's foliations of the second type.

Let $\pi : E \to B$ a continous projection

$S \neq \emptyset$ a subset of E such that for all x in B, $\pi^{-1}(x) \cap S$ has only isolated points in $\pi^{-1}(x)$.

$\mathfrak{F}$ is a foliation in $E - S$ of the same dimension as B.

Definition 3. $\mathfrak{F}$ is a Painlevé foliation of the second type if for each path $\{\ell, [0,1[\}$ in B and each point $m \in (\pi^{-1}(\ell(0)) - S \cap \pi^{-1}(\ell(0)))$ the path $\{\ell, [0,1[\}$ can be lifted in the leaf of m into a path starting at m and the path $\{\ell[0,1]\}$ (closed) in the closure of this leaf.

As a consequence we have for such a foliation : the restriction of π to the closure of a leaf is surjective.

Examples

1) $xdx + ydy = 0$ in $\mathbb{C}^2$ defines in $\mathbb{C}^2 - (0,0)$ a Painlevé foliation of the second type for the canonical projection on the x-plane.

2) $zdz + ydy + (z^2 + y^2)dx = 0$ defines in $\mathbb{C}^3 - \{y = z = 0\}$ a Painlevé foliation of the second type for the projection on the x-plane.

3) We shall see later that to many differential equations there is associated a Painlevé foliation of the second type.

2. LOCAL PROPERTIES OF SIMPLE FOLIATIONS

Let (E,π,B) be a complex analytic locally trivial fibration and denote by n the dimension of the complex analytic manifold B and by $n+p$ the dimension of the complex analytic manifold E. Consider now a complex analytic foliation of dimension n in E which is simple for the projection π.

Then we have :

Lemma 1. For each point $a \in E$, there exists a distinguished open subset $\Omega_a \ni a$ such that each "plaque" in Ω_a meets each fiber going through Ω_a.

Idea of the proof

(E,π,B) is a locally trivial fibration and the result of the le[mma] is a local result, so it is sufficient to proof the lemma in the following situation :

U a polydisk in $\mathbb{C}^n$

V a polydisk in $\mathbb{C}^p$

$\tilde{\mathfrak{F}}$ a complex analytic foliation in $U \times V$ which is simple for the projection p

$a = (o,o)$

Let Ω_o be distinguished open neighbourhood of (o,o) such that the fiber of (o,o) meets the "plaque" of (o,o) only in the point (o,o).
Now Ω_o is isomorphic to a product $U_1 \times V_1$ of two polydisks $U_1 \subset \mathbb{C}^n$ and $V_1 \subset \mathbb{C}^p$ such that

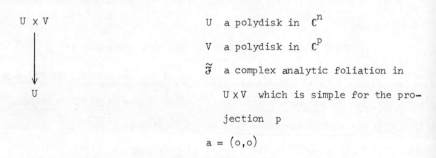

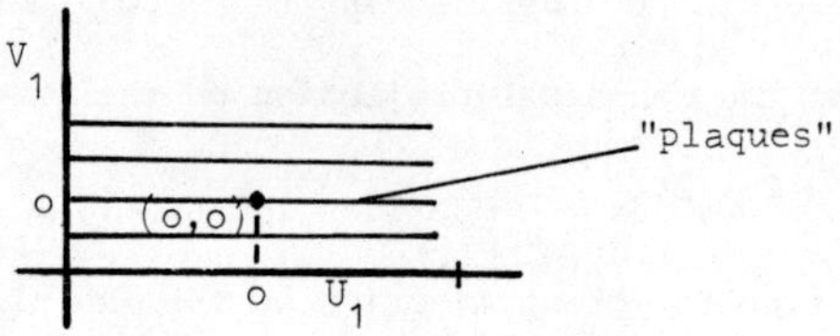

Denote by
$$h_o : U \times V \to U_1 \times V_1$$
$$(x,y) \to (x_1,y_1)$$

the application associated to the isomorphism.

Then h_o^{-1} has equations:

$$x = h(x_1,y_1) \qquad h_o^{-1}(o,o) = (o,o)$$
$$y = k(x_1,y_1)$$

A "plaque" in Ω_o is given by

$$x = h(x_1,y_1^o) \qquad y = k(x,y_1^o)$$

As $\tilde{\mathfrak{F}}$ is simple for the projection on U the system

$$o = h(x_1, o)$$
$$y = k(x_1, o)$$

has the unique solution $x_1 = o, y = o$.

Now look at the following product :

$$\begin{array}{c} U \times V_1 \times U_1 \times V \\ (x, y_1) \; \Big\downarrow \tilde{p} \; (x_1, y) \\ U \times V \\ (x, y_1) \end{array}$$

and the analytic set G defined in this product by

$$x = h(x, y_1)$$
$$y = k(x_1, y_1)$$

The system

$$o = h(x_1, o)$$
$$y = k(x_1, o)$$

aving the unique solution $x_1 = o, y = o$; $p^{-1}(o,o)$ intersects G only in (o,o) and then the theorem of Remmert-Stein [11] implies that there exist two polydisks $U' \subset U$, $V_1' \subset V_1$ such that $G \cap \tilde{p}^{-1}(U' \times V_1')$ is a finite ramified covering of $U' \times V_1'$. Then $h_o^{-1}(U' \times V_1')$ has the property announced in the lemma.

emma 2. If U is a polydisk in C^n , V a polydisk in C^p , $\mathfrak{F}$ a simple fo-liation in $U \times V \to U$ of dimension n , then every germ of paths at $0 \in U$ can e lifted into the leaf of (o,o) in $U \times V$.

f the leaf of (0,0) is transversal to the fiber of (0,0) then the result s trivial. In the other case the assertion follows immediately from the following acts :

1) in a distinguished neighbourhood of (0,0) the "plaque" of 0,0) is an analytic subset;

2) the projection $p : U \times V \to U$ is a "good" projection to apply

the theorem of Remmert-Stein to the "plaque" of (0,0) which is over a neighbourhood of $0 \in U$ a finite ramified covering.

3. THE MAIN THEOREMS ON PAINLEVE'S FOLIATIONS

The lemmas of §2 imply immediately the following theorem.

Theorem 1. Let (E,π,B) be a complex analytic locally trivial fibration, $\mathfrak{F}$ a simple foliation in E satisfying dim. $\mathfrak{F}$ = dim. B . Then for each germ of pa in B at a point $a \in B$ and each point $m \in \pi^{-1}(a)$, this germ is liftable ir the leaf of m and the number of possibilities is finite.

Now we are able to state a generalization of a theorem due to Paul Painlevé i the situation associated to a differential equation.

Theorem 2. If (E,π,B) is a locally trivial complex analytic fibration and $\mathfrak{F}$ a foliation in E which is simple and having the same dimension as B; an moreover the fibration has compact fibers, then $\mathfrak{F}$ is a Painlevé's foliation the first type.

Remark. This result remains true if π is only a proper submersion.

Proof. The compacticity of the fiber and Lemma 1 give us immediately the fol fact : if you can lift the path $\{\ell,[0,\varepsilon[\}$ then you can lift the closed path $\{\ell,[0,\varepsilon]\}$ in any leaf going through a point of $\pi^{-1}(\ell(0))$. Then with this pr perty and the local lifting theorem it is easy to conclude (for details see [12]).

As a consequence of this result we have :

"""" Theorem (P. Painlevé). Une intégrale $y(x)$ de $F(y',y,x) = 0$ polynôme en y',y à coefficients analytiques en x ne peut admettre comme points sir liers non algébriques que certains points fixes $x = \xi$ qui se mettent en évidence sur l'equation même.""""

Let us give the proof of this result for an equation of the following form :

$y' = \frac{P(x,y)}{Q(x,y)}$ where P and Q are polynomial relatively prime.

The general case is given as an exercise.

We write the associated pfaffian equation

$$Q(x,y)dy - P(x,y)dx = 0$$

and extend this equation to an equation $\tilde{\omega} = 0$

on $\mathbb{P}_1(\mathbb{C}) \times \mathbb{C}(x) \to \mathbb{C}(x)$

Denote by :

S_1 the set of singular points of $\tilde{\omega}$;

S_2 the subset of $\mathbb{P}_1(\mathbb{C}) \times \mathbb{C}$ which is the union of vertical leaves for the projection on $\mathbb{C}(x)$ (leaves of the foliation defined in $\mathbb{P}_1(\mathbb{C}) \times \mathbb{C} - S_1$ by $\tilde{\omega} = 0$) .

$\Xi = \pi(S_1) \cup \pi(S_2)$ (this is a discrete subset of $\mathbb{C}(x)$).

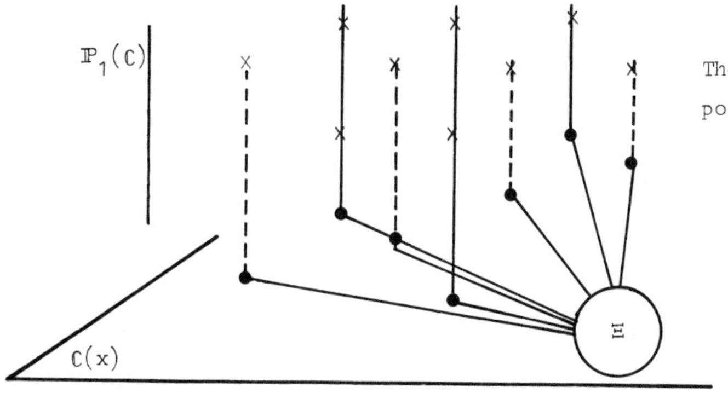

The points of Ξ are the points ξ of P. Painlevé.

Let $\mathfrak{F}$ be the foliation defined by $\tilde{\omega} = 0$ in

$$\mathbb{P}_1(\mathbb{C}) \times (\mathbb{C} - \Xi) \to \mathbb{C}(x) - \Xi$$

This foliation is simple for this trivial fibration (local existence and uniq ness theorem for a differential equation). So by theorem 2, $\mathfrak{F}$ is a Painle foliation of the first type and theorem 1 says that the movable singularities are at most algebraic ; each "plaque" is locally a finite analytic ramified covering of an open subset of the basis.

As a consequence of the theorems 1 and 2, the result of P. Painlevé is also t for certain pfaffian systems in which this general geometric situation arises In the general situation (E,π,B) with compact fiber and $\mathfrak{F}$ a simple foliat in E, it is also possible to give a generalization of:

""""Theorème (P. Painlevé). Soit y_o la valeur de $y(x)$ pour $x = x_o$ et so $y = \varphi(x,y_o,x_o)$ l'intégrale générale de $F(y',y,x) = 0$. Si $\bar{x}$ et $\bar{x}_o$ sont valeurs numériques quelconques distinctes des valeurs ξ la fonction $y = \varphi(\bar{x},y_o,$ ne présente dans le plan des y_o que des points singuliers algébriques.""""

Let (E,π,B) be a locally trivial complex analytic fibration with compact fiber , S an analytic subset of E which meets each fiber only in isolated points , $\mathfrak{F}$ a complex analytic foliation in $E - S$ which is simple f the projection π and whose dimension is the same as the dimension of B .

Theorem 3. $\mathfrak{F}$ is a Painlevé foliation of second type.

The only thing to prove is the following: if $\{\ell,[0,\varepsilon[\}$ $(\varepsilon < 1)$ is a path in B which is such that $\{\ell,[0,\varepsilon[\}$ can be lifted in the leaf of the point $m \in \pi^{-1}(\ell(0))$ into a path $\{\tilde{\ell},[0,\varepsilon[\}$ then $\{\ell,[0,\varepsilon]\}$ can be lifted in the c sure of the leaf of m ; that means $\hat{\ell}(\varepsilon) \in \overline{F(m)}$.

Using the compacticity of the fiber, there is a point $b \in \pi^{-1}(\ell(\varepsilon))$ and belongi to $\overline{\tilde{\ell}([0,\varepsilon[)}$.

Now three cases are possible :

1) $b \in E - S$; then the lemma 2 gives the conclusion

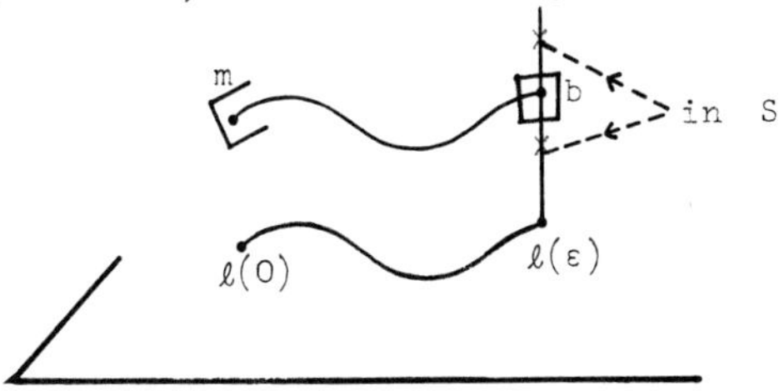

2) $b \in S$ and $\overline{\tilde{\ell}([0,\varepsilon[)} \cap \pi^{-1}(\ell(\varepsilon)) = \{b\}$

then the result is also true and

$\ell(\varepsilon) \in \overline{F(m)} - F(m)$, where $F(m)$ denotes the leaf of m

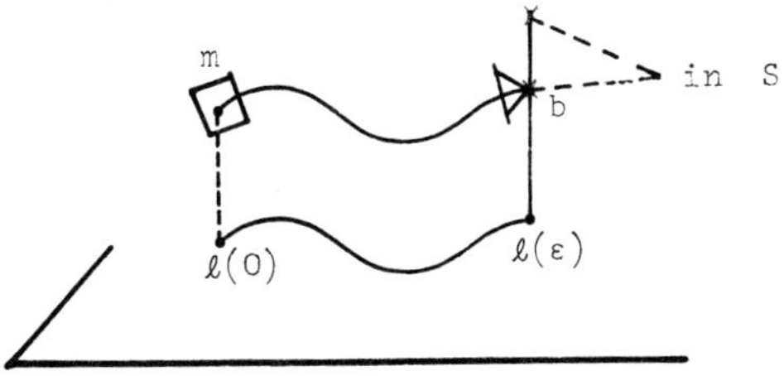

3) $b \in S$ and $\overline{\tilde{\ell}([0,\varepsilon[} \cap \pi^{-1}(\ell(\varepsilon)) \subset S$ has at least two points b and b' . But this hypothesis implies, as it is easy to see, that $S \wedge \pi^{-1}(\ell(\varepsilon))$ has non-isolated points (see the picture) .

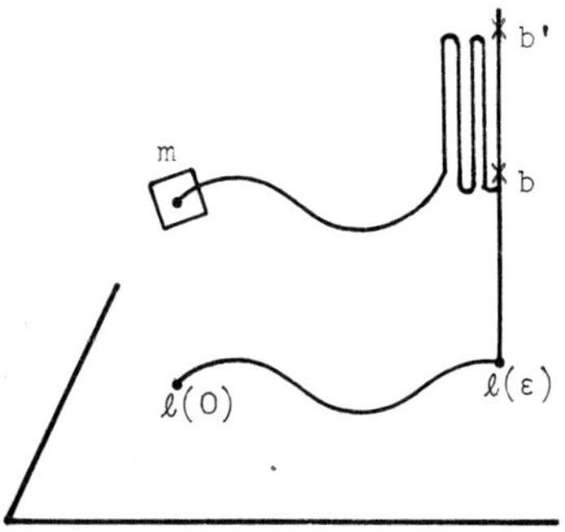

There are now infinite points of S in the fiber over $\ell(\varepsilon)$ because if a point $c \in \tilde{\ell}[0,\varepsilon[\cap \pi^{-1}(\ell(\varepsilon))$ is not in S we can apply 1), and b is not in the closure of $\tilde{\ell}[0,\varepsilon[$ which in contradiction with our hypothesis.

4. APPLICATIONS

4.1. Singularities of the solution of a differential equation

For simplicity we look at a differential equation of the following form :

$$\frac{dy}{dx} = \frac{P(x,y)}{Q(x,y)}$$ where P and Q are polynomials relatively prime

Let $\tilde{\omega} = 0$ be the extended associated pfaffian equation to $\mathbb{P}_1(\mathbb{C}) \times \mathbb{C}$.

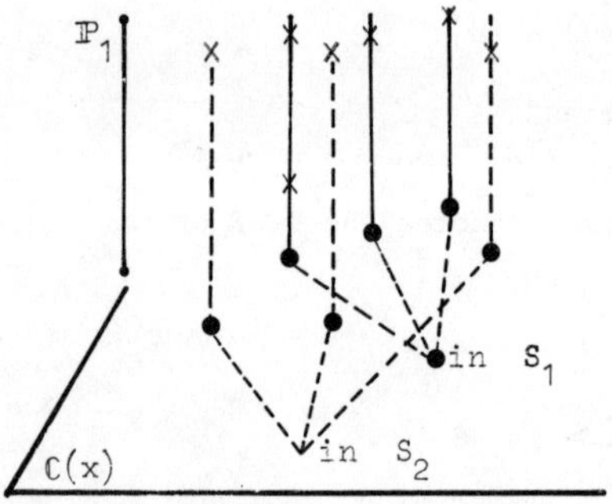

Denote by :

S the set of singular points of $\tilde{\omega}$;

S_1 the union of the projection of vertical solutions;

S_2 the set of projection of the singu points of $\tilde{\omega}$ which are not in S

In $\mathbb{P}_1(\mathbb{C}) \times (\mathbb{C}(x) - S_1) - S$, the pfaffian equation $\tilde{\omega} = 0$ gives us a foliation which is simple for the projection on $\mathbb{C}(x)$. As S ha the property given in theorem 4, the foliation is of the second type. This means analytically that if $y = \varphi(x)$ is a solution of our equation, th solution is in general a multivalued function which has a limit along each p going from a regular point x_o to a point $x_1 \in S_2$. This limit is the same all homotopic paths in $\mathbb{C}(x) - S_1 \cup S_2$. Such a point is called a transcendenta point by P. Painlevé. Now let us see what happens when we try to lift a pat in $\mathbb{C}(x) - S_1 \cup S_2$ with extremity in S_1 . There are two possibilities:

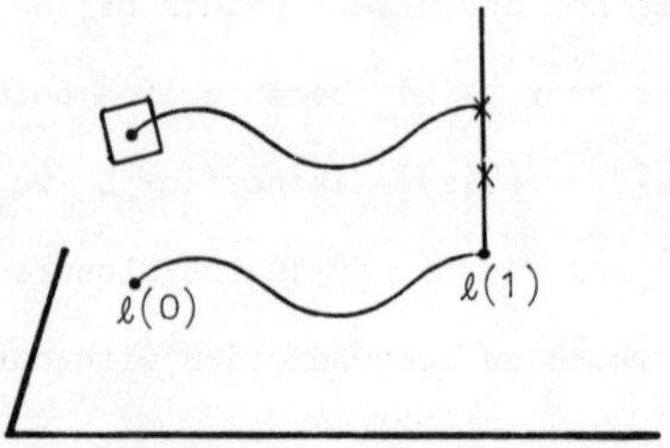

1) In $\overline{\ell[0,1[}$ there is a point of S the point x_1 is at most a transcenden singular point for the considered solut of our equation.

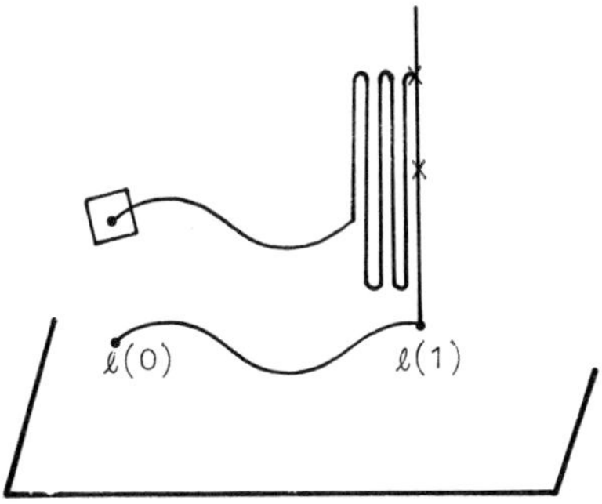

2) In $\overline{\mathcal{L}[0,1[}$ there is a point of $\pi^{-1}(x_1) - S$: then if we look at the foliation in $\mathbb{P}_1(\mathbb{C}) \times \mathbb{C}(x) - S$, $\pi^{-1}(x_1) - S \cap \pi^{-1}(x_1)$ is a leaf of this foliation and then the whole leaf is in the closure of the leaf of m .

This means that our considered solution has at the point x_1 an essential singular point.

As a corollary of the preceding considerations we have an analogy to Picard's big theorem on entire functions for the solutions of an algebraic differential equation : If a solution of the differential equation

$$y' = \frac{P(x,y)}{Q(x,y)}$$

has an essential singular point at x_1 then it takes near x_1 all values except perhaps the values y which are solutions of $P(x_1,y) = 0$.

For an analytic study of this question see [13] and [14].

4.2. Differential system

Consider a differential system of the following form

$$\frac{dy}{dx} = F_1(x,y,z) \qquad \frac{dz}{dx} = F_2(x,y,z)$$

where F_1 and F_2 are rational functions.

As it is done in the work of P. Painlevé, we consider the associated system in $\mathbb{P}_2(\mathbb{C}) \times \mathbb{C}(x)$:

$$(\Sigma)\ \frac{dx}{X} = \frac{tdy - ydt}{tA - yC} = \frac{tdz - zdt}{tB - zC}$$

where X,A,B,C are homogenous polynomials in y,z,t . Denote by q the degree of X; then A,B,C are of degree q + 1 . (In a little more general situation the coefficients of these polynomials are holomorphic in x) . In his lecture at Stockholm, Painlevé proves the following result :

""

I) Si les polynômes X, A, B, C sont les plus généraux de leur degré l'intégrale générale

$$y = \varphi(x, y_o, z_o, x_o) \ , \ z = \psi(x, y_o, z_o, x_o)$$

ne peut admettre de singularités mobiles non algébriques.

II) Pour que l'intégrale générale $y = \varphi$, $z = \psi$ admette des singularités transcendantes mobiles, il faut (mais il ne suffit pas) que les égalités

$$X = 0 \ , \ tA - yC = 0 \ , \ yB - zC = 0 \ , \ zC - tB = 0$$

soient compatibles (quelque soit x) pour les valeurs y, z, t qui ne soient pas toutes nulles.

III) Pour que l'intégrale générale $y = \varphi$, $z = \psi$ admette des singularités essentielles mobiles, il faut (mais il ne suffit pas que le polynôme $X(y,z,t,x)$ (ou l'un de ses diviseurs $X_1(y,z,t,$ définisse une intégrale première particularisée.

""

Now we are giving a geometric interpretation of these results.

Denote by

$\pi \ : \mathbb{P}_2(\mathbb{C}) \times \mathbb{C}(x) \to \mathbb{C}(x)$

S_1 : the set of singular points in $\mathbb{P}_2(\mathbb{C}) \times \mathbb{C}$ for the field of directions given by our system ;

S_1' : the set of points (y,z,t,x) S_1 such that $\pi^{-1}(x) \cap S_1$ has no isolated points ;

S_2 : the union of S_1' and of the integral manifolds which are in a fiber $x = Cte$.

The projections $\pi(S_1)$ and $\pi(S_2)$ are analytic subsets of $\mathbb{C}$. So we have three cases :

I) $\pi(S_1) \neq \mathbb{C}$ and $\pi(S_2) \neq \mathbb{C}$ (generic case)

II) $\pi(S_1) = \mathbb{C}$ and $\pi(S_2) \neq \mathbb{C}$

III) $\pi(S_2) = \mathbb{C}$ which implies $\pi(S_1) = \mathbb{C}$.

Geometrically these three cases give us :

I) The foliation associated to (Σ) in

$$(\mathbb{C} - \pi(S_1 \times S_2)) \times P_2(\mathbb{C}) \to \mathbb{C} - \pi(S_1 \cup S_2)$$

is a Painlevé foliation of the first type and as a consequence the movable singularities are at most algebraic.

II) The foliation associated to the system in

$$(\mathbb{C} - \pi(S_2)) \times \mathbb{P}_2(\mathbb{C}) - S_1 \to \mathbb{C} - \pi(S_2)$$

is a Painlevé foliation of the second type. This means that the movable singularities are at most transcendental singular points.

III) Movable essential singular points can occur.

4.3. As an exercise prove geometrically the following result of P. Painlevé on second-order differential equations of the form :

$$y'' = \frac{P(y',y,x)}{Q(y',y,x)} \tag{1}$$

where P and Q are polynomials in y', y with holomorphic coefficients in x and which are relatively prime. Denote by p the degree of P and q the degree of Q. Assume that $p \geq q + 2$ (this is in the case of the general situation).

The result of P. Painlevé:

"" I) Si $Q(y',y,x)$ n'admet pas de zéro $y = G(x)$ indépendant de et si $p > q+2$, l'intégrale générale $y(x)$ de (1) et sa dérivé $y'(x)$ ne peuvent présenter de singularités essentielles mobiles

II) Si $Q(y',y,x)$ admet au moins un zéro $y = G(x)$ indépendant y' et si $p > q+2$, l'intégrale générale $y(x)$ de (1) ne peut senter de singularités essentielles mobiles. Mais $y'(x)$ peut e présenter.

III) Si $Q(y',y,x)$ n'admet pas de zéro indépendant de y' mais $p = q+2$, la fonction $y'(x)$ peut présenter des singularités essentielles mobiles mais $y(x)$ n'en présente pas.

IV) Si $Q(y',y,x)$ admet au moins un zéro indépendant de y' et $p = q+2$, les fonctions $y(x)$ et $y'(x)$ peuvent présenter des singularités essentielles mobiles.

Hint: Look at the associated system in

$$\mathbb{P}_1(\mathbb{C}) \times P_1(\mathbb{C}) \times \mathbb{C}(x) \to \mathbb{C}(x)$$

It is easy to see how to generalize many results of Painlevé to pfaffian systems.

Part III

ALGEBRAIC PFAFFIAN EQUATIONS ON A PROJECTIVE SPACE

1. DEFINITIONS AND EXAMPLES

Consider in $\mathbb{C}^{n+1}$ $(n \geq 2)$ a differential form

$$\omega = \sum_{i=1}^{n+1} \omega^i \, dx_i$$

satisfying the following conditions :

1) for all i, ω^i is a homogenous polynomial of degree m ;

2) $$\sum_{i=1}^{n+1} x_i \omega^i = 0$$

The second condition means that the integral manifolds of $\omega = 0$ in $C^{n+1}\{0\}$ are cones with top at the origin of C^{n+1}. If we denote by X the vector field

$$\sum_{i=1}^{n+1} x_i \frac{\partial}{\partial x_i}$$

then

$$\sum_{i=1}^{n+1} x_i \omega^i = \omega \lrcorner X \quad \text{(interior product)}$$

An algebraic pfaffian equation on $\mathbb{P}_n(C)$ of degree m is a class modulo multiplication by elements of C^* of differential forms satisfying the two conditions 1) and 2).

The equation is called irreducible if the greatest common divisor of the ω^i's is 1 ; this means that the singular locus of ω in $P_n(C)$ contains no hypersurfaces. It is easy to see that if S is the singular set of ω (the set of common zeros of the ω^i) then $S \neq \emptyset$ unless $n = 2p+1$ and $m = 1$.

The pfaffian equation is called completely integrable if the equation $\omega = 0$ is completely integrable in C^{n+1}.

Example.

$$\omega = xdy - ydx + ydz - zdy + zdt - tdz = 0$$

is a completely integrable pfaffian equation on $\mathbb{P}_n(C)$ and $S = \emptyset$.

Proposition 1. *If $\omega = 0$ is a completely integrable algebraic pfaffian equation on $\mathbb{P}_n(C)$ then the singular set S contains always an irreducible component of dimension $n-2$.*

For the complete proof see [15]. In the general case the result is a conseq
of a theorem of Bott [16].

Proof in a particular case

It is sufficient to prove the result for the one-form ω in $\mathbb{C}^{n+1}$.
Denote by S the singular locus of ω. It is known that at each point $\xi \in$
at which $d\omega(\xi) \neq 0$ an irreducible component of S of codimension smaller
equal to two goes through this point. So it is sufficient to prove that $d\omega$
is not identically zero on S. If $d\omega$ is identically zero on S we must
the "Nullstellensatz" of Hilbert a relation of the form

$$\text{for all } i,j \qquad \left(\frac{\partial\omega^i}{\partial x_j} - \frac{\partial\omega^j}{\partial x_i}\right)^{n_{ij}} \in (\omega_1,\omega_2,\ldots,\omega_{n+1})$$

where, n_{ij} is for all i,j an integer greater or equal to 1 and
$(\omega_1,\omega_2,\ldots,\omega_{n+1})$ is the ideal generated by the ω^i's.
But in the case where S is reduced we must have

$$\frac{\partial\omega^i}{\partial x_j} = \frac{\partial\omega^j}{\partial x_i} \qquad \text{for all } i,j$$

So there exists a homogenous polynomial g of degree $m+1$ such that

$$\omega = dg$$

But this is wrong because of

$$0 = \omega \lrcorner X = dg \lrcorner X = (m+1)g \quad \text{(Euler's formula)}$$

An algebraic solution of codimension one of the algebraic equation $\omega = 0$
on $\mathbb{P}_n(\mathbb{C})$ is by definition an irreducible homogenous polynomial f such
that

$$df \wedge \omega = 0 \quad \text{on} \quad f = 0$$

Theorem 1. Assume $n \geq 2$. Then

a) The equation $\omega = 0$ in $\mathbb{P}_n(\mathbb{C}) - S$ has no compact integral manifold of codimension one.

b) If codim. $S \geq 2$ each algebraic solution Z of codimension one of $\omega = 0$ whose singular points are in S contains an irreducible component of codimension 2 of S . In particular, if codim. $S \geq 3$ the equation $\omega = 0$ has no algebraic solution of codimension one which is regular outside of S .

c) If codim. $S \geq 2$, each normal algebraic solution Z of $\omega = 0$ is of degree smaller or equal to the degree of ω .

Corollary. If ω is completely integrable then the foliation defined by $\omega = 0$ in $\mathbb{P}_n(\mathbb{C}) - S$ has no compact leaves.

Proof of the theorem

a) Let V be a compact integral manifold of $\omega = 0$; this is a connected regular algebraic submanifold of $\mathbb{P}_n(\mathbb{C})$ and so is defined by a homogenous polynomial f . Denote by γ the cone defined by $f = 0$ in $\mathbb{C}^{n+1}$ and $\gamma^* = \gamma - \{0\}$; γ is a complete intersection and is regular outside the top of the cone, which is of codimension ≥ 2 . This means that γ is normal.

As

$$df \neq 0 \quad \text{at each point of } V$$

$$\omega \neq 0 \quad \text{at each point of } V$$

and V being an integral manifold, we have

$$\overline{df} \wedge \overline{\omega} = 0$$

where $\quad \begin{matrix}\overline{df} \\ \overline{\omega}\end{matrix} : \mathcal{O}_{\gamma^*} \to \Omega^1_{\gamma^*} \otimes_{\mathcal{O}_{\mathbb{C}^{n+1}}} \mathcal{O}_{\gamma^*}$

are the induced morphisms.

This implies that there exists an invertible element $a \in \Gamma(\gamma^*, O_{\gamma^*})$ such that

$$\overline{\omega} = a\overline{df}$$

The cone γ being normal, the section a can be extended to an invertible ele also denoted by a of the $\mathbb{C}$ - algebra $\Gamma(\gamma, O_\gamma)$; that means to an element $a \in$ The irreducibility of f implies that

$$\omega = a\, df + h f \quad , \quad a \in \mathbb{C}^*$$

where h is a polynomial one-form ; but the degree of f is greater then th degree of df and this implies $h = 0$.
Now we have $\omega = a\, df$ and the Euler formula gives us the following contradiction :

$$0 = \omega \lrcorner X = a(\deg. f)f$$

b) The only remark we have is that

$$\text{codim.}(S \cap Z, Z) \geq 2$$

which implies that the cone associated to Z in $\mathbb{C}^{n+1}$ is also normal and the proof follows as in case a).

c) The hypothesis on S and Z implies easily that

$$\omega = a\, df + fh$$

where

$$a \in \mathbb{C}[x_1, x_2, \ldots, x_{n+1}]$$

and h is a polynomial one - form.

If we have deg. $f > m$ then fh has no homogenous components of degree m ; th implies that $\omega = a\, df$, which contradicts Euler's formula.
The following more general result is also true.

<u>Theorem 2.</u> <u>Let</u> Y <u>be a Stein manifold and</u> E <u>a complex analytic vector bundl</u> <u>on</u> Y <u>and</u> $\pi : P(E) \to Y$ <u>the associated projective bundle. Then every complex</u>

analytic foliation $\mathfrak{F}$ of positive dimension in an open subset U of $P(E)$ has no compact leaf.

The proof of this theorem follows from the following lemma.

Lemma 1 [17]. If X is a smooth compact submanifold of dimension $m > 0$ of $\mathbb{P}_n(\mathbb{C})$ then there does not exist on X a complex analytic vector bundle which is both ample and flat.

Idea of the proof of theorem 2

If X is a compact leaf, X is in a fiber because Y is a Stein manifold. Now we can restrict ourselves to the following situation :

$$\pi : \mathbb{P}_n(\mathbb{C}) \times V \to V$$

where V is an open subset of $\mathbb{C}^r$ containing 0, and moreover we can assume that $X \subset \mathbb{P}_n(\mathbb{C}) \times \{0\}$. If N is the normal bundle of X in $\mathbb{P}_n(\mathbb{C})$, since X is smooth and projective, N is ample (see [18] p. 105, example 2); then det.N is also ample (see [19] p. 68, corollary 2.6).

But a theorem of Ehresmann (see [1] p. 384, theorem 2.8) implies that the normal bundle $\bar{N}$ to X in $\mathbb{P}_n(\mathbb{C}) \times V$ is flat.

But the exact sequence

$$0 \to N \to \bar{N} \to X \times \mathbb{C}^r \to 0$$

implies that det. N is isomorphic to det. $\bar{N}$ and so det. N is also flat, which is a contradiction to the result of the lemma.

Definition. Let U be an open subset of a complex analytic manifold Z and $\mathfrak{F}$ a complex analytic foliation in U. A compact solution of $\mathfrak{F}$ is by definition a compact analytic subset A of Z such that $A \cap U$ is a leaf of the foliation.

Lemma 2. (for the proof see [17]). Let S be a closed analytic subset of $\mathbb{P}_n(\mathbb{C})$ and $U = \mathbb{P}_n(\mathbb{C}) - S$. Then there does not exist a closed algebraic submanifold X

of $\mathbb{P}_n(\mathbb{C})$ of positive dimension such that $V = U \cap X$ is smooth and satisfies the following two properties :

1) the normal bundle N to V in $\mathbb{P}_n(\mathbb{C})$ is flat and the restr[iction] to V of an ample bundle over $\mathbb{P}_n(\mathbb{C})$;

2) $\mathrm{Codim.}(S \cap X, X) \geq 2$.

From this lemma we deduce the following.

Theorem 3. In the situation of theorem 2 assume that $S = P(E) - U$ is a close[d] analytic subset and that $\mathfrak{F}$ is of codimension one. Then each compact solutio[n] X is in a fiber and

$$\mathrm{codim.}(S \cap X, X) \leq 1$$

In particular, a fiber P_y of a point $y \in Y$ does not contain a compact soluti[on] if $\mathrm{codim.}(S \cap P_y, P_y) \geq 3$.

Proof. We know that X is in a fiber P . Let N be the normal bundle of X in P; then from the exact sequence

$$0 \to N \to \overline{N} \to X \times \mathbb{C}^r \to 0$$

we deduce that N is the restriction of an ample bundle on $\mathbb{P}_n(\mathbb{C})$ and, moreo[ver] this bundle is flat, which contradicts lemma 2.

The theorems 2 and 3 can be generalized to pfaffian systems by using the same lemmas 1 and 2 (for details see [17]).

2. ALGEBRAIC SOLUTIONS OF A PFAFFIAN EQUATION ON $\mathbb{P}_n(\mathbb{C})$

The structure of the set of algebraic solutions of a pfaffian equation on a projective space was first given by Darboux [20] in the case $n = 2$.
Let us consider a pfaffian equation on $\mathbb{P}_n(\mathbb{C})$, i.e. an equation of the following form:

$$\omega = \sum_{i=1}^{n+1} \omega^i dx_i = 0$$

satisfying the conditions;

1) the ω^i's are homogenous polynomials all of the same degree m;

2)
$$\sum_{i=1}^{n+1} x_i \omega^i = \omega \lrcorner X = 0$$

where X is the vector field $\sum_{i=1}^{n+1} x_i \frac{\partial}{\partial x_i}$

We have :

Theorem [15]. For $\omega = 0$ there are only two possibilities:

1) All the solutions are algebraic;

2) The equation has only a finite number of algebraic solutions and this number is smaller than

$$N = \frac{1}{2} m(m-1)\binom{m-1+n}{n-2} + 2$$

A proof of this result was given by Darboux in [20] for the case $n = 2$.

Corollary. If we have N algebraic solutions $F_1, F_2, \ldots, F_N$ of $\omega = 0$ then there exist N complex numbers $\alpha_1, \alpha_2, \ldots, \alpha_N$ not all zero such that :

1)
$$\sum_{i=1}^{N} \alpha_i \deg.(F_i) = 0$$

2)
$$\sum_{i=1}^{N} \alpha_i \frac{dF_i}{F_i} \wedge \omega = 0$$

and in particular ω is completely integrable.

Notation

$$A = \mathbb{C}[x_1, x_2, \ldots, x_{n+1}]$$

K is the field of fraction of A;

$\Omega_a^1(\mathbb{C}^{n+1})$ the A-module of algebraic 1-differential forms on $\mathbb{C}^{n+1}$;

$\wedge(\Omega_a^1(\mathbb{C}^{n+1}))$ the exterior algebra associated to Ω_a^1.

I recall that the following complex of Koszul is acyclic:

$$0 \to \overset{n+1}{\wedge}(\Omega_a^1) \xrightarrow{\lrcorner X} \overset{n}{\wedge}(\Omega_a^1) \xrightarrow{\lrcorner X} \dots \to \overset{2}{\wedge}(\Omega_a^1) \xrightarrow{\lrcorner X} \overset{1}{\wedge}(\Omega_a^1) \xrightarrow{\lrcorner X} A \xrightarrow{\varepsilon} \mathbb{C} \to 0$$

where $\lrcorner X$ is the interior product by $X = \sum_{i=1}^{n+1} x_i \frac{\partial}{\partial x_i}$

$$\varepsilon(P) = P(0)$$

$[\overset{i}{\wedge}(\Omega_a^1)]_p$ is the set of elements of $\overset{i}{\wedge}(\Omega_a^1)$ whose coefficients are all of degree p.

A rational integrant factor of $\omega = 0$ is by definition a non-zero rational function $R = \frac{P}{Q}$ homogenous such that

$$dR = h\omega \text{ where } h \text{ is an other rational function.}$$

But the last relation implies that

$$dR \lrcorner X = 0 \text{ since } \omega \lrcorner X = 0 \text{ so } (\deg.R).R = 0$$

and R is homogenous of degree 0.

I recall that an algebraic solution of $\omega = 0$ is by definition an irreducibl homogenous polynomial F such that

$$dF = hF \qquad h \in \overset{2}{\wedge}(\Omega_a^1)$$

Denote by $\mathfrak{a}$ the set of all algebraic solutions of $\omega = 0$ and consider the $\mathbb{C}$-vector space with basis $\mathfrak{a}$: $\mathbb{C}^{\mathfrak{a}}$.

<u>Proposition 2. The $\mathbb{C}$-linear map</u>

$$\psi : \mathbb{C}^{\mathfrak{a}} \longrightarrow \Omega_a^1 \otimes_A K$$

$$u = \Sigma\, \alpha_i \rightsquigarrow \psi(u) = \Sigma\, \alpha_i \frac{dF_i}{F_i}$$

is injective.

We make use of the following lemma

Lemma 2. Let $C_i = \{F_i = 0\}$ $1 \leq i \leq p$ be p different algebraic hypersurfaces in $\mathbb{P}_n(\mathbb{C})$; then there exists a projective line D which cuts the surface C_1 in an isolated point and which is not contained in the other surfaces.

Proof of the proposition

Assume that

$$\Sigma\, \alpha_i \frac{dF_i}{F_i} = 0$$

for some α_i's and that, for example, $\alpha_1 \neq 0$.

Let D be a line given by the lemma for C_1 which has equation $F_1 = 0$, and θ a parametric representation of the line D . If we denote by $\overline{F}_i = F_i \circ \theta$ and z the coordinate on D, we have

$$\begin{aligned} \overline{F}_1 &= \mu_1 (z-a)^t \prod_{b_i \neq a} (z-b_i^1) \\ &\;\vdots \qquad\qquad\qquad\qquad \text{for all } s,\ b_i^s \neq a \\ \overline{F}_p &= \mu_p \prod (z-b_i^p) \end{aligned}$$

Then

$$\Sigma\, \alpha_i \frac{dF_i}{F_i} \circ \theta = \Sigma\, \alpha_i \frac{d\overline{F}_i}{\overline{F}_i} = t\alpha_1 \frac{dz}{z-a} + \left(\Sigma \frac{1}{z-c_k}\right) dz$$

and $\Sigma\, \alpha_i \dfrac{dF_i}{F_i} = 0$ implies $\alpha_1 = 0$, which is in contradiction with our hypothesis.

The proposition 2 allowed us to consider $\mathbb{C}^{\mathfrak{a}}$ as a subvector space of $\Omega^1_a \otimes_A K$. Denote by $e(F)$ the vector of the basis of $\mathbb{C}^{\mathfrak{a}}$ associated to the algebraic solution F and consider now the subvector space W of $\mathbb{C}^{\mathfrak{a}}$ defined by

$$W = \{\Sigma \alpha_i e(F_i) \mid \Sigma \alpha_i (\deg . F_i) = 0\}$$

The map

$$C^{G} \longrightarrow C$$

$$\Sigma\, \alpha_i F_i \rightsquigarrow \Sigma\, \alpha_i \deg.(F_i)$$

is a linear form on C^{G} and W is the kernel of this map and as a consequen
codim. $W = 1$.

Consider now the C - linear map :

$$\varphi : W \longrightarrow [\overset{2}{\wedge}(\Omega^1_a)]\, m-1$$

$$\Sigma\, \alpha_i F_i \rightsquigarrow \omega \wedge \Sigma\, \alpha_i \frac{dF_i}{F_i} = \Sigma\, \alpha_i\, \omega \wedge \frac{dF_i}{F_i}$$

Proposition 3

$\varphi(W) \subset [Z_2]_{m-1}$ (space of 2 - cycles of degree $m-1$ of the Koszul comple

It is easy to verify that

$$(\Sigma \alpha_i\, \omega \wedge \frac{dF_i}{F_i}) \,\lrcorner X = 0$$

Proposition 4. For the 1 - form ω there are only two possibilities :

1) it has a rational integrant factor;

or

2) Dim.(Ker. φ) = 1 .

Let $\Sigma\, \alpha_i F_i$ and $\Sigma\, \beta_i G_i$ be two elements of Ker. φ. By adding a finite numbe
of terms we can assume that these elements are of the following form :

$\sum_{i=1}^{s} \alpha_i F_i$ and $\sum_{i=1}^{s} \beta_i F_i$ (with some zero coefficients)

By assumption we have

$$\omega \wedge \Sigma\, \alpha_i \frac{dF_i}{F_i} = 0$$

$$\omega \wedge \Sigma\, \beta_i \frac{dF_i}{F_i} = 0$$

Denote $G = \prod_{i=1}^{s} F_i$ and $G_k = F_1.F_2 \ldots \hat{F}_k \ldots F_s$

Then we have :

$$\omega \wedge \sum_{i=1}^{s} \alpha_i G_i \, dF_i = 0$$

$$\omega \wedge \sum_{i=1}^{s} \beta_i G_i \, dF_i = 0$$

As $\omega \neq 0$ outside S, which is of codimension greater than or equal to two, there exist two polynomials k_α and k_β such that

$$\Sigma\, \alpha_i G_i \, dF_i = k_\alpha \omega \qquad k_\alpha \not\equiv 0$$

$$\Sigma\, \beta_i G_i \, dF_i = k_\beta \omega \qquad k_\beta \not\equiv 0$$

And then we have

$$\Sigma\, \alpha_i \frac{dF_i}{F_i} = \frac{k_\alpha}{G} \omega$$

$$\Sigma\, \beta_i \frac{dF_i}{F_i} = \frac{k_\beta}{G} \omega$$

and by differentiation

$$d\left(\frac{k_\alpha}{G}\right) \wedge \omega + \frac{k_\alpha}{G} \, d\omega = 0$$

$$d\left(\frac{k_\beta}{G}\right) \wedge \omega + \frac{k_\beta}{G} \, d\omega = 0$$

By taking the difference of this two expressions we obtain

$$d\left(\frac{k_\alpha}{k_\beta}\right) \wedge \omega = 0$$

And two cases appear :

1) $\frac{k_\alpha}{k_\beta} \neq$ Cte; then $\frac{k_\alpha}{k_\beta}$ is a rational integrant factor of ω and all

solutions of $\omega = 0$ are algebraic;

2) $\dfrac{k_\alpha}{k_\beta} = c$ (constant)

which implies easily:

$$\Sigma(\alpha_i - c\,\beta_i)\frac{dF_i}{F_i} = 0$$

That means

$$\alpha_i = c\,\beta_i \quad \text{for all} \quad i$$

and the dimension of $\ker.\varphi$ is 1 .

Proof of Darboux's theorem. We have only to study in more detail the case 2.
We have the exact sequence :

$$0 \to \mathrm{Ker}.\varphi \to W \xrightarrow{\varphi} V \to 0$$

where $$V = \varphi(W) \subset [Z_2]_{m-1}$$

Now $\mathrm{Dim}.[Z_2]_{m-1} < +\infty$, $\mathrm{Dim}.W < +\infty$ and since

Codim.W in $\mathbb{C}^{\mathfrak{G}}$ is 1, this implies that $\mathrm{Dim}.\mathbb{C}^{\mathfrak{G}} < +\infty$.

Moreover, $\mathrm{Dim}.\mathbb{C}^{\mathfrak{G}} = \mathrm{Dim}.W + 1$ and $\mathrm{Dim}.W = \mathrm{Dim}.V + 1$; that means $\mathrm{Dim}.\mathbb{C}^{\mathfrak{G}} =$ $\mathrm{Dim}.V + 2$.

But $V \subset [Z_2]_{m-1}$ so $\mathrm{Dim}.\mathbb{C}^{\mathfrak{G}} \leq \mathrm{Dim}\,[Z_2]_{m-1} + 2$

and an easy calculation gives us

$$\mathrm{Dim}.[Z_2]_{m-1} = \frac{1}{2}\,m(m-1)\binom{m-1+n}{n-2}$$

Which ends the proof of Darboux's theorem.

You can find in Darboux's original paper an example of pfaffian equation on $\mathbb{P}_2(\mathbb{C})$ which has exactly N algebraic solutions and no more.

Now let us give an other important result on the foliation defined by a pfaffian equation on $\mathbb{P}_n(\mathbb{C})$. Denote by $\omega = 0$ the pfaffian equation and S its singular set.

Theorem [21]. If all the leaves of the foliation $\mathfrak{F}$ defined in $\mathbb{P}_n(\mathbb{C}) - S$ by $\omega = 0$ are proper then $\omega = 0$ has an algebraic solution.

The proof of this result follows from the following lemma.

Lemma. The set

$$E = \{\overline{F} \mid F \text{ is a leaf of } \mathfrak{F}\}$$

is an inductive set.

The proof of this lemma is very long and make use of Painlevé's foliations.

Idea of the proof of the theorem

Let F be a leaf of the foliation $\mathfrak{F}$ and consider the set

$$E = \{\overline{F} \mid F \text{ is a leaf of } \mathfrak{F}\}$$

This set is inductive by the lemma. This implies that $\overline{F}$ for the considered leaf contains a minimal set, say K. Consider now the end $B(K) = \overline{\overline{K} - K}$ of K.

As K is proper we do not have $B(K) \supset K$.

As K is minimal we do not have

$$B(K) \cap K = \emptyset \quad \text{and} \quad B(K) \cap \mathbb{P}_n(\mathbb{C}) - S \neq \emptyset$$

So $B(K) \subset S$ and $\operatorname{codim} S \leq 2$ implies that K is an algebraic submanifold of $\mathbb{P}_n(\mathbb{C})$ and gives us an algebraic solution of the pfaffian equation $\omega = 0$.

In particular, we have also the following result :

Each leaf of the foliation $\mathfrak{F}$ is algebraic or contains in its closure an algebraic leaf.

BIBLIOGRAPHY

[1] HAEFLIGER, A., Variétés feuilletées, Ann. Sc. Norm. Super. Pisa, Sci. Fis. Mat. **16** (1964) 367–397.

[2] REEB, G., Sur certaines propriétés topologiques des variétés feuilletées, Actualités scientifiques et industrielles, Hermann, Paris (1952).

[3] SUZUKI, M., Sur les relations d'équivalence ouvertes dans les espaces analytiques, Ann. Sc. Ec. Norm. Sup. Paris, 4° série, t.7, fasc. 4 (1974).

[4] KAUPP, B., Über offene analytische Äquivalenzrelationen auf komplexen Räumen, Math. Ann. **183** (1969).

[5] PAINLEVE, P., Œuvres, Edition du CNRS, Tome I (1972).

[6] MARTINET, J., Doctorat de 3° cycle, Grenoble (1962).

[7] NARASIMHAN, R., Analysis on Real and Complex Manifolds, Masson (Paris) 1968.

[8] EHRESMANN, C., "Les connexions infinitésimales dans un espace fibré différentiable", Colloque de topol espaces fibrés, Librairie universitaire, Bruxelles (1950) 31.

[9] DIENER, F., Doctorat de 3° cycle, IRMA, Strasbourg (1974).

[10] PAINLEVE, P., Œuvres, tome III, Editions du centre nationale de la recherche scientifique, Paris (1975).

[11] REMMERT, R., STEIN, K., Über die wesentlichen Singularitäten analytischer Mengen, Math. Ann. **126** (1953) 263–306.

[12] GERARD, R., SEC, A., Feuilletages de Painlevé, Bull. Soc. Math. France **100** (1972) 47–72.

[13] KIMURA, T., Sur les points singuliers des équations différentielles ordinaires du premier ordre, Comment. Math. Univ. Sancti Pauli **2** (1953) 47–53.

[14] KIMURA, T., SIBUYA, Y., Essential singular points of solutions of an algebraic differential equation, Journal für Math. **272** (1973) 127–149.

[15] JOUANOLOU, J.P., Equations de Pfaff algébriques sur un espace projectif, IRMA preprint, Strasbourg (197

[16] BOTT, R., Lectures on characteristic classes and foliations, Lectures Notes in Mathematics 279, Springer (19

[17] GERARD, R., JOUANOLOU, J.P., Etude de l'existence de feuilles compactes pour certains feuilletages analytiques complexes, CR Acad. Sc. Paris **277** (1973).

[18] HARTSHORNE, R., Lecture Notes in Mathematics 156, Springer, Berlin (1970).

[19] HARTSHORNE, R., Publication No. 29, Institut des hautes études scientifiques, Bures-sur-Yvette (1966).

[20] DARBOUX, G., Mémoire sur les équations différentielles algébriques du premier ordre et du premier degré, Bull. Sc. Math. (1878) 60–96, 123–144, 151–200.

[21] TRAN HUY HOANG, Doctorat de 3° cycle IRMA, Strasbourg (1975).

SECRETARIAT OF THE COURSE

DIRECTORS

A. Andreotti	Department of Mathematics, School of Science, Oregon State University, Corwallis, Oregon 97331, United States of America and Istituto Matematico, Università di Pisa, Via Derna 1, Pisa, Italy
J. Eells	Mathematics Institute, University of Warwick, Coventry CV4 7AL, Warwickshire, United Kingdom
F. Gherardelli	Istituto Matematico "U. Dini", Università di Firenze, Viale Morgagni 67/A, Firenze, Italy

EDITOR

Miriam Lewis	Division of Publications, IAEA, Vienna, Austria

The following conversion table is provided for the convenience of readers and to encourage the use of SI units.

FACTORS FOR CONVERTING UNITS TO SI SYSTEM EQUIVALENTS*

SI base units are the metre (m), kilogram (kg), second (s), ampere (A), kelvin (K), candela (cd) and mole (mol).
[For further information, see International Standards ISO 1000 (1973), and ISO 31/0 (1974) and its several parts]

Multiply		*by*	*to obtain*
Mass			
pound mass (avoirdupois)	1 lbm	= 4.536×10^{-1}	kg
ounce mass (avoirdupois)	1 ozm	= 2.835×10^{1}	g
ton (long) (= 2240 lbm)	1 ton	= 1.016×10^{3}	kg
ton (short) (= 2000 lbm)	1 short ton	= 9.072×10^{2}	kg
tonne (= metric ton)	1 t	= 1.00×10^{3}	kg
Length			
statute mile	1 mile	= 1.609×10^{0}	km
yard	1 yd	= 9.144×10^{-1}	m
foot	1 ft	= 3.048×10^{-1}	m
inch	1 in	= 2.54×10^{-2}	m
mil (= 10^{-3} in)	1 mil	= 2.54×10^{-2}	mm
Area			
hectare	1 ha	= 1.00×10^{4}	m^2
(statute mile)^2	1 mile^2	= 2.590×10^{0}	km^2
acre	1 acre	= 4.047×10^{3}	m^2
yard^2	1 yd^2	= 8.361×10^{-1}	m^2
foot^2	1 ft^2	= 9.290×10^{-2}	m^2
inch^2	1 in^2	= 6.452×10^{2}	mm^2
Volume			
yard^3	1 yd^3	= 7.646×10^{-1}	m^3
foot^3	1 ft^3	= 2.832×10^{-2}	m^3
inch^3	1 in^3	= 1.639×10^{4}	mm^3
gallon (Brit. or Imp.)	1 gal (Brit)	= 4.546×10^{-3}	m^3
gallon (US liquid)	1 gal (US)	= 3.785×10^{-3}	m^3
litre	1 l	= 1.00×10^{-3}	m^3
Force			
dyne	1 dyn	= 1.00×10^{-5}	N
kilogram force	1 kgf	= 9.807×10^{0}	N
poundal	1 pdl	= 1.383×10^{-1}	N
pound force (avoirdupois)	1 lbf	= 4.448×10^{0}	N
ounce force (avoirdupois)	1 ozf	= 2.780×10^{-1}	N
Power			
British thermal unit/second	1 Btu/s	= 1.054×10^{3}	W
calorie/second	1 cal/s	= 4.184×10^{0}	W
foot-pound force/second	1 ft·lbf/s	= 1.356×10^{0}	W
horsepower (electric)	1 hp	= 7.46×10^{2}	W
horsepower (metric) (= ps)	1 ps	= 7.355×10^{2}	W
horsepower (550 ft·lbf/s)	1 hp	= 7.457×10^{2}	W

* Factors are given exactly or to a maximum of 4 significant figures

Multiply		*by*	*to obtain*
Density			
pound mass/inch^3	1 lbm/in^3	= 2.768×10^4	kg/m^3
pound mass/foot^3	1 lbm/ft^3	= 1.602×10^1	kg/m^3
Energy			
British thermal unit	1 Btu	= 1.054×10^3	J
calorie	1 cal	= 4.184×10^0	J
electron-volt	1 eV	$\simeq 1.602 \times 10^{-19}$	J
erg	1 erg	= 1.00×10^{-7}	J
foot-pound force	1 ft·lbf	= 1.356×10^0	J
kilowatt-hour	1 kW·h	= 3.60×10^6	J
Pressure			
newtons/metre^2	1 N/m^2	= 1.00	Pa
atmosphere[a]	1 atm	= 1.013×10^5	Pa
bar	1 bar	= 1.00×10^5	Pa
centimetres of mercury (0°C)	1 cmHg	= 1.333×10^3	Pa
dyne/centimetre^2	1 dyn/cm^2	= 1.00×10^{-1}	Pa
feet of water (4°C)	1 ftH_2O	= 2.989×10^3	Pa
inches of mercury (0°C)	1 inHg	= 3.386×10^3	Pa
inches of water (4°C)	1 inH_2O	= 2.491×10^2	Pa
kilogram force/centimetre^2	1 kgf/cm^2	= 9.807×10^4	Pa
pound force/foot^2	1 lbf/ft^2	= 4.788×10^1	Pa
pound force/inch^2 (= psi)[b]	1 lbf/in^2	= 6.895×10^3	Pa
torr (0°C) (= mmHg)	1 torr	= 1.333×10^2	Pa
Velocity, acceleration			
inch/second	1 in/s	= 2.54×10^1	mm/s
foot/second (= fps)	1 ft/s	= 3.048×10^{-1}	m/s
foot/minute	1 ft/min	= 5.08×10^{-3}	m/s
mile/hour (= mph)	1 mile/h	= 4.470×10^{-1}	m/s
		1.609×10^0	km/h
knot	1 knot	= 1.852×10^0	km/h
free fall, standard (= g)		= 9.807×10^0	m/s^2
foot/second^2	1 ft/s^2	= 3.048×10^{-1}	m/s^2
Temperature, thermal conductivity, energy/area·time			
Fahrenheit, degrees −32	°F − 32	$\frac{5}{9}$	°C
Rankine	°R	$\frac{5}{9}$	K
1 Btu·in/ft^2·s·°F		= 5.189×10^2	W/m·K
1 Btu/ft·s·°F		= 6.226×10^1	W/m·K
1 cal/cm·s·°C		= 4.184×10^2	W/m·K
1 Btu/ft^2·s		= 1.135×10^4	W/m^2
1 cal/cm^2·min		= 6.973×10^2	W/m^2
Miscellaneous			
foot^3/second	1 ft^3/s	= 2.832×10^{-2}	m^3/s
foot^3/minute	1 ft^3/min	= 4.719×10^{-4}	m^3/s
rad	rad	= 1.00×10^{-2}	J/kg
roentgen	R	= 2.580×10^{-4}	C/kg
curie	Ci	= 3.70×10^{10}	disintegration/s

[a] atm abs: atmospheres absolute;
atm (g): atmospheres gauge.

[b] lbf/in^2 (g) (= psig): gauge pressure;
lbf/in^2 abs (= psia): absolute pressure.

HOW TO ORDER IAEA PUBLICATIONS

An exclusive sales agent for IAEA publications, to whom all orders and inquiries should be addressed, has been appointed in the following country:

UNITED STATES OF AMERICA	UNIPUB, P.O. Box 433, Murray Hill Station, New York, N.Y. 10016

In the following countries IAEA publications may be purchased from the sales agents or booksellers listed or through your major local booksellers. Payment can be made in local currency or with UNESCO coupons.

ARGENTINA	Comisión Nacional de Energía Atómica, Avenida del Libertador 8250, Buenos Aires
AUSTRALIA	Hunter Publications, 58 A Gipps Street, Collingwood, Victoria 3066
BELGIUM	Service du Courrier de l'UNESCO, 112, Rue du Trône, B-1050 Brussels
CANADA	Information Canada, 171 Slater Street, Ottawa, Ont. K1A 0S9
C.S.S.R.	S.N.T.L., Spálená 51, CS-110 00 Prague Alfa, Publishers, Hurbanovo námestie 6, CS-800 00 Bratislava
FRANCE	Office International de Documentation et Librairie, 48, rue Gay-Lussac, F-75005 Paris
HUNGARY	Kultura, Hungarian Trading Company for Books and Newspapers, P.O. Box 149, H-1011 Budapest 62
INDIA	Oxford Book and Stationery Comp., 17, Park Street, Calcutta 16; Oxford Book and Stationery Comp., Scindia House, New Delhi-110001
ISRAEL	Heiliger and Co., 3, Nathan Strauss Str., Jerusalem
ITALY	Libreria Scientifica, Dott. de Biasio Lucio "aeiou", Via Meravigli 16, I-20123 Milan
JAPAN	Maruzen Company, Ltd., P.O.Box 5050, 100-31 Tokyo International
NETHERLANDS	Marinus Nijhoff N.V., Lange Voorhout 9-11, P.O. Box 269, The Hague
PAKISTAN	Mirza Book Agency, 65, The Mall, P.O.Box 729, Lahore-3
POLAND	Ars Polona, Centrala Handlu Zagranicznego, Krakowskie Przedmiescie 7, Warsaw
ROMANIA	Cartimex, 3-5 13 Decembrie Street, P.O.Box 134-135, Bucarest
SOUTH AFRICA	Van Schaik's Bookstore, P.O.Box 724, Pretoria Universitas Books (Pty) Ltd., P.O.Box 1557, Pretoria
SPAIN	Diaz de Santos, Lagasca 95, Madrid-6 Calle Francisco Navacerrada, 8, Madrid-28
SWEDEN	C.E. Fritzes Kungl. Hovbokhandel, Fredsgatan 2, S-103 07 Stockholm
UNITED KINGDOM	Her Majesty's Stationery Office, P.O. Box 569, London SE1 9NH
U.S.S.R.	Mezhdunarodnaya Kniga, Smolenskaya-Sennaya 32-34, Moscow G-200
YUGOSLAVIA	Jugoslovenska Knjiga, Terazije 27, YU-11000 Belgrade

---

Orders from countries where sales agents have not yet been appointed and requests for information should be addressed directly to:

Division of Publications
International Atomic Energy Agency
Kärntner Ring 11, P.O.Box 590, A-1011 Vienna, Austria

76- 08587